MW00356529

Making it FASTER II

THE INDIANAPOLIS AND GRAND PRIX CARS

Making it FASTER II

THE INDIANAPOLIS AND GRAND PRIX CARS
PRELIMINARY

NORM DEWITT

FOREWORD BY ROBIN MILLER

FASTER PUBLISHING
San Diego, California, USA

Cover photo - Neil Leifer of Sports Illustrated & Getty Images. 1969 Indianapolis 500 pace lap. AJ Foyt, Mario Andretti, Bobby Unser

Back cover photo - Brian Watson. 1967 Oulton Park Gold Cup qualifying – Sir Jackie Stewart in the H-16 BRM

Back cover small photo - Norm DeWitt. Force India in the monsoon, 2015 USGP

Title Page photo – Norm DeWitt. Helio Castroneves during one of his greatest drives, the 2017 Indianapolis 500

Cover design – Asa Wild / Norm DeWitt
Copy editing – Beth Bolwerk
Formatting and Layout – Yvonne Betancourt at ebook-format.com

Hardcover ISBN 978-1-948201-03-2
Paperback ISBN 978-1-948201-00-1
ePub ISBN 978-1-948201-01-8
Kindle ISBN 978-1-948201-02-5
Library of Congress Control Number 2018901565

Web – makingitfaster.com
Email – makingitfaster@cox.net
FASTER PUBLISHING
3779 Milan Street
San Diego, CA 92107, USA

Date of Publication: March 15, 2018

ACKNOWLEDGEMENTS

Due to space considerations, an enormous number of influential teams and manufacturers who had a significant impact on the sport of open-wheel racing world-wide do not have chapters in this book. Cooper Cars, Matra, Mickey Thompson, Shadow, Ferrari, March, Reynard, Surtees, Swift, Tyrrell, Ligier, Pat Patrick and his Wildcats, Galmer, Osella, Truesports, Minardi, Hesketh, and Jordan, were some of the more prominent omissions. Perhaps, someday in the future with a subsequent Making it FASTER book, I can endeavor to tell their stories as well.

There have been a number of references to the design details of various March, Ferrari, Hesketh, Tyrrell, Minardi, and Matra cars in the two Making it FASTER books and there is little doubt that each chapter could have formed the basis of a book in itself. My goal was to strike a balance in the size of the chapters included, with primarily first-hand stories by drivers, designers, engineers, mechanics, team managers, and owners, from these historic teams.

Thanks to all those racers who have helped me to understand and explain how it was to be inside these teams which made, or continue to make history.

For my parents, Ed and Marian,

who got a slot car set for me,

and everything changed

CONTENTS

Arie Luyendyk, Vince Granatelli, and Robin Miller –
Indianapolis 500 race day 2014 (photo - Norm DeWitt)

FOREWORD

When I think of Norm DeWitt, I'm instantly drawn to his essays and photos on motorcycle racing, now and then, because he's the keeper of records and memories for fifty years of two-wheel lore.

But, as you will soon learn while reading his new book, *Making It Faster II*, Norm is a racer first and foremost, and a student of history when it comes to Indy cars and Formula 1.

What he's managed to do so well here is capture the glory days of Formula 1 and the Indianapolis 500, when any idea was in play, innovation ruled and free thinking was around every corner.

Just his chapter on the brutish Novi makes us long for those days and having Bobby Unser as our tour guide illustrates why those days were so important in growing the "Greatest Spectacle in Racing."

Or the active-suspension revolution in Formula 1, another barometer for what made us care.

DeWitt talked to a lot of knowledgeable people in putting this time machine together and the reader is rewarded with first-person accounts of what really happened between Gasoline Alley and the world's most famous oval, or why a certain chassis could work at Monaco, but nowhere else.

From Dan Gurney to Colin Chapman to Mickey Thompson, it was a great ride and Norm was along for the whole trip.

Robin Miller
October, 2017

*My first race, Dan Gurney and Jim Clark – front row 1967 Rex Mays 300,
Riverside Raceway (Norm DeWitt)*

PREFACE

For everyone there is a moment of inspiration or a series of experiences that leads one down a certain irreversible path. For years I had been reading articles in slot-car magazines about incredible racing achievements by legends such as Fangio, Moss, Nuvolari, and Rosemeyer. In 1967, Grand Prix arrived at our local Cinerama, taking us into the world of Formula 1 racing, putting the viewer into the cars like never before. Technical wonders such as Turbine Cars and the H-16 BRM were racing, and it was a time of huge success for Americans and American manufacturers. Ford, along with American drivers Gurney and Foyt, had won Le Mans. Dan Gurney made history by winning the Belgian Grand Prix at Spa in his Eagle, and AJ Foyt won his third Indianapolis 500 driving the Coyote. Mario Andretti in the Brawner Hawk was the reigning two-time Champion in USAC Championship Car racing, also having won both the 12 Hours of Sebring and the Daytona 500 that year. The 1967 Championship would be decided in the final race between Mario Andretti and AJ Foyt.

1967 First Annual Rex Mays 300 ticket stub with Dan Gurney's Eagle
(Norm DeWitt)

The 1967 Rex Mays 300 at Riverside was the first USAC Championship Car race held in the Los Angeles area in thirty-one years. Riverside native Rex Mays won that race in January 1936, with riding mechanic Chickie Hirashima (legendary Indianapolis engine builder from the 40s-60s). This race followed on the heels of the Mexican Grand Prix, so it boasted a superstar lineup of drivers. Costa Mesa's Dan Gurney was the local favorite, making the home-track debut of his All American Racers Eagle. Gurney took pole position, followed by Jim Clark, Bobby Unser, John Surtees, AJ Foyt, and Mario Andretti in qualifying.

Gurney's Eagle had a flat tire, which required an unexpected pit stop. With Andretti moving into the lead, and Foyt having to take over Roger McCluskey's Eagle at mid-race, the championship swung in Andretti's favor. When Andretti stopped for a splash of fuel with six laps remaining, the championship swung back to Foyt, and the race lead swung to Bobby Unser's Eagle. In the final turn, coming to the white flag, Gurney swept past Unser to take the lead as the roar from the crowd drowned out the engines. It's what Gurney has described as a "Hollywood ending." He had won at his home track, with his Eagles finishing 1-2, and another Eagle carrying Foyt to the championship, all this to close out a year during which his Formula 1 Eagle had won the Belgian Grand Prix.

1971 Ferrari 312B. My first Formula 1 race – 1971 Questor Grand Prix, Mario Andretti winner (Norm DeWitt)

One could hardly overestimate the impression left upon this particular 13-year-old fan in the Turn six grandstand. Over the next two years I got to meet and chat with Mario Andretti, Bobby Unser, Al Unser, Johnny Rutherford, Art Pollard, Chris Amon, Denny Hulme, and the Granatelli brothers, along with dozens of others who had come to race at Riverside Raceway in the Indianapolis or Can-Am races. Age 16 saw my introduction to Formula 1 at Ontario Motor Speedway in the Questor Grand Prix. Jackie Stewart put the new Tyrrell 001 on pole for the race, which saw Andretti in the flat-12 Ferrari 312B moving up through the field to battle Stewart for the lead, while Pedro Rodriguez set the track record in the new BRM P160. Combined with Amon's wailing Matra V-12, the entire experience is best described as being in sonic and technology heaven. Those races from 1967-1971 had opened my eyes to a world of wonder.

Sadly, BMX and off-road motorcycles were far more within my budget than Formula 1 or Indianapolis cars.

1983 Runoffs sponsored by Dick Barbour's Performance Datsun

*Leading overall at the 1985 Riverside Memorial Day National
(Don Hodgdon)*

While at university studying Architecture, I entered a National Solo II Autocross in Tucson, and won. After success at Autocross and an encouraging outing at the Jim Russell School in a Formula Ford coached by Dennis Firestone, I fabricated and developed a road-racing car. My cars always carried #36, an homage to the greatest of Southern California's racers, as Richie Ginther and Phil Hill had carried 36 on their Ferraris, along with Dan Gurney's winning Eagle at Spa.

After a promising National debut, Dick Barbour provided critical sponsorship through Performance Datsun in San Diego. The Runoffs in that era were populated by rising stars destined for Indianapolis, such as the brothers Andretti, Groff, and Bren, along with Parker Johnstone, Chip Ganassi, Scott Sharp, and Jimmy Vasser. Development on the engine and chassis brought track records and National wins in the West, along with three years at the Road Atlanta Runoffs in D-Production and GT-3, the highlight being fifth place in 1983. Soon after, I shifted over to sailboat racing. Stars & Stripes had won the America's Cup in Australia and Point Loma became the world's primary venue for sailboat racing. My decision was reinforced by the demise of Riverside Raceway, Carlsbad Raceway, and Holtville Airport, three local tracks where I had won Nationals.

Eventually, writing for magazines led to renewing racing contacts and publishing stories with fellow Runoff racers, such as Rahal, Vasser, Sharp, Ganassi, and the Andrettis. That in turn, led to decades of writing for motorsport and aviation magazines all around the world, and eventual book offers. Thanks to all the racers, mechanics, engineers, and designers who helped make this and my past efforts a reality. *Making it FASTER II* is my third book, and I hope you enjoy it.

Norm DeWitt

Norm DeWitt, 2014

Gian Paolo Dallara 2016. For a decade, every car in the Indianapolis 500 has been a Dallara (Norm DeWitt)

INTRODUCTION

For generations the top series for open-wheel racing have been Indianapolis car racing (in its various guises from AAA, USAC, CART, IRL/Champ Car, Indycar) in America, and Grand Prix racing primarily in Europe. These two worlds often found themselves dominated by the same designs beginning in 1913 with the Indianapolis 500 won by Jules Goux in the Peugeot Grand Prix car, which featured 4 valves per cylinder and twin overhead camshafts. The following year, the race was won by Rene Thomas in another French Grand Prix racer, this one made by Delage. There were two Delages and two Peugeots in the top four, and pole position was taken by Georges Boillot in yet another Peugeot.

During this time in hyper-nationalistic Europe, there could hardly have been a more important race than the French Grand Prix that followed soon after Indianapolis on the Fourth of July, 1914. It proved to be a battle royale between Germany and France, the two greatest automotive powers of the day. It was the Mercedes factory cars versus Boillot in his four-wheel-braked Peugeot. The Mercedes led early, before giving way to Boillot for most of the race. In the end however, Mercedes cars finished 1-2-3 with Goux fourth for Peugeot.

Jim Nabors about to sing 'Back Home Again in Indiana' for the last time, 2014 (Norm DeWitt)

That race was held the week after the assassination of Austria-Hungary's Archduke Ferdinand and is often seen in the context of being the opening act for the Great War (the First World War) which erupted three weeks later. It was likely the most consequential motor race ever held. Ironically, the winning Mercedes Grand Prix car was on a victory tour in England when war broke out, and studying the cars gave the English important insight into improving their engines courtesy of Mercedes.

Twenty-five years later it was Wilbur Shaw in the Italian Maserati 8CTF, the Boyle Special, winning the race back to back in 1939-40, then was leading the 1941 race until a wheel failed. Alberto Ascari brought over the Ferrari team in 1952 for the 500, and nine years later, World Champion Jack Brabham brought a specially built Cooper-Climax to restart the rear-engine revolution (the rear-engine Miller of the late 30s and 40s predated that revolution, see *Making it FASTER*). Lotus arrived soon after, followed by racing car manufacturers Brabham, Lola, McLaren, AAR (Eagle), Penske, March, and Reynard, which similarly competed in the Indianapolis 500 along with competing in Formula 1, most of them finding success at both the Speedway and in Grand Prix.

Louis Meyer, Inc.

Ford Racing Engine
SALES
and...

SERVICE

We Have All The Modern Facilities To Keep Your Equipment in Top Running Condition

• • •

In 1967, Indianapolis race programs carried ads for buying 4-cam Ford engines. Build your best chassis and race.

OFFENHAUSER

manufactured by

Drake Engineering & Sales Corp.

If you prefer, buy a Supercharged Offenhauser instead.
Radical innovation was on display throughout the grid.

Emerson Fittipaldi's 1974 World Championship
winning M23 McLaren (Norm DeWitt)

The rich history of these two different types of racing being intertwined continues into the current generation. In 1995 Jacques Villeneuve won the Indianapolis 500 and CART championship, followed soon after by winning the Formula 1 World Championship for Williams-Renault in 1997. In 2017, Fernando Alonso skipped the Monaco Grand Prix to compete at the Indianapolis 500 and become the most recent example, as former Formula 1 driver Takuma Sato won the race. It was reminiscent of 1993 when Nigel Mansell drove his first oval race, dueling to the finish with Emerson Fittipaldi and Arie Luyendyk,

Emerson Fittipaldi on the Long Beach pitlane in 1978
with the Copersucar (Norm DeWitt)

or 1965 when Jim Clark won ahead of Parnelli Jones and Mario Andretti. The Indianapolis 500 is truly a Race of Two Worlds.

My previous book, *Making it FASTER*, was a collection of first-hand stories from drivers, designers, engineers, and mechanics about how technical innovations were developed, or in some cases, failed. This book continues that theme with the focus limited to these forever-linked premier classes of open-wheel racing, both in America and Europe, Indianapolis and Grand Prix cars.

Two-time Formula 1 World Champion and two-time Indianapolis 500 winner Emerson Fittipaldi put it well when I asked him if he had regrets leaving Team McLaren and their highly successful M23 in favor of racing the Copersucar for 1976. Emerson explains, "No, because since that I came to Indianapolis. Teddy Mayer [an American formerly with Emerson at Team McLaren] and I won Indianapolis together, and Teddy was crew chief on my car. I would do the same again."

Norm DeWitt

Orange Juice? Emerson Fittipaldi won the '93 Indy 500,
earning On Track's 'Drive of the Decade' (Penske Racing)

Chuck Sprague – "I was the voice on the radio that always asked "Tudo bem?" ("All good?") just before the green flag at every race. His reply would be "Tudo otimo" ("All is great")."

CHAPTER ONE

RACING FOR ENGLAND

The BRM V-16 cars were England's great hope for postwar Grand Prix glory (Norm DeWitt)

In postwar England, the ambitious BRM V-16 project had been a source of great national pride. The top manufacturers in England came together to build a car that could put England foremost upon the world stage in Grand Prix racing, which was currently dominated by the Italian teams of Alfa-Romeo and Ferrari.

British racing car manufacturers had struggled to compete at the highest level of motorsport in the prewar era, the glowing exceptions being Seagrave at the 1923 French Grand Prix and the ERA voiturettes (Formula 2) of the mid-1930s, driven to great success by Dick Seaman, B. Bira, and Raymond Mays. After the war, Mays headed the effort to get British automotive and aviation industries behind an effort to build an all-British racing car that would dominate Grand Prix racing.

The formula of the day was 1.5 litre supercharged (4.5 litre normally aspirated) and, given the enormous experience of Rolls Royce with their supercharged aero engines such as the Merlin, a two-stage supercharged V-16 of 91 cubic inches was pursued. Looking at the resultant engine, it was essentially two V-8s joined at the center to minimize the twisting forces. Each cylinder head was essentially a compact 4-cylinder, twin-cam design. Despite these considerations, the resulting engines were hand grenades. At the car's public debut at Silverstone in August 1950, the car broke on the grid at the drop of the flag. Quoted power figures

Cutaway supercharged V-16 BRM (Norm DeWitt)

were literally all over the map. It was a PR nightmare as the wide gulf between expectation and reality began to sink in.

Former Norton racer and postwar Norton director, Tony Vandervell, had been part of this failing BRM effort. Despite some early efforts with different versions of a supercharged Ferrari 125, he was now campaigning an extensively modified 375 with a 4.5 litre V-12 Ferrari under the banner of "Thinwall Special" to promote Vandervell's thin wall bearing company which had made him a wealthy man. BRM factory driver, Reg Parnell, with little to do given the endless problems with their V-16 car, found success with the Thinwall Special, winning a number of nonchampionship events and taking fourth place at the 1951 French Grand Prix.

By the time the BRM was somewhat sorted, it was 1953 and the Formula 1 World Championship was no longer for 1.5 litre supercharged/4.5 litre normally aspirated engines. As a result, this V-16 technological marvel/nightmare was consigned to exhibitions and Formula Libre races. The BRM effort to dislodge the Italian firms of Alfa-Romeo and Ferrari from the pinnacle of Grand Prix racing had failed. The World Championship had shifted to being based upon Formula 2 spec cars (2 litre normally aspirated) and with the Alfa-Romeo factory's departure, it was a field day for Alberto Ascari and Ferrari. The Ferrari effort at the 1952 Indianapolis 500 had been humbling, as Ascari found

Continued success by the Norton Manx inspired a 4-cylinder version, the Vanwall (Norm DeWitt)

his car to be seriously lacking in speed (the 500 counted towards the World Championship in the 1950s), but once back in Europe racing the 2 litre Ferrari, Ascari won all remaining six races that season. 1953 brought more of the same as Ascari won five of the eight Grand Prix (not counting Indianapolis). Other than one Grand Prix victory by Mike Hawthorn in France, the British had little to celebrate.

Meanwhile, for 1953 Vandervell continued to develop the Thinwall Special, and the car eventually sported many innovations such as disc brakes, but his eyes were again upon Grand Prix racing. The Norton Manx had been the dominant engine in motorcycle Grand Prix racing prewar as well as postwar, and was now being extensively used in the new Formula 500 car racing to great success. Emerging talent Sir Stirling Moss was a dominant force in Formula 500 racing from 1948 through the early 1950s. Stirling recalls:

> If you could get it, what you wanted was known as a 'double knocker Norton,' which was a 500cc racing engine, and you'd quite often have to buy the entire bike and then throw the other bit away as they were like hen's teeth. From my point of view you bought the best engine. It was hard to find state-of-the-art engines from motorcycle manufacturers. The designer of the Norton Manx engine, Vandervell commissioned him to do a 4-cylinder of that as the Formula 1 of that era was 2.5 litres.

In America, Warren Olson had entered a Cooper Mk IX for emerging star Bruce Kessler. Warren explains:

> The Norton was a proper racing engine. It had twin overhead cams, great cooling, there were no ifs, ands, or buts about it. I don't think there was any rule about the number of cylinders in Formula 500, and I remember a number of people running Triumphs. The Kieft [chassis] were very good but it ended up that the Cooper Norton was the formula. The JAP or the Norton were the ones to run. Getting parts for the JAP wasn't bad as they were very popular for short-track motorcycles.

Bruce Kessler was a local talent in the Los Angeles area and was fortunate enough to be in the right place at the right time. Bruce says:

4

Where I got started was Warren Olson had his eye on me; he had a shop on La Cienega Boulevard [Los Angeles] and he felt that I was a real up-and-coming driver and he wanted to get involved with Cooper, an unknown chassis in this country. If he would get a chassis, I would get an engine and a drive. It would be a good way for me to come up, as I was only 18 or 19 years old. There was a problem… when we got the Mk IX Cooper it had a JAP engine in it, but the first race we went out and won. Everybody was switching to Norton and we were forced to switch and put a Manx Norton in the car.

The availability of the Norton Manx engine in America was a similar situation to what it was in England. Warren says, "We were lucky in that we got the Norton without having to buy the bike. We had a spare engine, but it was a JAP. The stuff that we used a lot of were sprockets and chain, they really suffered. I'd run a primary chain once, a secondary chain twice. You had to do that, as they weren't very reliable. It wasn't unusual at all to have a chain come apart during a race, but as for problems that was about it." Warren was to become the American distributor for Cooper Cars. The Norton Manx had become the dominant engine in both English Formula 500 and motorcycle Grand Prix, and often was the engine of choice in AMA road racing (four straight wins in the Daytona 200 between 1949-52), and SCCA Formula 500 racing.

Warren was to later create the greatest American sports racer of the 1950s, the Scarab. Warren recalls, "I don't remember precisely where I ran across the name, but we did. It seemed to work out alright and everybody seemed to like it." It's probably just as well that nobody mentioned that the Scarab was a dung beetle.

There was no getting around the basic goodness of the Norton Manx racing engine. Given Tony Vandervell's position at Norton, having access to Manx engines and engineering wizard Leo Kuzmicki was a given. With their results in motorcycle Grand Prix and Formula 500, it seemed that Vandervell's ambition plus Norton Manx technology was a match made in heaven for either the current 2,000cc Grand Prix formula, or for the new 2,500cc formula announced for 1954. Work soon began on the 4-cylinder Vanwall Special, initially in 2 litre form, for the 1954 World Championship season.

Cylinder head of the Manx-inspired 4-cylinder Vanwall (Norm DeWitt)

Results were few and far between initially, as Grand Prix racing was dominated by the might of Mercedes-Benz and their 8-cylinder desmodromic valve train racers for 1954. Fangio had started the season with two wins in a Maserati 250f while awaiting his new W196 Mercedes, and the championship was a Fangio runaway. Stirling Moss had shown promise from the start, finishing on the podium in third with his "Equipe Moss" privateer Maserati 250f for its first race in the Belgian Grand Prix at Spa. For comparison, the best result for the Vanwall Special that season was a seventh place by the talented English racer Peter Collins in the Italian Grand Prix at Monza.

For the following year, Stirling Moss joined the Mercedes team alongside Juan Manuel Fangio, and 1955 was to be another Silver Arrow runaway with Fangio winning four of the six European rounds, and Moss getting his first World Championship win in the British Grand Prix at Aintree. In comparison, Vanwall had truly dismal results with the latest 2.5 litre car. Hawthorn did not finish (DNF) at both Monaco and Spa, and later in the season Harry Schell got DNFs at both the British and Italian rounds while sharing Ken Wharton's second Vanwall to a ninth place finish at Aintree. At this point the Vanwall effort could only be described as a complete and utter failure. All the Kuzmicki-Norton engine technology in the world was of little use without a chassis to match. However, work was afoot to build a new car for 1956, one that would change the team's fortunes and eventually lead to them becoming a World Championship Constructor.

The new Vanwall was a clean sheet design for the chassis and bodywork. Relative unknown Colin Chapman had been hired by Vandervell to design a space frame chassis. Colin was a talented young engineer who had already founded a company specializing in kit cars and small displacement sports racers – Lotus. Their state-of-the-art space frame Mk VIII and Mk IX with Frank Costin bodywork had impressed all when released and the Mk IX debuted at Le Mans in 1955. Chapman's passion for low weight combined with chassis rigidity made him a reasonable and inspired choice to bring into the Vanwall team. That Chapman would have de Havilland Aircraft's Frank Costin to continue their successful relationship completed the puzzle. The new Vanwall was far more rigid than the typical ladder-framed Grand Prix car of the day. It was highly aerodynamic, looking like a flattened blimp in cross-section, with the NACA duct on the hood feeding the air intakes.

Vandervell did not suffer from the "England-only" mindset of the BRM V-16 effort and had ditched the Amal carburetors for Bosch fuel injection on the latest version of the 4-cylinder Manx Norton. The new car was powerful, quick, and handled well. At last Vandervell had his contender for World Championship glory. Moss was brought onto the team for a nonchampionship race at Silverstone and won, showing the new car's potential.

With the Le Mans disaster leading to the withdrawal of Mercedes-Benz for 1956, Moss cast his lot with Maserati, a prudent and proven choice. Stirling says, "The Maserati 250f was the most user friendly of all Formula 1 cars. It had really beautiful balance." Fangio started with a win in his native Argentina, driving for Ferrari. Moss countered to win at Monaco, proving the wisdom of his decision. Collins, teammate to Fangio at Ferrari, countered with wins at Spa and Reims. The championship appeared to be wide open, but there was a notable increase in pace by this latest Vanwall. Harry Schell finished fourth at Spa and then at Reims had relieved the ill Mike Hawthorn and battled for the lead. In their third season of Grand Prix racing, Vanwall was now a potential race winner if the various reliability gremlins of the latest car could be resolved.

The 1956 season closed out at Monza with Fangio as champion, thanks to Collins turning his car over to his teammate, and Stirling Moss winning the race in a dominant run. It was looking like 1957 would be a Maserati steamroller year, but Stirling had always wished to drive for an English team if possible. Stirling understood that he was walking away from a stellar car in switching to the Vanwall from the new lightweight 250f, but there were other issues at play. Moss explains, "Frankly, I did everything I could to drive a British car, but one didn't arrive until the Vanwall. When that came along I signed up along with Stuart Lewis-Evans and Tony Brooks. Stuart was a very fast driver, I'll tell you that. The whole team was very fast... obviously Tony Brooks and myself were at the top at that time."

Heading into 1957, Surry-based Connaught was having as much or more success, Tony Brooks in 1955 winning the non-championship Syracuse Grand Prix, with the 1956 highlight seeing their cars finishing third and fifth in the last race of the season, at the Italian Grand Prix. The third-place Connaught was a lap down to Moss' Maserati, whereas the Vanwall of Schell had battled in the lead pack until it struck trouble. At that moment, Connaught appeared to be as reasonable of a choice as the fast and fragile Vanwall for 1957, but with thirteen DNFs at Monza, it flattered to deceive.

The Vanwall Collection at Donington Park's Grand Prix Collection (Norm DeWitt)

Bernie Ecclestone was the agent for his friend Stuart Lewis-Evans, and Bernie would initially have Stuart driving a Connaught for 1957. Lewis-Evans had continued racing his 500cc Cooper-Norton to great effect, winning at Crystal Palace again in May of 1957, but what really got everyone's attention was when he finished a shocking fourth in his first Grand Prix (at Monaco) with the Connaught. The car didn't really deserve of that type of finish.

American Formula 500 champion Bruce Kessler (Cooper-Norton) was to drive at Monaco the following year in a similar Type B Connaught for entrant Bernie Ecclestone. Kessler recalls:

> When I went there everybody said that there was no freaking way that the Connaught was going to qualify. I had never had used a preselector gearbox, it was very awkward and strange to me. You select the gear and when you wanted to shift you'd hit the clutch and it would change. All I was thinking about was, 'How do I drive this freaking car with this weird transmission in it?' 1958 was also the beginning of the development of the Cooper and they got a bigger engine, a bored out Coventry Climax, and entered it for Maurice Trintignant, who was considered to be over-the-hill, on the downhill side of his career. Well, Moss comes to me and says, Trintignant is not going to be able to qualify and they are going to put you in Trintignant's car. Of course, Trintignant qualified and ended up winning the race. I was plan B, but it just didn't happen. At least I was comfortable in a Cooper, it was something I was used to.

Kessler's experience puts Lewis-Evans' stellar performance at Monaco the previous year into perspective. Stuart was quickly drafted onto the 1957 Vanwall team by Vandervell to join his all-British driver lineup of Stirling Moss and Tony Brooks in their massively improved cars. Leading the French Grand Prix at Reims secured Lewis-Evans' position on the team.

Bernie Ecclestone experienced the patriotic support given to the British teams of the day. When you think back on the "Racing for England" slogan from the Hesketh–James Hunt era of 1973-75, it simply pales in comparison to the success achieved by Vanwall in trying to bring home the World Championship with their new car from 1957-58. Bernie explains, "That or BRM, but they weren't as successful as Vanwall. You had to support them... it was one of those things I've been involved with. I've still got a Vanwall racer, so I still collect that sort of thing." Not something you hear every day – that somebody has a Vanwall.

With all three drivers at Vanwall being British, as were two of the drivers at Ferrari (Mike Hawthorn and Peter Collins), these were glory days for British racers in Formula 1. No doubt in 1957 the highlight for the home crowd had to be Vanwall winning the British Grand Prix at Aintree, driven by Tony Brooks and Stirling Moss (who took over Brooks' car when his failed). Moss says, "It was the first time that a British car had ever won the British Grand Prix. It was a big step forward in motor racing." *Was Aintree 1957 the biggest win of your career?* "Up to that point, yes. I had been waiting so long to get a [British] car that was really competitive."

Vanwalls 1-2-3, Stuart Lewis-Evans pole, Stirling Moss, and Tony Brooks, 1957 Italian Grand Prix (Bernard Cahier)

At the end of the season, Lewis-Evans was to start from pole position for the Italian Grand Prix, the top three slots on the grid being the three British drivers in their Vanwalls. However, 1957 was another Fangio year, having won four of the first five races (not counting the Indianapolis 500, which was part of the championship). Stirling Moss was to win the last two races of the season at Pescara and Monza in the Vanwall, and between Moss and Brooks they had set fastest lap at four of the European season's seven rounds, so things looked promising for the British team in 1958.

However, there were issues with the mandated changes from wild, unregulated combinations of alcohol- and benzene-based fuels to standard 130 octane aviation fuel, which Vanwall struggled to adapt to their large-bore 4-cylinder engines versus the smaller-bore V-6 Ferrari or the in-line 6-cylinder Maseratis. These were serious problems, and

Vanwall missed the opening Grand Prix of 1958 in Argentina. Did Vanwall not show up because they hadn't got it sorted with the new AvGas requirement? Stirling Moss weighs in, "Possibly. They had quite bad flat spots that got sorted out as the year went past. Certainly the engines did improve, not in power, but in the way you drove it. The power curve and everything was improved considerably. Before that [pre-1958] you could run anything you'd like. I did in fact run on nitro methane, which was virtually an explosive. But all the normal cars were set up with alcohol to start with." Moss was to win the Grand Prix of Argentina in a Cooper, the first Formula 1 win for a rear-engine car (not ignoring the prewar exploits of Stuck, Rosemeyer, and Nuvolari in their Auto-Unions).

The Norton Manx-based Vanwall still had one season remaining in its last and greatest challenge. Originally the great strength of the package upon the introduction of the Vanwall in 1954, the Manx-based engine was now the weak link. Stirling explains:

> Yes, probably. But one thing you've got to remember is that our races were three hours minimum. The races were a tremendous test not only of the driver, but of the machinery. You've got to realize that in those days, cars were not reliable like they are today. Winning a race was quite an achievement. The engines, gearboxes... they were nothing like it is now. Now most cars finish, in those days most cars didn't.

What the Vanwall team didn't know was that Stuart Lewis-Evans had struggled with a health issue, although it certainly didn't seem to affect his sheer speed or endurance in the cars. Bruce Kessler says:

> In 1958, after I won the GT class at the 12 Hours of Sebring, John Cooper brought in Bruce McLaren, and also talked to me about coming to England on a scholarship to England to drive Formula 2. However, John only had one car so I drove for Rob Walker in Formula 2. When I got to England, I moved in with Stuart Lewis-Evans' family, we knew each other from when he came to the USA and we got to be friends. He was a wonderful person. Stuart was really concerned, as he had a bleeding ulcer and he was afraid that they would find out and he wouldn't be able to drive for Vanwall. I never said a word about it.

11

Tony Brooks had won at Spa, Moss broke when he over-revved the engine, Lewis-Evans finishing third. Ferrari's Peter Collins won at Silverstone, leading home teammate Hawthorn, while Moss had retired with a broken valve while in second (Lewis-Evans finishing fourth). To quote Road & Track magazine's race coverage from the British Grand Prix, "With the recent Ferrari superiority, we do not see, except for a miracle in Vandervell's shops, how Moss' talents can overcome the combination of Ferrari and Hawthorn." There was something to this, as the engine reliability issues remained a continual problem. In the following race, the German Grand Prix, Moss shot away from pole, well in the lead when the magneto broke on lap 4. As at Spa, Tony Brooks won after Moss' retirement. Any chances at winning the Driver's World Championship indeed appeared to be slipping away from Moss. After the penultimate round at Monza, Brooks had three victories, Moss had three victories, and Mike Hawthorn only one. But the reliability of the Ferrari had made the difference and Hawthorn was ahead in points.

Commemorative stamp issued for Sir Stirling Moss and the Vanwall

The World Championship came down to its climax at the final event in Morocco between Stirling Moss and Mike Hawthorn. Indicative of the carnage of that era, both of Hawthorn's Ferrari teammates had by then died in Grand Prix, Luigi Musso at Reims, and Peter Collins at the Nurburgring. Sadly the nightmare was now shifting to Vanwall in what should have been their greatest hour.

Moss had to win, and take the fastest lap. He adds, "And have Mike not be second." Vanwall desperately needed a 1-2 finish. Moss was leading, but Tony Brooks was unable to help, his Vanwall having blown up halfway through the race. In the midst of all this Championship drama, Stuart Lewis-Evans was now Vanwall's last hope. Having qualified third, he had the pace but needed to move past the Ferraris into second place to ensure the title for Moss. No doubt the pressure was heavy upon Stuart to catch the Ferraris. As Stuart pushed the car to the limit, with twelve laps to go his engine locked up, which threw the car off the track and it erupted into flames.

In the end, Hawthorn was World Champion by one point over Moss as Hawthorn wrung the neck of his Ferrari to deny Moss the fastest lap, and Phil Hill fell back moving his Ferrari teammate Hawthorn up into that critical second place. Stirling reflects, "Everything fell into place,

Alma Hill (Phil's wife) shows up with a cake for
Stirling Moss' 85th birthday (Norm DeWitt)

and then Phil quite correctly and understandably backed off and gave him the position he needed and that gave him [Hawthorn] the title."

Vanwall had won the 1958 World Manufacturer's Championship, the first time that title had been offered, but it was a hollow and somber achievement in the wake of Stuart's accident. Stuart Lewis-Evans succumbed to his burns six days later. Vandervell by all accounts was shattered at the loss of his young driver and there were only half-hearted efforts to continue the team, with Brooks being entered at the British Grand Prix for 1959, the only race Vanwall contested that year. World Champion Mike Hawthorn was soon lost as well, killed in January 1959 when street racing against Rob Walker, the Formula 1 team owner. The entire 1958 Ferrari Grand Prix team was gone, with only replacement driver Phil Hill surviving to tell the tale.

Onto Ferrari's 1959 team of Dino V-6 cars came Tony Brooks and California's Dan Gurney, making his debut at the French Grand Prix, the race where Jean Behra was dropped from the team after getting into a dustup with Ferrari Team Manager Tavoni. Gurney's second race was the German Grand Prix at Avus and, ironically, Behra was killed in a support race for that Grand Prix, driving his Porsche over the top of the banking.

Grand Prix racing of that era truly lived down to its reputation of being a death sport, and the loss of Collins, Lewis-Evans, and Hawthorn within six months left only Moss and Brooks surviving from Ferrari and Vanwall to represent Britain the following year. It was the end of an era, and the end of Vanwall, the only front-engine World Manufacturer's Championship-winning car.

CHAPTER 2

THE NOVI

The prewar supercharged Winfield V-8 in the 1941 Indianapolis car of Ralph Hepburn (Norm DeWitt)

The Novi, named for a town in Michigan, actually has its roots in prewar Southern California. The engine, initially known as the Winfield 8, was the brainchild of Los Angeles mechanical genius and speed guru Ed Winfield. That car first ran at the Indianapolis 500 in 1941, finishing fourth as the Bowes Seal Fast Special driven by another Los Angeles-area legend, Ralph Hepburn, previously a motorcycle racing champion for both Harley-Davidson and Indian. Postwar there were the front-wheel drive Frank Kurtis-built chassis that became the best known examples from this lineage of cars, with the name now changed to the Novi Governor Special. In 1946 new team owner Lew Welch (from Novi, Michigan) returned with veteran Hepburn, who turned a staggering lap at 134.449 mph in qualifying, the only car to lap at over 130 mph. Due to the oddities of Indianapolis qualifying, Hepburn was to start from the seventh row, but this latest Novi had thrown down the gauntlet. The racing world had truly returned to the Speedway as qualifiers included Luigi Villoresi, while non-qualifiers included Rudi Carraciola (grievously injured in practice), Zora Arkus-Duntov, and Achille Varzi. The Novi shot to the front and led the race before mechanical gremlins hit, and when the car stalled at lap 121, its race was done.

The Novis returned in 1947, Cliff Bergere putting one in second on the grid, Herb Artinger in fourth. On race day, Bergere led handily until stopping for tires, eventually dropping out with a failed piston after fighting his way back into contention. Artinger handed over to Bergere in the second Novi, who then finished fourth, but the cars had clearly shown they had speed. Starting at the back was the next great driver of the Novi era, as Duke Nalon ran the 1947 Indianapolis 500 with a Grand Prix Mercedes. Patrick Nolan, Duke's son, says:

> A lot of things in Dad's career was being in the right place at the right time. When he first started he was stooging for a couple of guys. Somebody didn't show up one night and the guys he'd been helping told the owner that 'this Nalon kid goes really good out west.' Nalon had never been 'out west,' but they convinced the car owner to put him in the Midget. He spun in his heat race, but came back to win the feature, his first race.

In 1948, when Cliff Bergere walked off the team after struggling with the heavily modified Novi, Ralph Hepburn returned to drive it again at the age of 52. The legendary racer on two wheels and the one person who could get the maximum from any Novi was back. Shockingly, Hepburn was killed in practice. With Bergere having walked away and Hepburn dead, most considered the 1948 Novi to be a jinxed car. Patrick Nolan tells the backstory as to how Duke Nalon took the Novi and turned the fastest lap in qualifying:

> When Dad got the opportunity to drive one of the Novis, he found out pretty quickly that you didn't need all of the accelerator. He would hold his fingers about 3/8" apart and say that there was that much of the throttle that he didn't even need. Of course the tires were narrow and very hard, and certainly not up to today's standards. Once the tires did evolve for Indy cars, he would always say, 'Give me a set of today's biscuits [tires], put 'em on the Novi, and I'll show you the quick way around this place.' Some of the guys that tried to drive that car found out the hard way that you couldn't drive that front-drive Novi like the normal Indycar of its time. There weren't a lot of takers from what I understood. In his first few laps in the car, Dad spun it and got it down into the infield. That was enough to get his attention.

After being a contender for the lead all race long, he finished third after a late-race pit stop, the result of not getting a full fuel load during an earlier pit stop in the race. Starting in 1949, Novi brought dominance to the Speedway, with the two cars of Duke Nalon and Rex Mays starting side by side on the front row, having qualified 1-2. Patrick Nalon continues:

> With today's racing there are team orders, and there were team orders that day in 1949 too. Dad was on the pole and Rex Mays was in the middle of the front row. The owner of the Novis said right before they went out, 'Whoever comes across the start-finish line ahead at the end of the first lap, that's how you are going to run all day. You will run 1-2, there will not be any racing, as we've got the cars to beat.' Well, Dad's car had a different gear in it, it was a little lazier than Rex's car, so Rex had a better gear for starting. Going down the back straightaway headed towards the green, Dad feigned making the move into high gear and Rex saw Dad do that. They came down the front

straightaway and actually that start should have been pulled back as Dad was just gone. Coming out of two and going down the back straightaway on lap 1, he finally shifted into high gear. I don't remember the numbers, but I know that when they did the math on what that thing was turning for rpm in that gear at that speed, it was lucky it stayed together right there. After coming so close in 1948 with the fuel stop, he was going to be in front all day.

Duke Nalon was not interested in finishing second at Indianapolis. Patrick Nalon says, "He was in front, came in and got fueled up, went back out, and then the rear axle broke." The broken axle put him backwards into the wall in turn 3, resulting in a fiery crash that made every newsreel across the country, one from which Nalon was lucky to escape with his life. Patrick recounts the harrowing event:

He said that he was lucky that he wasn't knocked unconscious. He hit the wall and there was a big ball of flame. People want to send me pictures, and quite frankly I've seen enough of it. I'd seen the car after the blaze, and thank God he lived. He bailed over the wall and crawled over to a yellow shirt [track worker] who was sitting there eating lunch. Dad said 'can you help me, can you get me help.' The guy looked at him, vomited, and passed out. Dad was one of the few that wore a cotton uniform, made by Hinchman here in Indianapolis. Dad wanted to look a little better, but everything was burned badly except for around where his name was embroidered onto the uniform.

By lap 50, the sister car of Mays was out with a broken magneto. Never again would the Novis show such dominance. Patrick says, "Later on that night Rex Mays came to the hospital to see how Dad was doing. Rex said, 'Duke, there was no way you could have outdid me that way going that quick.' Dad told him 'Well I wasn't in high gear.' Rex said, 'Duke, you must be on a lot of meds, as I saw you go into high gear.' Dad said, 'No, you think you saw me put it into high gear.'"

1950 was a fiasco of broken cars and neither car (Duke Nalon and Chet Miller) qualified. Patrick continues, "He had a good year of rehab, came back in 1951 and put it on the pole again. He still had that drive. In 1983, it was the second year he drove the Buick Indy Pace Car, and

The Novi of the late 1940/50s later benefitted from the twin magneto from the W196 Mercedes (Norm DeWitt)

when he went out to take a few test laps he asked if I wanted to go. He was as smooth as smooth could be, even at an advanced age he still had that feel for it." When Nalon put the Novi on the pole again, Miller qualified at nearly the same pace but started twenty-eighth. Both cars failed in the race, again.

1952 saw Chet Miller setting the fastest speed, fully 1 mph faster than pole-sitting Agabashian in the Cummins Diesel. Miller's time being set on the last day of qualifying meant he lined up thirtieth, just ahead of Alberto Ascari's Ferrari in thirty-first. Both Novis retired with supercharger failure. Sadly, Chet Miller was to lose his life in the Novi during practice for the Indianapolis 500 in 1953, a race where Nalon finished eleventh. On and on, the cars had always been blazingly fast, and equally snake-bit when it came to results.

Due to their legendary speed and distinctive sound, there are few cars more iconic and fondly remembered than the Novis. Tom Malloy has had a lifetime fascination with the Novi, fueled by having seen the cars back in their day. "My dad's car ran at the Speedway in '54 and '55, he brought me to the race in 1954 when I was 15 years old. I couldn't believe I was at the Indianapolis Motor Speedway. The Novis were big at the time, Troy Ruttman took one of the cars out on the track for testing and I was right there. Ruttman drove for my dad, we knew each other and he was only eight or nine years older than I was. I was mesmerized by the car and that engine."

Appropriate that one of the Novi chassis would find its way to Tom

Malloy later in life, and it was the first one to be raced at Indianapolis. It had a bizarre history as yet another big idea combining Preston Tucker and Harry Miller (see the rear-engine Miller saga in *Making it FASTER*), this venture being with Edsel Ford. Tom explains:

> A chassis was up for sale at the McPherson sale here in California about six or seven years ago. They were like hens teeth to find and I was going to own one, so I bought the car. That chassis is one of the ten Miller-Fords that Harry Miller built for Ford to run the 1935 race. Only four of the ten cars made the race and all four of them dropped out as the steering gear box was too close to the exhaust manifold and the bushings all froze up. Henry Ford was so upset that he put them all away after the 1935 race, embarrassed by all the money he had spent. In 1937 Lew Welch had built parts for Ford and they were very good friends. Lew talked him into taking one out of storage and he put an Offenhauser engine in it, running it at Indy in 1937, 1938, and 1939. When they took out the flathead Ford he had to rebuild all the gears in the front of it for the Offenhauser. They were also building the Winfield V-8 so they took that car and put that V-8 into it for the 1941 race. They had to rebuild all the gears again as the original gearbox was built for 80 hp and the V-8 had 400. By 1941 they didn't have riding mechanics, so they put the oil tank where the riding mechanic previously sat next to the driver. They had to make a different cowl, side panels, and hood to accommodate the width of the Winfield V-8. It qualified and finished fourth, and then the races were over because of the war.

Soon after Tom had acquired this original car, opportunity knocked again:

> It's been four or five years ago when somebody called me and said, 'Are you aware that the balance of the Novi engines and components are for sale?' He gave me the information that was on some website. The City of Novi had put it up for sale and they had a buyer that was going to give them $6,900. Somebody got ahold of them and said, 'Do you know what you have there? Oh my gosh, you can't do that.' So they yanked it off and put it

on this website. I had my mechanic and grandsons stay on top of the auction. We started bidding on it and eventually ended up with the whole deal. It was a lot more money than the $6,900 that they were about ready to accept, but for a hell of a lot less than you would pay for this. It was a semi-truck load of stuff. Five blocks and crankshafts, pistons, the original stand they used when they built the engine. You look in the Novi books and there is this stand they used to put the engine together. The Ferguson four-wheel-drive car that Unser drove, we got one of the transfer cases in this conglomeration of pieces and parts. I think we have the molds for the uprights on that car. I bought all the patterns, there were four boxes of those. It is my understanding that they had a car that they didn't want to get rid of, but they were willing to let go of all these miscellaneous original pieces. When this stuff came up for sale I had acquired Ed Pink Racing Engines, and if I hadn't had that, I probably would have let it go. I had the vehicle to put something together, and I had the facilities.

Frank Honsowetz, formerly of Nissan/Datsun Competition, is currently an engine guru at Ed Pink racing engines. He was tasked by Malloy to build the Novi engines for vintage racing. These include the later twin-spark-plug/cylinder versions as well as single-plug versions. He says:

One engine is in a car, one in on an original stand and we are completing another one. His car is technically the very first one, technically a Winfield and in a 1935 Ford. That's the only Novi he has that is running. Then he has a Novi engine on the original stand and that engine is configured for a 1958 car. All the engines are 1950s through 1964. Most of the stuff we have comes from when Granatelli folded up the deal and probably for the sake of taxes he gave it to a museum in Novi for their motorsport hall of fame.

The heads are integral with the cylinders. It's a Leo Goosen engine, essentially a V-8 Offy or Miller, kind of an early Offy. The block and head are cast iron integral. It's got bolt-on cam carriers, all sorts of stuff like the Offy has. It is interesting that there is an aluminum case that holds the crank, and we also do have a magnesium case [likely the final variant]. The big side cover plate on the crankshaft is cast iron. The case is what holds the crankshaft and that bolts to the block, a separate casting. It's the same way on the Offy – surrounding the cranskshaft there are big windows in the side of the case. When you put it together you take the cover off, and you put the piston in the cast iron block assembly from the bottom and lower it down onto the case, pulling it down onto the crankshaft. Then you go in through the side windows to torque up the rods, then put the cover back on. It's a nice engine, but I wouldn't be surprised if the engine weighed 700-800 pounds. It's a pain in the ass to work in, but you have some access. When you put the crankshaft into an Offy case, you put the crank on its nose and lower the crankshaft in from the rear, so that the flywheel end of the engine is pointed at the sky. All the Goosen stuff is like that, same with the Novi.

The power numbers on a Winfield/Novi V-8 are jaw dropping. In 1946 it was supposedly putting out 510 hp. Frank continues:

I believe that, the parts are huge. In the '50s there are photos of the front-wheel drive Novi cars with blue smoke boiling off the front tires on the straights. There are some guys that couldn't hang on to it because of the power and the front-wheel drive. The speed of the supercharger is probably the most interesting thing… the big supercharger on the front. The impellor in the supercharger is 10-inch diameter and runs through compound gears at the back of the engine and a shaft runs the length of the Vee, and there are compound gears at the front. The supercharger runs at four times engine speed, although there are probably different gear ratios. If they really ran 8,000 rpm in those things, now you've got 32,000 to 36,000 rpm supercharger speed. When you take an impellor that is a 10-inch diameter impellor, the tips are going way supersonic which is why they are famous for the noise. Anybody you've ever talked to that heard the noise at full supercharger speed, they are always talking about the noise.

The efficiency for a fan falls off radically when it goes into supersonic tip speed, so it's not at all a performance benefit in making that distinctive sound, but one supposes that the Novi would be a footnote to history versus an Indianapolis legend without that wail. One also has to wonder why, given the team members involved with the initial Winfield V-8 design, that it wouldn't be essentially a 4-valve per cylinder supercharged Offy V-8? Frank ponders:

> I have no idea why they did a 4-valve Offy and a 2-valve Novi. As far as I know the Novis only ran two events outside of the Speedway. They ran Langhorne and they ran an oval in Europe [The Race of 2 Worlds, Tony Bettenhausen setting a track record over 177 mph at Monza in 1957]. The story is that they went to twin sparkplugs because at the race in Europe they met the Bosch guys who told them to use twin plugs and use the magneto out of the W196 Mercedes Grand Prix car [a twin-plug straight 8 design]. That's where the mag came from that was in the twin-plug Novi, a mag setup for sixteen sparkplugs.

This story dovetails with Andy Granatelli's tale of how the twin magneto cars of 1959-60 had done so poorly at the Brickyard, as the mags were preset by Bosch to provide the required 30 degrees advance, and the crews unwittingly then installed them at the 30 degrees BTDC that the previous Novi single-plug magnetos required. At speed the resultant 60 degrees of advance had destroyed both parts and performance.

Tom Malloy's 1941 Bowes Seal Fast Special of Ralph Hepburn, Winfield Supercharged V-8. (Norm DeWitt)

There were also different fuel systems across the Novi era. Frank explains, "From 1961 on [the Granatelli years] they are fuel injected with Hilborn stuff. The one we did for 1958 has a variable venturi Holley aircraft carburetor. There's a big black square box in front of the supercharger, it sucks the air through the carburetor. Tom's [Winfield V-8 1941] car has three carburetors that are under the cowl and the mixture sucks through the carburetors for the supercharger."

Fragility does not begin to describe what having a Novi/Winfield V-8 engine does to a drivetrain. Frank says, "The car that Malloy has, the 1935 Ford front-wheel drive car, has the transmission behind the differential. He has to baby it unbelievably in first gear. The diff is first and then the axle housing contains the transmission, so it's a weird deal. It's not any good for the Novi. He's broken transmission gears in front of the shop just trying to load it."

There is still another car pending installation of a Novi engine, a 1958 roadster that Tom Malloy also owns. Tom recalls, "I bought a fake Novi with a Maserati engine in it. I was going to use that and take this fake chassis and put a real Novi in it, putting the [twin-plug] engine that is on the stand into it. A fake car with a Novi engine in it is better than a fake car with a Maserati engine. The engines I have would fit in that car." Frank agrees:

> He has what he believed was a 1958 car that ran the Speedway, but it turns out that it's not a real car… it's a real body, but not a real car from what he thought it was. It's like the case of the twenty-five Nissan GTPs when they built seven of them. We built this engine and configured the outside systems for '58. We had several blocks from the 1961-64 era where they changed the features around the spark plug. It was the same twin plug deal but the hole around the spark plug is a D shape instead of a round circle. The spark plugs are counter-bored into the top of the block. When they went to the twin plug, from then until '60 it's a round counterbore. In approximately '61 they went to a D shape with increased material between the two counterbores.

As with most racing machinery, there was constant revision and evolution, but just knowing that there are potentially more Winfield/Novi in the pipeline is very good news indeed.

Tom says, "I also know where there is a real Novi car, but I can't seem to get them talked into getting rid of it. The two engines I have would fit right into that car. I went over and I opened up the hood... it's a wooden replica of a Novi engine, but everything else is there, original just like they rolled it off the track. The Speedway also has one of the original ones that was built in 1946, and they've got the same thing... they've got a wooden replica engine in their car as well."

It's absolutely amazing that the owners of these original Novi chassis weren't all over that engine sale. Tom agrees, "That's the thing and you wonder about it sometimes. I've got a half a dozen crankshafts, some are cracked so maybe I'll clean them up and make lamps out of them?" Stranger things have happened, as at Kenny Roberts' ranch, he has the crankshaft from a V-5 Moto GP engine configured into a toilet paper holder. Hopefully the remaining unusable Novi bits find a reasonable final use. Cutaway engine anyone?

The later Novi efforts of the 1950s were notable, but without significant results to show for it. The new Kurtis rear-wheel drive Novis debuted at the brickyard in 1956, where Paul Russo battled for the lead until a blown tire and subsequent fire led to the car only being classified thirty-third, in last place. The following year Russo was to give the Novi one of its best finishes in years, taking fourth in 1957, ahead of Tony Bettenhausen's Novi in fifteenth, which finished five laps down. The Novis continued to qualify through 1958, with thirty-third starting Bill Cheesbourg finishing tenth, and teammate Russo a DNF in eighteenth place, but it appeared to be the end of the trail. In 1959 and 1960 every qualifier was powered by an Offenhauser and the race was dominated by either the lightweight Watson offset chassis (and the clones), or the Epperly laydown cars that had won in 1957-58.

Eventually the Granatelli brothers bought the remains from the previous Novi efforts and armed with their obsolete Kurtis chassis, returned to the Indianapolis 500 race in 1961. The 1956 rear-drive Kurtis chassis Novis were far heavier than the Watson and the clones. Andy Granatelli's engineering mentor was Ed Winfield, so the effort had a personal aspect as well.

For the Granatellis, 1961 is best described as "musical chairs," as the exhausted team with their old Kurtis-Novi had no driver signed. In the end, Granatelli withdrew the Novi entry. The following year brought more disappointment as neither of the brand new Kurtis-Novi model KK500K cars qualified, although Jim "Herk" Hurtubise showed that the

Novi still had the speed to compete at the front with a few blistering laps. In 1963 Herk was to prove that it was also capable of leading the race.

Bobby Unser's Kurtis-Novi. Jim (Herk) Hurtubise led the
1963 Indianapolis 500 with the sister car (Norm DeWitt)

Bobby Unser is one of the famous Unsers of Pikes Peak International Hill Climb fame. His resume includes three wins in the Indianapolis 500, but it all started with a Novi. Bobby recalls:

1963 was my first year to go to Indianapolis as a racing driver. The whole car was heavy and everybody knew that. The engine was heavy and the car was heavy, there was nothing modern about that. With the Kurtis cars in the beginning, I was spinning the tires on the straightaway coming off the number 4 and 2 turns. I could come around and see my tire marks from the previous lap. That son-of-a-gun… you can't do that without a ton of freaking horsepower. One day I was going down the front straightaway. I'd changed the weight jacker in it and broke the goddamn rear tires loose at almost 200 mph. It was freaking powerful and it was smooth power with 8 cylinders. It was a nice engine to run, but it didn't last. The engine was a big engine, it took up a lot of room in the engine compartment. I qualified fifth fastest with the yellow Novi, but I didn't qualify until the third day and they line you up on the grid behind the cars that qualified the first two days. I spun out on the second lap of the race and hit the wall. That car I drove in 1963 is in Al's museum [Al Unser] at the Unser Racing Museum.

Ryan Falconer is a legendary engine builder, but in the early 1960s, he was a mechanic with the Granatelli Novi team at Indianapolis. He recounts:

Bobby was a hell of a driver. In 1963 the only thing that killed us was that Bobby was driving faster than the car on the first lap, otherwise he'd have been fine. He just got too anxious or he'd have been cool. When you think that the engine was designed and built in 1941... it was really an archaic engine at the time, yet it still made over 840 horsepower. It was only 167 cubic inches against the Offys that were 270. The Offy was a 4-valve engine whereas the Novi was only a 2-valve engine, so there were a lot of things. It just didn't have the low-end torque because of the centrifugal supercharger, which has to get going. So it would come off the corners too slow, off-boost. Then when it comes on-boost, it was going too fast down the straightaway. Even when he drove the earlier Kurtis chassis, Bobby knew what he had. They weren't state of the art, but Kurtis built us two new chassis in 1962 and we went to the Speedway and didn't qualify... there was a reason for it but that's a long story. It was mainly because they didn't know what they were doing. I'm sure at the time the Kurtis was a lot cheaper than the Watson was. We went back in 1963 and qualified all three cars [Herk, Bobby, and Art Malone], and Malone drove the old car that was built in 1956.

Herk qualified in the middle of the front row and led the first lap after swapping the lead with a determined Parnelli Jones. Ironically Herk was disqualified for an oil leak, which was likely no worse than the one that was overlooked in Parnelli's leading Watson-Offy, later in the race. What was certainly a potential top-three finisher had to be parked.

Andy Granatelli knew that despite that success in 1963, he needed a more modern chassis to have any sort of reasonable chance of success in 1964. He took note of how impressed Stirling Moss had been when racing a four-wheel-drive Formula 1 car, the Ferguson P99. It was his only experience in a four-wheel-drive car and Stirling had first driven the car in the 1961 British Empire Trophy, a formula libre non-championship event. Moss exclaims, "It was incredible! I remember it very well when I took it over at Aintree. A Cooper was in the lead and

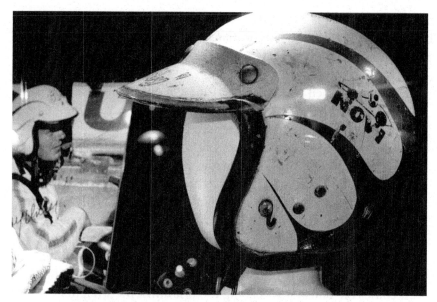

*Bobby Unser's helmet from the 1964 Ferguson P104 4wd Novi at
the Indianapolis 500 (Norm DeWitt)*

my car [a Rob Walker-entered Lotus] had broken. It was so fast compared to anything else in the rain. I got in the thing and I remember
passing Phil Hill on the outside of a corner and giving him the thumbs
up sign. I was catching up to the Cooper and John went to the organizers and said I shouldn't be allowed to race as I hadn't qualified the car.
It was the most satisfying car to drive and in the wet it was ridiculous."

Did the Ferguson take grip to another level? Moss says, "Absolutely.
I think it was the balance that comes with four-wheel drive. It had a 2.7
maybe, or a 2.5 engine. It had a couple of differentials in the thing and
it had the best balance I can remember, which was of course increased
with the wet. But it had it even in the dry. I know where the P99 is. Tony
Rolt bought that at a very realistic price and it's worth a bloody fortune
now."

Stirling went on to win the wet Oulton Park Gold Cup race with
the P99. Moss would have loved more drives in the Ferguson but the
new 1.5 litre engine formula put an end to the experiment given the
additional weight of the various hardware in the Ferguson P99. The car
went on to win the 1964 British Hillclimb Championship, showing the
benefits of 4wd on an entirely new stage.

After a promising test with the old P99, a new car was ordered to be
designed for the Novi engine, known as the Ferguson P104, and it was

a monster by any definition. Bobby Unser says, "The all-wheel drive [1964 car], you could have different amount of pressure on the front and rear. Torque split it was called, they didn't have the same amount of torque. The thing carried about 74 gallons, and multiply that out at about roughly 7 pounds a gallon. I used to see it on the scales and it was 2,500 pounds with very little fuel in it. Nobody would ever really know unless you talked to Big Vince or Little Vince, and they probably wouldn't tell you unless you got them half-drunk or something. The Granatellis kept that stuff really secret you see."

Ryan Falconer has said that car weighed almost 3,000 pounds. Unser agrees, "Full of fuel he's probably right. Andy used to hide that. You wouldn't believe how much time he spent hiding that. We knew it was heavy, but what were you going to do about it? You can't cut off half the car and throw it away. Andy would set that chassis up on the scales for setup, and he would always covered up the scales before you'd leave so nobody could read what they were." In Granatelli's book, *They Call Me Mister 500*, Andy goes into great detail about how the frame only weighed ninety-five pounds, undoubtedly a diversionary statement about the real weight issue of the car. Bobby Unser laughs, "That's Andy. I never read that book, but that totally tells you the way he operated."

Unser continues, "The cars were fast. When we got to the AWD car [1964], if we could have ever gotten the thing past loose oil lines, things breaking, and all the things that went wrong with the Novis, I had a chance to win the race with those suckers. It was fast enough to win the

The later twin plug Novi V-8 with Bosch twin magnetos was used throughout the Granatelli era (Norm DeWitt)

race. It had lots of horsepower, that engine really produced. It probably did have more than 800 horsepower, I would have no doubt to that in the world. The first AWD car I had was the Ferguson and it was really good."

Bobby Unser continues, "You couldn't do a one-stop race unless it had gasoline. The Novi never ran gasoline, it ran alcohol. Not methanol, alcohol... with gasoline you could make mileage. The Offys were putting out 500 hp, as soon as they went from the 270s to the 250s, they could get more RPM. The smaller engine made the Offys go faster, and they also got better gas mileage. But with the Novi, hell, we never worried about fuel mileage. If that son-of-a-bitch had run far enough to worry about the fuel, that would have been a nice thing. I have no idea what the gas mileage would have been. That guy just had a little trail of bad luck that followed him."

Ryan Falconer says, "In 1964, Carl Taylor was the head guy from Ferguson and he was just an obnoxious asshole. I couldn't deal with him and told him if he crossed a line in the garage I was out of there. So, he'd come up to it and put his feet up to it, but he didn't cross it." The best guess is that Taylor had paid heed to the stories about Falconer from the previous year. Ryan recounts, "In 1963, Andy tells me to go get another barrel of fuel. So I get back to the garage and it's about two-thirds full, we are running out of time so... no bulls**t... I grab it by the end and picked it up, a two-thirds-full 55-gallon drum, and put it on my shoulders and run all the way to the start/finish line. Parnelli tells the story that I've got a 55-gallon drum under each arm!" [laughs]

Falconer continues, "The Ferguson car was overweight as Andy had it built for all this horsepower and torque, so they built this giant transmission and differential that was way overkill, it was way too much. They just went berserk. The car was so goddamned heavy that the car was just laboring all the way around the track. How Bobby ran as quick as he did is amazing, as he was running 154 mph laps. They were trying to run those big-ass Dunlop tires that were just going to chunk on us. Dunlop was really overconfident and they hadn't that much experience at the Speedway. There is nothing wrong with having a little bit of experience around that place."

Considering that Jim Clark had to retire when the rear suspension failed on his Lotus due to chunking Dunlop tires, the Novi team wasn't alone in their tire struggles. Ryan had seen enough. "I quit after 1964 as I couldn't take any more of the bulls**t." In the end, Unser ended up involved with the front-stretch nightmare crash of 1964 that took the lives of both Dave MacDonald and Eddie Sachs.

For 1965, there was a lighter version of the Ferguson Novi, although it was still using the iron block Novi powerplant. Unser ponders:

Well, I don't think that would have made enough difference. They built a new car for me in 1965 and it was faster. It was quite a bit lighter, but by how much how in the hell would you know how much lighter if you were driving for Granatelli? It drove better and it handled like a race car instead of driving a choo-choo train. We didn't drive it very much because it got wrecked when Ebb Rose from Texas spun his car and cut across the track. He was going to hit the cement wall head on, but I hit him before he hit the wall, which saved his life. The problem is that it ruined my car and I had to drive the '64 model of the car in the race.

Bobby qualified eighth for the race with the older version of the Ferguson Novi, which makes one wonder what he could have done with the lightweight one, as none of the trick parts made it from the '65 car to the older chassis. Unser explains:

There was nothing changed. It was a day or two before qualifying, so there was no time. They thought I was hurt too bad in the wreck to be able to qualify the next day anyway, so they didn't do anything to get the old car ready for me, it was in the truck somewhere. But I had a little piece of paper called a contract. I told Andy, 'You remember that little piece of paper that we drew up? It says you are going to have a car for me in qualifying. Well, I wrecked the good car, so what are you going to do? I need a car.' So he worked that poor crew solid, they had to work all night and did a tremendous job to get it ready for me. But don't think that we changed over parts as there was no time for any redesigning. It was one hundred percent the same as the car from the year before. It was still a good car, but just too heavy. But that goddamn thing had so much power you just couldn't believe it. All the press decided, even with how little experience he had, that Bobby Unser had a fairly good chance of winning the race. If the car happened to keep running, that four-wheel drive was going to be way better than a two-wheel drive car when there was oil and rubber on the track. It gets really bad and that would have given me a big advantage.

The car made it about a third of the way through the race before it broke, the official cause of the retirement was an "oil fitting." Bobby says:

I don't remember what broke, but you've got to remember that Andy Granatelli said whatever he wanted to say, and that wasn't necessarily what went wrong. An 'oil fitting' could have been a broken rod in the engine. Who knows? But it's not likely that an oil fitting broke. With an Offenhauser, I'd believe anything. The Offys would shake the crap out of you and you would lose feeling in your hands. People would put hardwood on the steering wheel spokes, mainly with the 270s – the drivers would do anything to slow down the vibration. But a Novi V-8 wouldn't break an oil fitting as it was so smooth I could set a glass of water on the hood and it wouldn't spill.

Ryan Falconer agrees, "Now I could surpass what the Novi did easier than s**t, but it was amazing in its time. When you think about it, we made 840 horsepower with 147 inches of boost." Regardless of what put the car out, 1965 marked the end of the road for the Novi as a contender at Indianapolis, although Greg Weld did attempt to practice the STP Novi in 1966, wrecking it before qualifying.

Patrick Nalon says, "When we lived in Phoenix, Andy Granatelli paid for us to come meet with him in California to discuss putting the Novi in a rear-engine chassis. He wanted to meet with Dad, but my mom and I were in the meeting. It was pretty close to happening, but never came to fruition because the weight of the engine was too much. I'd heard that the first idea was that the ['67] turbine car was going to be a Novi, he couldn't let loose of that." The idea was abandoned and Granatelli turned his attention to building the revolutionary STP turbine car that Parnelli Jones drove to nearly win the 1967 Indianapolis 500. (For the story of that car, see *Making it FASTER.*)

An author's apology: In 1968, I had approached Vince Granatelli in the STP garage at Riverside where they were working on their Lotus 56 Turbine cars. Asking for an autograph from Vince by name, he was clearly pleased as Andy's brothers had worked in relative obscurity

Andy Granatelli slowly moving along the grid shaking hands with everyone
2012 Indianapolis 500 (Norm DeWitt)

compared to their showman sibling. When Andy Granatelli walked up, I wheeled around to him, handing Andy my pen and autograph book. I could see the disappointment in Vince's face. Sorry Vince, I was just a dumb kid being a knob.

Patrick Nalon continues, "The Novis were unbelievable cars. I got to drive the Novi that Ralph and Chet got killed in, at the Speedway and at Goodwood. Even at low speed it was a pretty impressive race car. I work in the IMS museum every once in a while, and people who know

33

racing usually want to see three race cars. One is the 1911 Marmon Wasp, one is the 1967 Parnelli Jones turbine car, or a wedge turbine car… that's fine too. The third is the Novi, a car that never won the race. They were always known for their brute horsepower and the sound. The Novi that is in the museum, that is missing engine parts, is the number 54 car, the car that Dad drove."

It speaks volumes when you arrive at the museum, there is a limestone Novi in front of the building. Nalon explains, "The people who made that were from Bedford, Indiana, the limestone capital of the world, and they were huge Duke Nalon and Novi fans. Dad actually saw that before he passed away, and he told them, 'This would be really good for my headstone.' They donated it to the Speedway who then decided to put it right out front. It's pretty cool to have it there."

The Limestone Novi of Duke Nalon at the entrance to the Indianapolis Motor Speedway Museum (Norm DeWitt)

*Duke Nalon on pole – 1951 Indianapolis 500. The Kurtis-Kraft Novi
was a spaceship vs the competition (Patrick Nalon)*

*Duke Nalon on the grid before the 1951 Indianapolis 500,
as the mechanics make last minute adjustments (Patrick Nalon)*

The 1963 World Champion, Jim Clark, was showing most everybody his heels at Indianapolis in the Lotus by 1964-65. Stirling Moss says, "Of course he was such a bloody good driver as well." Which leads us back to Europe, away from the era of front engine Grand Prix cars and Indianapolis Roadsters, to the saga of the team that is synonymous with Jim Clark, Team Lotus.

1964-65 Ferguson STP Novi of Bobby Unser on display in the City of Novi Public Library (April Stevenson)

CHAPTER 3

TEAM LOTUS IN THE PRE-JPS YEARS

Jim Clark's 1964 Indianapolis 500 leading Lotus 34
(Norm DeWitt)

In 1958, during the final season for Colin Chapman's Vanwall chassis, Colin had entered Formula 1 with a tiny front-engine car, the first Grand Prix car to carry the name "Lotus," the Model 12. This was the design that got Team Lotus their first points, as Cliff Allison finished sixth at both Monaco and Zandvoort. This was followed by fourth at Spa, about a minute behind third-place Stuart Lewis-Evans' Vanwall. That first Formula 1 Lotus was soon replaced by the much sleeker Lotus 16 midseason that teammate Graham Hill used through the end of 1959, taking a sixth at Monza. That the Lotus 16 resembled a three-quarter-scale Vanwall was unmistakable.

By 1960, Lotus had switched to the rear-engine layout, with Stirling Moss winning Lotus' first Formula 1 victory while driving the Rob Walker Lotus 18 at Monaco. The following race at Zandvoort saw the first start for Jim Clark at Team Lotus, a combination that in future years would become legendary.

Dick Scammell had joined Team Lotus as a mechanic for the 1960 season, and advanced to being assigned to Jim Clark's car for the 1962 season. The Lotus 25 monocoque car changed the Formula 1 game forever when it was introduced at Zandvoort for the season opening Dutch Grand Prix in 1962. Undoubtedly there was a jaw-dropping paddock reaction something akin to first seeing a picture of the six-wheel Tyrrell in 1976. Dick Scammell says, "I built the monocoque. It was completely different and we tried to keep it a secret as long as possible. People knew we were up to something, but they weren't quite sure what, as everyone was banned from the racing shop. It was a rather nice surprise from everybody except for the customers who had bought their cars from the other side of the business. There had been somebody back in the 30s who had built a monocoque car, so it wasn't the first ever and there were airplanes built like it. Colin being Colin he thought, 'Ah! There is something in that, isn't there?'"

This was the beginning of the aeronautical engineer trade in Formula 1, as when Gian Paolo Dallara had started as an aerodynamicist at Ferrari in 1959. Scammell recalls, "People were beginning to catch on. Let's face it, Colin had his rules that weight was a terrible thing, which it is. You have to accelerate it, you have to stop it, and you have to get it around the corner. Frontal area was not a good thing to have, hence he started laying drivers down, which they didn't like."

The 1.5 litre cars didn't have much power, so one could see how that formula would lead to efficiency above all, in aero and everything else.

It was the perfect sandbox for an innovator like Chapman, and his trademark obsession with weight was on full display.

Dario Franchitti driving Clark's 1965 winning Lotus 38 in pre-race festivities, 2016 Indianapolis 500 (Norm DeWitt)

Having been a garage mechanic in Salisbury, David "Beaky" Sims was hired by Team Lotus at the end of 1965, which had to be one of the most coveted jobs in racing given their dominance that season in both Formula 1 and the Indianapolis 500. As Beaky explains:

They were North London then, in Chesnut. Basically it was just a garage with a couple of lathes, some stands, and wooden benches. The transporter was outside. The first day I was there they started up a Ford V-8 Indycar on stands and as a youngster you thought, 'this is the business.' They accepted me. Bob Dance is still around, still in Norrich. Dick Scammell was there then too. Bob was the guy who nicknamed me 'Beaky.' He said, 'That kid that just joined us has got a hell of a nose on him, a hell of beak,' so I've been 'Beaky' ever since. It is what it is. Colin Chapman, Dick Scammell, Jim Engelrod, they put me on the Cortinas for the British Saloon Car Championship for Peter Arundell and Jimmy Clark. It had a Lotus engine sent to BRM, it was specifically built by BRM for Lotus only, and it was a piece of work. It was a totally different engine from the Rudd

Big Valve, anybody who has got one of those BRMs has really got something.

1965 marked the beginning of the 16-cylinder adventures, or perhaps more accurately put, misadventures. There was the Lotus 39 built for the flat-16 Coventry climax engine that never ran in the Lotus chassis. Dick says, "The engine never really arrived in any form that we could use." That car ended up with a V-8 in it and ran in the Tasman Series.

For 1966's 3 litre Formula 1, there emerged the Lotus 43 designed to take the BRM 16-cylinder engine. Nobody really knows what was in the water of England at that time, but one supposes there were 16 cylinders of it. Dick continues, "Well, it was really that the car constructors had little to do with it as they had to take the engines that they could get, Coventry Climax and so on. BRM was saying how this engine version they were developing was wonderful and that they were also going to do an Indy version of it as well, and it was a disaster."

The concept of the H-16 was taking advantage of the existing 1.5 litre V-8 BRM technology and parts by stacking two flat 8s. Of course the initial concept was to have shared camshafts, but the more compact version ran into crankshaft interference problems and they ended up with essentially two separate flat 8s sitting on top of each other. Somehow the engineering staff at BRM must either have not realized what that would do to the center of gravity, or had begun to actually believe their optimistic power output projections.

Beaky recounts, "The H-16 was being developed and at the same time as in Formula 1, they were also making one for Indycar. You could never keep it running, it used to grenade itself on the brake at BRM all the time and in the end it was a no go. If you've ever seen the inside of the Formula 1 H-16 BRM, and you see the engineering, the gears, the gearing, the crankshafts, the camshafts, it is phenomenal engineering. It is just unbelievable."

There were tests at Snetterton for the Indianapolis 4.2 litre H-16 and "disaster" really doesn't begin to describe the experience. Scammell recalls:

I was in charge of building the car. It arrived late and we couldn't start it to start with. When we did get it started, I rang Tony Rudd and told him that I thought we'd wrecked it because the

*Jim Clark at Monza with the H-16 powered Lotus 43 in its first race,
the 1966 Italian Grand Prix (Bernard Cahier)*

oil pressure was so low. He asked, 'What is the oil pressure?'
I told him it was at about 15 psi, and he said, 'Oh, you've got
a good one' [laughs]. We then went and tried to run it. I was
there to drive the car for the first three outings, because I was
just making sure that it changed gear and the clutch worked and
all that sort of thing before the drivers showed up. I refused to
drive it after that because I was getting a bad name. The engine
always failed while I was driving it. Anyway, the fourth time we
took it, it failed as well. It was a complete drama from beginning
to end. They never really produced the Indy engine at all, in any
shape or form.

The H-16 was a DNF at Spa, then Clark drove it at Monza and it
broke again. There were zero expectations for the H-16 by that point
late in 1966. Actually, there weren't any expectations once the Lotus
team first took delivery of the engine and realized just what they had
signed up for. As Dick tells it:

We didn't. By the time it arrived late and once we saw it, I had
the engine and gearbox weighed, which were all BRM parts as
we provided the chassis. We put the scales with its mounting

and exhaust system on it and we had 50 pounds left for the rest of the car to the minimum weight limit. It was hugely overweight. It used to catch fire. The gears weren't good enough as teeth came off the crank gear that put together the two crankshafts. They would explode out the top of the engine and run oil down the driver's neck. I always tell everybody that it was my worst year in motor racing. Of course they decided to make the standard Formula 1 H-16 even more complicated by doing four valves per cylinder [a much later version that arrived in 1968]. Because the little intakes were horizontal, if the fuel ran back into the air-box – which it usually did because, in fact, it injected fuel against the slides – of course the whole lot caught alight. We had many occasions where the car was alight in the pit lane. You couldn't start it often without flooding the barrels. Strangely enough there is a chap over here who built an H-16 BRM car. It is beautiful, I've seen it and seen it run. It is absolutely a fantastic job and a few things have been altered to make it more sensible. I helped a bit with that, and he rang me up one evening and said, 'We just started it, but it caught fire.' I said, 'You've got a proper one then, haven't you,' as they all used to catch fire. The design of the horizontal trumpets, fuel was bound to run out of them. The ignition system was rubbish and the clutch was rubbish on it. They did in fact put the flywheel on the layshaft of the gearbox, and put the clutch plate on the end of the crank, which then meant you couldn't change gear. It was a complete and utter disaster, but the one marvelous thing we did with it was to win the Watkins Glen Grand Prix.

Watkins Glen was one of the largest purses in racing and could go a long way to pay the bills. In 1966, Lotus managed a win there with the Lotus 43 using a heavily cobbled up H-16, which had previously blown up and was patched back together with bits of metal plates. So ironic that of all things, this would be the engine that got to the finish line. Dick agrees, "Exactly, sometimes you make some of your luck and sometimes you get some luck. Overnight on the Saturday night we put another battery in parallel as they were very difficult to start and the car had to start on the starter. Fortunately we did that, Jimmy drove it brilliantly and he won the Watkins Glen Grand Prix. The Prize Money – mechanics used to get 10% of the prize money. It didn't work quite that way as there was no starting money, so that start money was taken out first, but it was still a very nice payday for the mechanics. We always wanted to win that one."

Everybody talks about the subsequent Lotus 49, but the 43 was an innovative Lotus design that had the engine as a stressed member in the chassis the previous year. Dick says, "On the chassis end of it, for sure. We didn't have any problem with the structure of the car and the engine, it was just that the engine was rubbish."

The Cosworth DFV powered Lotus 49 won in its first race at the 1967 Dutch Grand Prix (Norm DeWitt)

It was a learning curve as sometimes the suspension failed. Such as at the Nurburgring during the German Grand Prix of 1967. Dick recalls, "The rear suspension mountings were mounted to the gearbox and it wasn't a very wonderful design. It had some very short bolts in it and it was very difficult to get it done up, and we didn't on that occasion. We lost one of the bolts out of the suspension. One of the glorious benefits of hindsight of course, it was not a very good design and it got altered."

The Lotus 49 also came apart when Clark won at Watkins Glen with a car that resembled a banana, which somehow Jimmy nursed to the finish. Dick reminisces, "One year at Watkins Glen, Jimmy drove the last two laps very carefully with the rear suspension collapsed on one side, but being Mr. Clark, he managed to get it around the corners when the top link was off, by being clever – by being Jimmy Clark. He was a complete and utter natural. You could see the difference between his gearboxes and other people's gearboxes after the race. He nicely put it into gear where other people got it into gear."

Beaky was moving up within the Lotus team and had been assigned to the Formula 2 campaign. He recounts:

After the Cortinas they put me on F2, in mid-1967 they put me onto Jimmy's car. It was Jimmy and Graham. By then we were at Norrich, in a new factory adjacent to the production factory where they built the Lotus Elan. It was a nice workshop, beautifully done. It was just me and another guy, Mike Gregory was on Graham's car, and that was it – only two of us with one of us on each car. We loaded the Lotus van and drove it, just the two of us, all the way to Barcelona. The first race [of 1968] was Barcelona, Jacky Ickx in the Dino put in new brake pads on the grid. First corner – no brakes – and Jimmy was out. So that was that, and Graham's engine blew, so that night we loaded up and headed for Hockenheim. We drove all night, all day, got there to the Hotel Luxof on the River Rhine just inside Hockenheim. So we started an engine change on Graham's and he helped me do the repairs on Jimmy's car.

It was freezing cold, minus 5 degrees, and the metering units would seize up. You'd have to boil up a kettle and pour it all over the metering unit just to stop it from seizing the bearings. It was terrible, and of course Jimmy crashed in the race. Jimmy and Chris Amon were on the same row, about the fourth row. Amon and Jimmy said, 'Well, don't expect anything from us.' In the damp conditions you couldn't get temperature in the tires at all, nothing. We tried everything with shockers and pressures, in those days technology wasn't what it is now. It was instant deflation of the right rear Firestone tire that caused the accident. Jim Clark was a guy who never crashed, he never spun off. It was horrible.

Clark was killed as his Lotus flew into the forest at high speed, tearing the car to shreds. Paul Butler was the Dunlop tire representative for Formula 2, (later to become Moto GP series Race Director), and was there at Hockenheim that sad day. Paul remembers:

I still have got the newspaper somewhere. It was really extraordinary as the whole place went completely silent – not just in the pit lane but everywhere – when the news got out. It was literally chilling, because it was unbelievable. It was a race that he didn't need to be doing. It was combined with the World Sidecar Championship and alongside the F2 there were these sidecars. It was the first time I'd ever seen sidecar racing with all these blokes that looked like they'd come up out of the ground, all in their black leather. It was really quite macabre, and to lose the greatest racing driver of his era in that environment was chilling and an extraordinarily emotional occasion for everybody. It struck the entire event. It was dumbfounding, really.

Beaky reflects:

We had to get back [because] if a driver got killed in Germany it was manslaughter, a capital offense. So there was a little VW police car there with the transporter. Colin Chapman said, 'Get it back.' We told him, 'We can't, there's a police car there.' But Colin said, 'I don't care, get it back.' At about eleven o'clock at night, suddenly the police car went off and so we decided, 'Let's go.' We kept off the Autobahns and went all through the villages and eventually headed for Belgium, towards Spa, up over the Ardennes. We came to a border where it said 'Frontiere' and there was nobody there, it was up in a forest. So we lifted the barrier and headed into no-man's land. We didn't know where we were, and then we looked back and it said Belgique! We were in Belgium, so we made our way to Zeebruge, there is a ferry there. We asked if there was an open ticket for carriage there that we could have? 'Yeah, yeah, you have one hour. Please, we need your passports and for... Ah! Jimmy Clark! I want to see, take pictures.' We said no you can't and he said, 'Okay, no, you stay here, no boat, no ferry, no... sorry.' Christ, we had to let him in and open the door for him to take pictures of some bits and pieces, whatever. We got on the ferry and when we got to England the police were waiting for us and escorted us all the way to Lotus no problem and that was it, they were good.

When you are that young and you just lost the best driver in the world... You know how it was. And all the newspapers got on it: 'A mechanic left a wheel loose,' or 'A mechanic didn't do the engine,' all this stupid media stuff. So, I told Chapman, 'What do I do?' Colin said 'What do you mean, what do I do? What do you think you are going to do? Well, Graham's new chassis, the new 49B, has got to go to Jarama tomorrow. You are towing it with the van on the open trailer. Get it loaded, the boys will help you get it loaded.' I had never been to Jarama, so me and another kid drove all the way to Spain towing this Lotus with no cover on it, and thankfully it didn't rain. It was Graham's elongated chassis for this race. Because he was 6 feet tall it had changes in the pedals and the seat back area. I was in Formula 1 at Lotus until 1971.

It was during this time that Hughie Absalom of Lotus got to work on his favorite car. He reminisces:

The one that has a soft spot would have been the 56, the Turbine. The concept and achievement, it was just an amazing bit of kit. We were on Spence's car. But then Spence went down to Granatelli's and he never came back. We didn't know he'd even gone down there. We were changing gear ratios on his car and then the lights came on, and then we found out Mike had been driving their car and got killed in it. So we had to reboot from there. We had four turbines. Three for Team Lotus and Granatelli had one. Joe Leonard was in the other one [the old 67 Parnelli Jones car]. Pollard replaced Spence in the Team Lotus car, which became Graham Hill and Art Pollard.

We had two cars that had the upgraded fuel pump drive because the normal one for the engine in the helicopter had a shear drive, meaning if it did seize up, it would just break the shaft. The whole month, the whole time we were there, we had been running on whatever it was. And then the American Oil deal that Granatelli had stipulated, we had to run gasoline. On Carburetion Day they came along and said, 'Okay, this is what you are going to run.' We had the wrong shaft in the wrong car, really. I can't remember, but I think maybe two of our cars had them in, and somewhere along the line they took one of our cars – the spare after Mike smashed one up. There were originally two [upgraded] shafts out of the four cars.

Graham Hill's car was the only Lotus that crashed during the race and it carried one of the upgraded pump shafts. Neither Joe Leonard (leading the race), nor Art Pollard's cars had this upgrade, and both cars rolled to a stop with nine laps remaining.

Clive Chapman with the Lotus 56 Turbine cars in pre-race festivities for the 2014 Indianapolis 500 (Norm DeWitt)

The Formula 1 race in May at Monaco was the first track where the dive planes showed up on the nose of the Lotus 49, and the rear engine cover became a wing-like tray. Beaky explains, "Downforce. We had to make the whole thing. Then came the tall Lotus 49 wing, and where the bottom of the wing strut went into the upright, it broke off. Too much downforce versus the structure. It wasn't wind-tunnel tested or anything."

The wing failure described above was the result of the first attempt by Team Lotus at the French Grand Prix in 1968. Jackie Oliver's car had a massive crash following a wing failure. Dick says, "I think I was chief mechanic by that time. The wings were really difficult things, weren't they? They were so tall." The Lotus wings were far higher and larger than anything anyone else was running, when most were putting little wings on the engine covers. Dick continues:

That was Colin, wasn't it? That's why Colin was so successful, he didn't mess around. He [Colin] said, 'Right, if you've got a wing, where do you put it? You've got to put it out of the influence of the air over the car. It would be in clean air, so you've got to put it reasonably high. Where do you want the load? Well, you don't want it on the chassis as it just pushes the chassis down. You want it straight onto the tires, so you put the wing mountings on the uprights don't you?' Of course it's difficult because the wheels want to go up and down, it's quite a difficult job. When you take it to the sprung part of the chassis, the car goes down the higher the load gets. Therefore it's not really what you want all the time.

At that point one would have to wonder what the mindset was at Lotus, having lost Clark at Hockenheim, Spence at Indianapolis, and then nearly lost Oliver with the wing failure in France when the car was scattered the length of the straight. Beaky recalls, "It was busy. They were like fighter pilots in the Second World War. The Luftwaffe came over and the pilots scrambled. It was, 'Johnny didn't come back, oh well better get somebody else in.' It wasn't changed until people like Max Mosley and Ecclestone got on the program years later. When I was in Formula 1, eighteen people got killed from malfunctions, badly designed circuits, and badly designed cars."

The drivers were considerably more affected than most of them would admit at the time. David Hobbs remembers:

It happened about once a month, but you never ever thought the next one would be me. Jimmy Clark's death was a major shock for all of us. We all thought, 'S**t, if it can happen to him, I suppose it could happen to me.' I was at the BOAC 500 race, and when I was told I was out at Clearways watching Paul Hawkins who was driving the car [Hobbs-Hawkins Gulf Ford GT-40], I thought, 'You've got to be kidding.' It was pretty bad. I went to his funeral with Michael Spence and he and I were coming out just gutted, both of us a bit teary-eyed as there had been some good speeches. Then it seemed like the next week I was at Mike Spence's funeral. That got the missus' attention.

Late in 1968 was when a young American driver had his first experience racing for Lotus, eventually to become a World Champion at the wheel of their cars. Beaky recalls:

They took this guy on named Mario Andretti. He was driving a Lotus 49B in 1968, he didn't even touch the car and took pole position at Watkins Glen. He just got on with it and I thought, I like this guy... amazing. When Jimmy Clark came back from America after all his times at Indianapolis, and he said, 'There is a guy called Mario Andretti and his car control is incredible. If he ever goes into Formula 1, I guarantee you he will be World Champion.' Everybody was going, 'Right, he's a roundy roundy guy, he's never going to be any good.' Jimmy was right, he was unbelievable, and he's my hero. Dirt cars, anything – his car control was phenomenal. You've got to look at who is left: Dan Gurney, Jackie Stewart, Mario, John Surtees [RIP]. They are few and far between now.

By the end of 1968, the high-wing 49B was more reliable and Graham Hill won the World Championship in winning the season finale in Mexico City. Then they added a high front wing and larger nosedive planes as well for 1969. Beaky says, "It was a test experiment for Oulton Park because it was a non-championship race, and nobody seemed to know, 'Was it working or not?' It was a different chassis geometry with different front uprights. I remember the engineer couldn't balance the front to the rear. It seemed to be a question mark. We never really did figure it out. In the end we went back to the conventional 49 front wings, radiator exit in the nose, and a proper rear wing, and started to get a grip on it all."

Joe Leonard in the final version of the Lotus 56 turbine Indianapolis car in the 1968 Rex Mays 300 (Gary Hartman)

In 1969 that aerodynamic reliability with the wings went away with a rash of wing failures in Barcelona from an extended rear wing and front dive planes. Beaky explains:

The reason being was, again, Colin Chapman said we haven't enough downforce, the Ferrari is too quick. We've got to get more downforce on the rear. So we have to put 6 inches of rear wing on each end, adding another foot of rear wing, overlapping the rear wheels. [Jackie] Stewart came up and said, 'What are you guys doing? Jesus...' It's what we've got to do. Doing it without testing, no bloody testing... to do it in the race. At the Monjuich circuit the cars used to leave the ground going over the hump and we heard this 'bang' and the wing collapsed in the middle. The 49 was like a horseshoe and the marshals couldn't get him [Jochen Rindt] out of the car. The battery was fizzing away under his legs, still shorting. We rushed up there. Graham had crashed in the exact same place, wing gone – same wing failure, but not as bad. We said to Graham, 'What should we do with him? He's still semi-conscious.' Graham said, 'Get him out, get him out, it's going to catch fire!' Graham straddled the car with his two legs, and grabbed him by his lapels and pulled him out. Jochen didn't wear a full-face then, he had an open-face motorcycle helmet and it had split his skull and was bleeding quite a lot.

It was the last race for the high-wing Lotus 49, banned shortly thereafter mid-practice at Monaco. Later in the season Rindt was to finally achieve his first Grand Prix win at Watkins Glen. But a new car was afoot, one that was more similar to Maurice Philippe's Indianapolis Lotus 56 Turbine Car of 1968 than it was to the model 49, and that car was the Lotus 72.

Philippe's 72 set the design trend in Formula 1 for a decade. Dick Scammell agrees, "It really did, but the original design of the 72 was a bit of a disaster because it had anti-squat and anti-dive on it and the drivers said you couldn't feel the car, they didn't know what the car was doing. So when you put on the brakes the front of the car didn't go down and the drivers didn't like it. In the end we had to come back, throw the car away and design another chassis for it." The revised 72 discussed above was the car with which Jochen Rindt later dominated the 1970 season.

The 1970 Lotus 72 continued Maurice Philippe's design philosophy from the Lotus 56 of 1968 (Norm DeWitt)

The drivers wanted nothing to do with the car until revisions were completed. Dick explains, "Yep, originally they didn't want to drive it. The drivers just said it was horrible and that they couldn't feel the car at all and they would say, 'bring me a 49.' The last race win for the Lotus 49 was at Monaco 1970. At Monaco in 1970 they ran the Lotus 72 triple element rear wing on the 49.

Beaky says, "That was a definite improvement, the rear wing did work. It was a test piece, it was what Philippe wanted to test, and it was adjustable too, each wing plane could adjust. It was very successful."

There was a key issue with the car, and that was the hollow front brake shafts to the inboard front brakes. Beaky recounts:

> In Austria the brake shaft broke up the outer CV joint, which then broke the shaft splines and the shaft failed. There was a modification made to the brake shafts in Yarmouth, made of special material. Chapman insisted they were to be hollow, it was a hollow brake shaft. You couldn't just chuck it into the truck because if you scratched it, it would break. To this day it's never 100% proven that it was a brake shaft that failed in Italy. What people don't realize is that Chapman told us to take the wings off because Ickx was a half second quicker. It was the no-chicane Monza and it was quick. 'Take the mirrors off, take the front wings off, take the rear wings off,' but the car was designed with wings. Jochen came back in and he said 'Ahhhhh, scary but it's absolutely quicker, and we have to go up two gears in ratios.' We went up two gears on ratio and it was phenomenally quick. Jackie Stewart came up again and said 'Colin, what in the f**king hell are you doing? You know this is just f**king ridiculous. This has got to stop, all this bulls**t.' Chapman said, 'Beaky, John Miles, get them wings off, this is the way to go. John Miles took the wings off, came in after one lap white as a sheet, and said 'I'm not driving, that is impossible.' White as a sheet, he had scared himself.

> It was undriveable. How Jochen got that around in the times he did was just amazing. Denny Hulme said about five laps before the accident Jochen went past Denny into the Parabolica absolutely all over the place, flat out like Jochen would, and Denny thought 'Jesus, that's scary.' I probably might upset history, but we got to the crash, his foot was facing me, it was off. The 72 was a wedge, and it went under the Armco and twisted around, took the front end and mashed whatever was left – pedals, etc. – with it. The seat belt… he never wore a crotch strap and he went down and the buckle slit his throat as well when he went down underneath. They opened up his chest and put a big tube in, but you were going 'he's dead.' No way was it survivable. When you look at the crash, there was nothing left. There was no way anybody could say it was the front brakeshaft. It was like an aircraft crash. There was nothing.

The concept behind the hollow brake shaft was certainly that a hollow bar or tube can have almost the same torsional value as a solid bar, yet at a much reduced weight. In practice, it turned out to be a sketchy proposition. Herbie Blash concurs, "Correct, yeah it was. I was Jochen Rindt's mechanic. The problem in those days was obviously drilling. The idea was you had to come in drilling from both ends, and if it wasn't absolutely spot on then that was ready to crack. The only crack testing we used to do was just external. They used to twist as well, they had a lot of twist in them."

The Achilles heel of the Gold Leaf Lotus 72 was hollow brake shafts, later updated to solid shafts (Norm DeWitt)

The Rob Walker Lotus 72 team of Graham Hill went their own way and created their own solid shafts after inspecting the wreckage and made the following race at Canada, a race skipped by Team Lotus. Dick Scammell ponders, "Was this another time when the obsession with weight had tragic consequence? No, I don't think so. It was obviously very different and we had problems with the odd shaft. What actually happened with Jochen I don't think anybody knows, really. They all assume something broke because it just turned left. We got on with the shafts, and I approached it rather than just making it thicker would make it better, trying to make it the best we could with surface finishes and everything else on the shafts. It was trying to get rid of unsprung weight."

Inboard brakes had a bad rap forever as a result, and Amon refused to drive the inboard brake Ensign of 1975-76 until it was modified. Scammell recalls, "That's the trouble isn't it, they obviously have a very bad name. The drivers didn't like it, but if you do it right, it was the right answer. It was just a question of doing it well enough to making it work, then everybody loves it. It's like an unstable fighter isn't it?"

Once again the concept of Formula 1 driver as test pilot comes to mind. Scammell continues, "Yeah, and you are, because you want to make progress as fast as you possibly can, so you always make it to the next race meeting with something new on the car. We did it as safely as we could, and I think you have to say that. We never said, 'Whoa, let's see if we can get away with that.' We really thought we were going with something better and good, we never hung it out that way, but you didn't always get it right."

Two races later at Watkins Glen, Lotus returned with a factory modified version of the 72. Beaky describes:

Emerson Fittipaldi and our Swedish kid, Reine Wisell, we finished first and third. The car was good, still kept the torsion bar suspension, but there were modified front brake shafts, we made them solid. After that, things started to get fragile on the 72, they started breaking rear suspension pickup points. At Ontario [the Questor Grand Prix of 1971], the left rear wishbone front pickup point broke again, so I said to Chapman, 'Why don't we make it bigger bolts, why don't we do something?' Colin said, 'Do as you are told.' We had to fabricate and build the cars ourselves, and we had to weigh every nut and bolt and washer that you put on the car with a grocery scale. One day, I was a bit of a rebel when I was doing this and I said, 'What is the point of all this?' Colin said, 'I'll show you what the point is. I want you and you [another employee] here tomorrow.' The next day they strapped to me half a hundred weight hanging on my back, at least fifty pounds, believe me. Then the other guy got nothing, and Colin said, 'I want you to start the race when I say, up to the guardhouse. I said, 'Of course I can't with all this weight on me.' Colin said, 'Right, just do as you are f**king told. Do you f**king realize what weight does?' I thought, 'Oh, okay.' Carrying excess weight obviously wasn't good. I left in 1971 and went to March Engineering to look after Niki [Lauda] and

Ronnie [Peterson]. The Lotus later went from torsion bars to coil springs when the car went black [JPS livery].

Dick Scammell left the team as well in 1971. "I left Lotus after eleven years because I needed a holiday, and if you worked for Colin you couldn't have a holiday. I set up the Museum at Donington Circuit." Everyone seems to have had a different assessment of Chapman. For Nigel Bennett he was the "White Tornado," for Beaky he was "The Evil Genius." Beaky explains, "Late nights, if you had to work all night to do it, that's what you'd do. You had to work all the time. Working at Team Lotus was far more than a full-time commitment, and no doubt the result was eventual burn-out and a lack of team continuity."

Emerson Fittipaldi's four seasons at Team Lotus had equally spanned the Gold Leaf and JPS Lotus 72 eras. He is clear about what he thinks of those cars when recently asked if he preferred the Lotus 72 or his following two seasons with the McLaren M23, both legendary cars having carried Fittipaldi to World Championships. Emerson says, "The Lotus was a fantastic car. The Lotus was better, but McLaren was a better team than Colin's. Alistair Caldwell was the team manager and he was fantastic." McLaren's less hostile work environment paid dividends versus the burnout and subsequent revolving door for those employed at Team Lotus.

*The tight footbox of the 4WD Lotus 56 turbine car.
Note the differential tube above the pedals. (Norm DeWitt)*

55

The end of the turbines - This Lotus 56B was raced by
Emerson Fittipaldi in the 1971 Italian Grand Prix (Norm DeWitt)

Chapter 4

BRM

The BRM H-16 engine was overweight, underpowered, and complex.
It won one race, in a Lotus 43 (Norm DeWitt)

As noted in Chapter 1, the initial BRM efforts with the V-16 engine were examples of complexity combined with disastrous unreliability. From the ashes of that effort, Sir Alfred Owen took the reins of the remaining company in Bourne, Lincolnshire. In contrast to the previous V-16 efforts, a 2.5 litre, 4-cylinder highly conventional engine program was pursued resulting in the P25 BRM. The less said about the seasons of 1956-57 the better, but by 1958 the car was capable of picking up podium finishes, Schell and Behra finishing 2-3 behind Stirling Moss' winning Vanwall at Zandvoort. The following year at Zandvoort, they achieved their first World Championship points-paying Grand Prix win with Jo Bonnier at the wheel. That the car had started from pole position and that Bonnier had finished ahead of future champion Jack Brabham's Cooper-Climax showed that it was no fluke.

British team participation at the 1959 Grand Prix of Holland was, as BRM's Louis Stanley pointed out, "Ten out of fourteen was a fair sized proportion of green cars." The British manufacturers had reached a point of dominance in Formula 1 even with the recent departure of Vanwall from the scene, although Vandervell did run a single car for Brooks at Aintree, resulting in an early DNF. Stirling Moss drove a BRM at the British Grand Prix that day, narrowly beating out Bruce McLaren for second place at the line.

Although championship contender Tony Brooks and the Dino Ferrari won at Reims and Avus, the 1959 season was the swan's song for the front-engine Grand Prix car and the days of the BRM P25 were soon over. It was replaced by the rear-engine P48 of 1960, which never achieved much other than a single podium by Graham Hill at Zandvoort, offset on that same day when a brake failure on Dan Gurney's P48 sent him crashing into a spectator area, killing one person. Dan has stated that in the wake of that accident in the BRM, he could never quite get beyond giving the brakes a precautionary tap before applying them. Although you could never tell it from the results, he felt it compromised his future capabilities of attaining the absolute fastest lap time possible with a car, to say nothing of destroying his confidence in the BRM.

The year 1961 brought the new 1.5 litre formula, and the very future of BRM hung in the balance as they developed a new space-frame car with a V-8 engine. Tony Rudd was advanced to the position of chief engineer. The intakes being within the Vee, the initial versions of the engine had carburetion, but the championship-winning 1962 version incorporated Lucas Fuel injection along with a curious stovepipe exhaust system coming out the sides and pointing skywards at an angle. It became the trademark feature of Graham Hill's championship winning P57 in 1962. A surprising feature of the Formula 1 engines of this era, including Peter Berthon's new BRM V-8, was that they were 2 valve per cylinder engines. This was also a characteristic of their chief opponent, Coventry-Climax. Both companies eventually built 4-valve versions for comparison, but stayed with the 2-valve. Ferrari had made the same decision with their 1964 Type 158, yet another 2-valve engine.

Graham Hill won the '62 World Manufacturer and Driver's Championships in the Stovepipe V-8 BRM P57 (Norm DeWitt)

For 1964, designer Tony Rudd got the memo and debuted the monocoque P261, a slender cigar-shaped car that was very much in the mix along with Lotus and Ferrari. Oddly enough Graham Hill won both the Monaco and United States Grand Prix – the same two Grand Prix he had won in 1963 with the P57, and the same two Grand Prix he was to win in 1965 with the P261. The one year that the P261 could have, or should have, won the World Championship was 1964. Jim Clark had to win the final race in Mexico and hope that neither Surtees nor Hill had a decent finish. The odds against that were high, and in the race Clark had his engine fail with just over a lap to go (eventually was classified fifth). Surtees needed to finish well ahead of Hill to win the title, and he was certainly helped in his quest when his Ferrari teammate Bandini rammed Graham's BRM out of third place. When Clark's engine failed, that put Bandini up to second behind eventual winner Gurney. Ferrari team orders came into play, and when the Bandini Ferrari slowed to let his teammate through into second, it swung the title to John Surtees. Graham Hill limped his damaged BRM home in eleventh, and although he had more points than Surtees (41 to 40), the BRM's reliability meant that they had to drop two points from a fifth place finish at Spa and the title went to Surtees by a single point.

Never before and never again was the Formula 1 title to swing in favor of three different drivers in the final two laps of the season. Hill and Clark could both be justified in feeling robbed of the championship, but with Hill it had been the actions of Surtees' teammate that sealed his loss. Rarely has the race winner been reduced to a footnote in history, but that was very much the case with Gurney's win for Brabham, although it showed that there were four manufacturers that could contend for the title in 1965. Graham's teammate Richie Ginther had finished fifth in points with his BRM and signed to race for Honda.

Scotsman Sir Jackie Stewart was added to the BRM Formula 1 team alongside Graham Hill for 1965. Jackie says:

> In 1965 I finished second to Jim Clark three times: at the Belgian Grand Prix, the French Grand Prix, and the Dutch Grand Prix. Scotland was a very small country with about 5 million people in it, but we produced some wonderful drivers, and with the two of us on the podium that was a pretty big deal for Scotland. They called us 'Batman and Robin' and there was no doubt who was Batman and who was Robin. It was a great apprenticeship. Following Jim Clark taught me everything about driving – the smoothness, the cleanness. In those days you couldn't push cars around, you couldn't move over the edge of the racetrack. There were trees, or telegraph poles, walls, or whatever. It was a very dangerous business at that time. I look at today's racing drivers and today's tracks and it's hard for them to understand how treacherous it was. In my day motor racing was dangerous, and sex was safe.

Stewart was to get his first of his 27 Grand Prix victories in the 1965 Italian Grand Prix at Monza. With the advent of the 3 litre Formula 1, 1966 brought a level of complexity that hadn't been seen at BRM since the V-16 days. That was due to Tony Rudd's new baby, the H-16 engine. There was a decision to make at BRM, between engine designers Peter

Berthon and Aubrey Woods' 4-valve per cylinder V-12, and Tony Rudd, who favored a stacked pair of flat 8s, integrated into an 2-valve per cylinder H-16 layout. "H" configurations were used with success in aero engines, and Rudd, coming from Rolls Royce, was certainly influenced by the success of the Napier version. Knowing there was certainly an argument to be made for using the successful 1.5 litre V-8 as the basis of a 16-cylinder 3 litre powerplant, Tony Rudd carried the day. The Berthon and Woods design became the basis of the Westlake V-12 and that engine was to appear late in 1966 powering Dan Gurney's Eagle, ironically at the same race that saw the race debut of the H-16.

The 1966 season started with Stewart dominating the Tasman series in his P261, winning four of the eight races. Graham Hill was to win two, with Dickie Attwood winning one, all in their P261s, while Jim Clark was limited to a single win for Lotus. There was certainly a synergy between the P261 and the 2 litre version of the BRM V-8, and that was to be the car that BRM used for the start of the 1966 Grand Prix season.

Sir Jackie Stewart recalls, "We didn't use it [the H-16] to begin with. I won the Monaco Grand Prix and was in pole position, won the race and did the fastest lap with the 2 litre [V-8]. I then had my accident at Spa with the 2 litre. Tony Rudd, who was the team manager and would be today called a team principal, was an old Spitfire man. He was involved in doing the Spitfire engines, an engineer on the Rolls Royce Merlins. He was a high-level engineer and he really and truly believed that doing the H-16 BRM was going to be revolutionary."

The massive engine/gearbox/clutch assembly of the BRM H-16
seen from the rear (Norm DeWitt)

Although the car did run in practice at Reims, shocking all with its straight-line speed, the first race for the H-16 came at Monza and the car did not qualify well. Jackie explains, "It carried more water than the Queen Mary. It carried so much fuel, so much oil, and so much coolant, the weight of all the liquids alone was horrendous. It sounded beautiful, but it just had no beef."

One would assume that stacking two flat 8's on top of each other probably didn't do much for the car's center of gravity. For Stewart, that wasn't the major concern. "The center of gravity didn't seem to be too much of a problem, it was just the lack of performance of the engine. It sounded good, but it took a long time to get its revs up."

The BRM P83 chassis was one of the first rear-engine cars in Formula 1 to have the engine as a stressed member and the tub ended at the front of the engine. Stewart was more impressed when looking at the Lotus 43 H-16 racer also in the paddock. He says, "That Lotus was a better handling car. Jimmy won the only Grand Prix that the car won, but it was a peculiar victory as it didn't lead from start to finish, if you know what I mean."

Stewart gave the H-16 a few good outings in 1967, starting with his pole position at both non-championship rounds in England – the spring race at Oulton Park (see back cover photo) and also at Silverstone. Following his dominant run (until the differential broke) at Monaco in the V-8 car, Stewart led the Belgian Grand Prix at Spa in the H-16, the scene of his massive accident the previous year. He remembers, "The car was really a dog. We didn't win anything with it. I was having to hold it in gear all the time, but Spa was an unusual place, very few people drove around Spa all that fast. It was one of these old sayings, 'That's when men were men.' Driving around Spa at the limit was not everybody's first choice, it was a long circuit, with doing Masta flat, Burneville, and those sorts of things. Never mind Eau Rouge, as Eau Rouge wasn't considered by far to be the most exciting corner in the world at that time." Stewart eventually finished second to Dan Gurney's Eagle-Westlake at Spa, the only points-paying Grand Prix that the Eagle won. The great irony of course being that the H-16 on its best outing was defeated by the V-12 that had been previously rejected by BRM. He who laughs last...

The new lightweight P115 H-16 powered car was first raced at the Nurburgring in 1967 instead of using the 2 litre V-8 from Monaco. This car had magnesium replacing the previously used duraluminum, surrounded by fuel, which must have been a joy to contemplate driving around the Green Hell with only the occasional fire extinguisher stationed around the 14-mile circuit. One would have thought the 2 litre V-8 car could have done quite well around the Ring. Stewart tells it, "By then the BRM people were getting pretty nervous about seeing the cars as a failure, and I would say that they decided they just had to run with the car. I wouldn't have had any choice, but the V-8 around there would have been a dandy little car, with a 2 litre around there I would have done quite a decent time." Formula 2 cars often had strong performances even though they started behind the 3 litre cars in the German Grand Prix (Ickx in the 1967 Grand Prix with a F2 Matra is one example, running as high as fourth overall). Stewart adds, "I've got a Ringmeister ring… There is a ring that you get with the Nurburgring on it, and I was given it not for when I won the Formula 1 race by four minutes or more [1968]. I got it for the previous year, in 1967, when I won that Formula 2 race with the Matra."

The special lightweight BRM P115 H-16 car which Sir Jackie used at the 1967 German Grand Prix (Norm DeWitt)

Shockingly, Stewart had put the new P115 H-16 car on the front row, behind only Jim Clark and championship leader, Denny Hulme. Stewart was one of the best drivers ever to race at the Nurburgring, and his time in the H-16 was probably deceptive as to the true capabilities of the P115. One would imagine sticking the landings over the Nurburgring's jumps with an H-16 car would have been a back-breaking experience. There had to be concerns about landing from the jumps, knowing the entire back half of the car was only bolted to the back of the tub, versus sitting in a cradle. Sir Jackie recounts, "It was almost too heavy to jump. You always took off because it was in fourth or fifth gear."

The Lotus 49s with their engines similarly installed as stressed members suffered rear suspension failures during the race. Stewart explains, "Chapman never made anything all that robust. That's why I went to BRM in the very first place. I didn't want to drive a Lotus, I wanted to drive a BRM. It was a much stronger car. Chapman was brilliant but everything was made too inside the margin." Jackie is very clear in never having any desire to drive even the greatest of their cars, the Lotus 49. "I had absolutely no intentions of ever getting into it." Stewart decided to throw in with the new Matra MS-10 Cosworth run by Ken Tyrrell, his long-time entrant in Formula 3 and Formula 2. Ferrari was less than impressed as they had fancied signing the rapid Scot. That move came back to haunt Stewart in 1978 when he compared all the current Formula 1 cars, actually driving each of the cars on-track. All that is, other than the Ferrari flat 12. Sir Jackie recalls, "This is actually what Enzo Ferrari said, "Our car wasn't good enough for Stewart in 1968, and it's not good enough now." Those were his exact words! [laughing]"

With Stewart's departure came the end of the H-16 program, although there were still some developments with a 4-valve-per-cylinder version in 1968 that would break surface from time to time along with reports of outrageous HP numbers, but it never returned to any Grand Prix beyond the first race of the season in South Africa, where the new V-12 car wasn't ready for Spence (in his last Grand Prix). For a few races during 1967, Bruce McLaren's car was running a 2-valve version of the new BRM V-12 and starting with 1968, the V-12 was to be the engine layout for BRM that was to take them through the end of their years as a Grand Prix manufacturer. The first to score points for the new BRM V-12 was current World Champion Denny Hulme in his only race with the McLaren-BRM, at Kyalami.

In the non-championship races that followed in England, both Spence and Rodriguez looked to be a potential race-winning combination with Len Terry's new P133 V-12 cars, Rodriquez finishing second to McLaren at Brands Hatch. Then Spence was killed in an accident at Indianapolis, replaced at Monaco by Dickie Attwood, who finished a charging second place to winner Graham Hill. At Belgium, McLaren came out on top of Pedro Rodriguez and their scrap for second became the race decider when leader Stewart's Matra MS-10 ran out of fuel on the last lap. The rest of the season became a slow slide back into mid-field for the V-12 cars, the best subsequent finishes for the BRM drivers being fourth at the Italian Grand Prix by Piers Courage, and a third by Rodriguez in Canada.

Bobby Unser tested and raced for the team in 1968 and was less than impressed with the overall situation:

I was with BRM as you know, and I did go over there and do a lot of testing, but we just didn't get along good. Their goddamn brains, they were so secretive and they weren't really good people. Now, the mechanics and I got along super good, but the BRM was a s**tbox [the 1968 V-12 car]. I was running their car in testing at Snetterton in the rain, and it was stupid to be out there in the rain. Why the hell would I be at a new racetrack with a Formula 1 car in heavy rain? I'm going around that race track and all of a sudden there was a little MG running around the track. I couldn't believe what I saw that day. People would just drive through the gate and run around the race track. No firetruck at the track, no ambulance at the track, and they are telling me that it's normal.

I'm going around there and say the engine is running at 80 degrees centigrade. At high rpm there was a little flagger shack off to my right and I'd see where I'm lined up with that to see what gear I was in. If I had to go past that shack and then shift into fifth, where before I was shifting into fifth say 50 yards before that shack. That was a way to see if I'd improved or lost my speed [in that segment]. The temperature was still good, but I was losing horsepower bad. I'd come in and tell them that, 'Guys, when I first go out it's got say 400 hp and then I'd go around 10 laps and maybe it's got 300 hp, it's losing horsepower really bad.' The boss would say, 'You know Bobby, we don't do things like you guys do in the States in getting a flash reading and then shut it down. Over here we'll leave the throttle wide open on the dyno, and we're liable to go to lunch and be gone for an hour and a half.' Okay, I accept that, but I didn't give a s**t as the car was still losing power.

Now you've got the radiator in the front and the engine in the back. I told them that the water pump is cavitating. They said, 'The temperature is always good and if it says 80C, it's 80C and it's not going up too high.' I figured that it's a goddang good bet that the water is not getting to the sensor. In the old BRM garages they used to have old rear-wheel dynos that were permanently in the cement. I looked at that and said, 'Let's chain the back of the car to the wall, and put the car up onto the chassis dyno. The water pipes would go outside the chassis so I said, 'Let's take this pressure pipe right here and whack out some of the aluminum pipe and replace it with clear plastic tubing with hose clamps because we could then see what the water was doing.' So they did it for me. You know what? It was cavitating at 5,000rpm just like I told them. The water flow turned into a garbage disposal and makes more goddamn noise than you can imagine. That's where the horsepower was going. The mechanics were afraid the car would shoot through the wall as the dyno wasn't designed for that era of car. They shut everything down and then they were nice to me. I was just getting smarter then, but first of all they think Americans are really dumb people, and maybe we are, I don't doubt that. But you had to have the water pump push instead of suck, as water pumps have too much gaps in them between the impellor and the housing. When I was with Gurney, he installed the entire cooling system of an Indycar on the dynamometer, using the pipes and actual radiators at the actual heights.

There was a "skunk works" project going on at BRM under the radar in 1969, to develop a ground effects car with Peter Wright. BRM team driver from 1971-72, Howden Ganley, recounts:

That was done back in the 1960s when Tony Rudd was there. So then the thing got canned. Peter Wright left BRM and he then went to Specialized Mouldings, who were doing the March bodywork, so I think that's probably where that aerofoil sidepod [on the March 701] came from. What a lost opportunity... there are so many opportunities with things that get missed. Everyone is desperate to run the cars and you think you will get back to things later and you run out of time and never do. With Tony

Rudd's departure they kind of swept everything out the door and started again.

I was fortunate when I came at the beginning of 1971, as they were on the upswing. I was delighted at being in Formula 1. Being at BRM with all that history, they made their own cars, their own engines, their own gearbox and all of that. It was a proper race car. The P153 had been improved, they only had 2-P160s initially, and so I had to drive the 153. The BRM made good horsepower compared to the Cosworths then. The DFV cars had the advantage of using a lot less fuel, the BRM suffered from that disadvantage like Ferrari, as did the Matra.

There was an opportunity for Howden to do a proper back-to-back comparison between his P153 and the "tea tray" March of Ronnie Peterson in 1971. Howden continues:

Tony Southgate's P-153 BRM of 1970, and his last BRM, the P-180.
Rodriguez won Spa with the P153 (Norm DeWitt)

The tea tray car was really good. We were testing at Silverstone one day with BRM and at lunchtime the team said, 'We've done everything we needed to do, we need to go back now.' I wandered down to talk to talk to Robin [Herd] and Ronnie. They asked what I was doing and I told them that I'm going home now as BRM has packed out. About five minutes later

Robin comes running up to me saying, 'Ronnie wants to go back to Sweden, could you run our car this afternoon for some tire testing?' Tim Parnell had gone back to Bourne so I couldn't ask him, so I got to spend all morning in the BRM, and all afternoon in the March, a direct comparison. The March was a quicker car.

I drove the 153, 160, and the 180. The P180 was very unloved to start with and then it turned out to be the best of the three, right at the end. Tony Southgate left and Louis Stanley being a little cuckoo, didn't campaign the car even though it was better than the earlier cars, it was a vindictive thing. It was Tony's last design and Stanley was going to stop it. I thought I was taking Tony to Williams with me, as I was getting out of BRM, but somehow Don Nichols got in there and took Tony to Shadow. Unfortunately we didn't finish up with the right designer, but that's racing.

They were well-engineered cars generally, properly made. The P153 was the 1970 car that Pedro won at Spa with. The 153 you could drive sideways. A short-wheelbase car like the 153 you could slide all about. The '71 car, the P160, was that much better again, but it was harder to drive. It responded well to the really high-speed corners, and it had good aerodynamics on it as long as you didn't let it get too far out of shape. I was talking to Tony Southgate just last week about it and someone told him how the steering of the 160 was heavy. I told him, 'Remember Tony, that if you got the P160 too far around, you couldn't pull it back'. Tony said, 'That's because of the geometry on the front. If you got the wheels too far around it sort of went over a cam and you physically couldn't pull it back. At that point you'd have to put your foot on the brake and let it go around.

There's one proper reason not to flat track a P160 as for all intents and purposes the suspension geometry gives the effect of seizing the steering rack.

Vic Elford drove the P160 at the Nurburgring in 1971, stepping to replace the recently lost Pedro Rodriguez who had been killed in a Ferrari 512 at Norisring. Vic remembers, "It should have been great, and it would have been, but all the way through practice and qualifying the engine never ran on 12 cylinders, it was on 11 cylinders always. The car was great, it handled well as it had been Pedro's car, which he had gotten set up nicely. Pedro wasn't like Jo [Siffert], who could drive anything versus Pedro who used to set cars up nicely. I'd have had a great race except for that idiotic engine problem."

Howden Ganley won the Peter Bryant Challenger Award, with the First Lady of Racing, Linda Vaughn (Norm DeWitt)

Even a properly set up P160 could bite if you weren't careful. Elford says, "Yeah, but once you knew what it was doing it was a nice car. It was like driving a 911, once you know it."

Howden gives his summary of the BRM P160 and the challenge of getting the best out of what could be a scary proposition:

The P160 had a longer wheelbase and was a narrower track, it was more of a knife-edge car. You had to be a lot more precise with it. Pedro Rodriquez and Siffert were big rivals from Gulf [Porsche] and their rivalry continued on at BRM [sadly both killed during the 1971 season]. The P180 you could throw around and it was quicker than the P160 as well. It was really coming right and had Southgate stayed on, that would have been a great car, but by then the engines were running out of steam. Beltoise preferred the P180 because of his gamy arm, as it had much lighter steering. He won Monaco with the P160, but that was in the rain and so was without the heavy steering. When he stayed on, they took his favorite car away. The end started at the beginning of 1972 when Louis Stanley tried to run two teams. He didn't have enough facility to run one team of three cars, and then he was trying to run two three-car teams. Mostly he was only able to run five cars, because he didn't have enough engines.

With the recent World Championships being won by Tyrrell and Lotus, BRM would have been better served by investing in lighter and more powerful engine developments. To enter five cars sounds like madness. Howden continues:

Madness with a Capital 'M'. At Watkins Glen [1971] we had the three regular works cars and two extras, as he had rented a car to John Cannon, so there were five BRMs in the race. There were thirty cars entered and fifteen cars finished the race. Siffert was second, and I was fourth, so of all the fourteen finishers, five of them were BRMs – fully one-third of the cars that finished the race. Stanley thought he had motor racing in the palm of his hand thinking this was easy, he'd just bring five cars all the time. Of such things dreams are made.

Another factor was that the BRM engine was pretty good on the high-speed tracks where you didn't have to slow down. The car didn't have a lot of torque, but once you got it up there and you kept it there, it was good. If you came down to a hairpin and then had to drag out of it, then you were struggling a bit. Over the winter on 1971 the Cosworths took a big leap up. BRM got rid of Aubrey Woods, the guy who had sorted it out, and they brought in new people and it seemed to me that they lost horsepower. Unfortunately the 'British Ferrari' went down the drain.

The P180 had all the weight location problems to start with, but eventually that all got sorted and it was brilliant. It had too much weight in the back and not enough in the front as it had the radiators behind the rear wheels. Southgate was a fantastic designer. The Can-Am car, the 154, wasn't good, but he redesigned it into the P167 and it was brilliant – way better than that McLaren [Mk 8F]. It was Lauda and Reggazoni in 1973, and because Louis Stanley canned the 180, when Mike Pilbeam took over as designer he got the P160, so he had to go and update that. He obviously had to do something quickly to update the P160 and in the meantime he was able to design the P201, which was all his and it was a really good car as well.

As discussed in *Making it FASTER*, the pyramid-shaped 201 was similar to the tidy trapezoidal monocoque Gordon Murray designed Brabham BT-42/44 cars that had preceded it, the BT-44 winning its third Grand Prix, at South Africa in 1974. Pilbeam's P201 had its debut at that same race, and Beltoise followed Reutemann across the line to finish second. Despite having this stellar introduction, it flattered to deceive and this was the last podium for both the P201, and for BRM as a manufacturer.

The P201 was an improvement over the Niki Lauda/Mike Pilbeam developed P160, but it wasn't enough. Mike Wilds was to drive the P201 at a couple of Grand Prix in 1975 and had this assessment:

The chassis was quite good, but the engines were sadly lacking in power. I went testing with BRM at the end of 1974 at Snetterton with the sadly missed Chris Amon. Initially I was given a P160 to drive whilst Chris drove the P201. I really liked the P160, a docile, very well-balanced chassis. At the end of the day I was entrusted with the P201. My impression was that the car handled very well, the inboard front brakes were good, although they did try to fry my ankles! As was to be the story of my short time with BRM, the engine failed after a handful of laps, the engine ventilating its block dramatically on the start/finish straight, showering Jean and Louis Stanley with oil as they were watching on the pit wall. When I arrived in Argentina for the first race of the 1975 season I noticed that the car was fitted with the same engine that had blown up at Snetterton with the engine having been repaired with a welded block! Needless to say, I didn't finish the race due to an engine malfunction causing a fire fairly early in the race. The engine felt as though it was about 100 horsepower down on an 'off the shelf' DFV. Because of this power deficit, it was unfair to the chassis as I was over-driving the car in the corners to try and overcome the lack of power. So, in summary, it was a good chassis let down by the use of old engines and a lack of budget to run the car properly.

Although the team limped along from 1975-77, the death of Sir Alfred Owen in late 1974 meant that "Owen Racing Organization" was replaced with "Stanley-BRM" on the P201, which was never again close to being a competitive proposition. BRM had been reduced to an object of ridicule as the cars were often referred to as "Stanley Steamers." The end came in 1977 at the Italian Grand Prix with the hopelessly off-pace P207, where Teddy Pilette was unable to muster enough speed out of it to qualify. It was a sad end to a legendary team.

CHAPTER 5

BRABHAM AND THE CLONES

Mario Andretti in his '66 pole winning Brawner Hawk at the 100th running of the Indianapolis 500. (Norm DeWitt)

Jack Brabham restarted the rear engine revolution at Indianapolis with a Cooper-Climax in 1961, and when he returned in 1964 it was with a car of his own – Ron Tauranac's Brabham BT-12. Sir Jack Brabham's first efforts with his own Formula 1 car (BT-3) came in mid-1962, achieving a number of reasonable finishes and a non-championship race win in 1963. There were no World Championship wins until the debut of the new BT-7 racer, and both wins were by team driver Dan Gurney in the 1964 French and Mexican Grand Prix. The new and successful BT-11 Formula 1 car was introduced in 1964 along with Brabham's first Indianapolis car, the BT-12. Powered by an Offenhauser, the tube frame BT-12 was known as the John Zink Trackburner Special and entered for the 1964 Indianapolis 500. There was nothing remarkable about the car's performance, and Brabham (a two-time World Champion) was only able to qualify twenty-fifth on the grid, a full 6 mph off the pace of Jim Clark and his Lotus-Ford. Starting that far back in the field put Jack squarely into the debris field of the nightmare crash that claimed the lives of both Dave MacDonald and Eddie Sachs, and the BT-12 retired after seventy-seven laps with fuel tank damage, a result of the earlier carnage.

The car was then driven by Jim McElreath for a number of races during the 1964 season and was improved, taking pole position at Trenton. The chassis was comprehensively written off in tire testing later in the season, but was to inspire a car that became the most successful chassis of the following two seasons, the Brawner Hawk.

During 1964, newcomer Mario Andretti got the ride in the Dean Van Lines Watson-Offenhauser, working with chief mechanic Clint Brawner. That team had been one of the great teams of the roadster era, and that ride must have been at the top of any driver's dream list. It was an incredible break for a rookie to get. Mario recalls, "Oh for sure, it was the break that I was waiting and hoping for in my career, no question about it. I knew how fortunate I was at the time. It was the perfect combination between Clint Brawner and Jim McGee – there was the old solid and the new thinking mixed in there for my benefit. The timing was so correct for me to take advantage of what I had, so again, those are the things you dream about and hope for and it definitely worked out."

Mario had achieved a podium in his first partial season of 1964 with the Watson Roadster. The decision was made for the team to build the Hawk, a rear-engine design. Mario says, "The handwriting was on the wall, with the influx of the Brabham and then Colin [Chapman] coming on. You knew that had to be the direction to go, seeing how well the Brabham tube chassis worked on the car that McElreath was driving for Zink."

When the Brabham BT-12 was torn apart in a crash during tire testing, Brawner got the remains of the BT-12 from John Zink to reverse engineer for the new Hawk. Mario explains, "They made a deal and out of the pieces they could put together, they made a jig to build a car. I had a little something to do with it, I told them, 'Don't change a thing! I mean not even by one-quarter inch. Put exactly the same pickups, the same suspension, the same uprights.' I had followed Formula 1 and knew Ron Tauranac and Jack Brabham. Their reputation was that they always had really great handling cars. We are lucky that we could go in that direction."

The author with his daughter and Brabham designer
Ron Tauranac at Laguna Seca in 1993.

Brabham was also known for building stout cars that didn't fall to bits with wheels coming off and the like. Mario agrees:

Exactly. Well, I think the one guy that was helpful on behalf of Zink was Kuzma, who had a great relationship with Clint as they did dirt cars. Here I am a rookie in 1965, doing the first part of the season with the roadster. Of course the objective was to have the rear-engine car at Indy, I didn't want to take the driver's test in the roadster at Indy. It would have been pointless to do that because I could not apply anything to the rear-engine car. So I sat it out, watching everybody else run for a whole week, which was the toughest time of my life, to do that.

I'd never driven around it in a passenger car, the first time I'd ever set foot on that track all the way around was in the race car [the Hawk] and it was a week and two days late. The car arrived the second week of practice, arriving on Tuesday, and the last day I could pass the driver's test was Wednesday. Talk about biting my fingernails to the bone.

There was never a question about the engine as the decision had been made to commit to the 4-cam Ford from the word go, as Andretti recalls:

It was a foregone conclusion that it was going that way. In qualifying we were so nervous about everything, the reliability, you know that these guys qualified me on 10% Tolulene with alcohol, a very light mixture that reduced the compression of the engine and made it run easier, when most of the guys were running 30-40% Nitro. If I'd have qualified with even 10% Nitro in that car, we would definitely have been on the front row. The shortcoming of the car, as we found in the race, was that we didn't have time to think it all through. If you look at the car, the fuel tanks were big saddle tanks on each side with a crossover so the fuel would flow, but we didn't think about having a check valve to preclude the fuel from moving from the inertia and going into the right side. From mid-fuel on, all the fuel was in the right tank. It held 40 gallons per side. The car felt the best when it was full, when it was balanced. Mid-fuel it was bicycling through the corners. There was a period where I was really struggling and we didn't realize it until after the race. We didn't have time to do anything for Milwaukee, but for after that we added two electric pumps, which turned out to be an ace in the hole for me. Then I could transfer fuel as I wanted to balance the car from there on and that became a nice little secret. In the McElreath car, there was a valve, but with that you don't have the leeway to transfer it like with the pumps. You think back at what might have been, 1965 in the race at Indy. If we'd had that it could have been so much better.

Mario was Rookie of the Year for the Indianapolis 500 having started from fourth on the grid and ultimately finishing third. Andretti and the Hawk were to take their first pole position soon after Indianapolis at Langhorne. Mario again had pole on the road course at IRP, and then took their first win. Although AJ Foyt was to take ten poles and win five races, Andretti's consistency with ten top-three finishes had allowed him to secure the title despite only having that single win. Andretti in the Hawk was the 1965 USAC Champion, which was a stunning result for any driver in his first full season.

1966 started for Mario with six straight poles, winning eight of the twelve pavement races through the season. Obviously there was nothing in Indycar racing with the pace of Mario and his Brawner Hawk. Mario says:

We were blessed with being one of the favorite teams to do tire development for Firestone. By doing so much running, obviously you learn a lot about the car, all the idiosyncrasies, to learn the dynamics of the car by having all that seat time. I would change springs by 25 to 50 pound differences, I knew the car so well. We knew we needed a stiff chassis, and that was the only mod that was made to our chassis that was different from what the original Brabham frame was. We knew how important that was so if we put cross-weight in the car, it could respond.

McElreath had won the opening race in Phoenix with his new Brabham-Ford. Both the Brabham and Brabham clones were looking dominant headed into race three at Indianapolis. In qualifying, Mario took pole position over Jim Clark's Lotus, the defending race champion.

Things were looking promising for the young American to win the biggest race in the world. It was not to be and the car retired early in the race with a bad valve.

Mario in this Brawner Hawk started from pole in the 50th 500 and dominated the 1966 season (Norm DeWitt)

Third place in the 500 went to McElreath in his Brabham-Ford, behind Graham Hill's Lola and Jim Clark's Lotus. As to which of the first two cars was the actual winner, that both cars headed to Victory Lane speaks volumes.

Mario was to take nine poles and win eight races that season on his way to a second consecutive USAC Championship Car title. It was unsurprising that the second place car in the points by season's end was the similar Brabham-Ford of Jim McElreath. Johncock and Leonard were a distant third and fourth in points.

Andretti reflects upon his early Indianapolis luck that could have seen him in the winner's circle, "When I look back to 1966 or 1967, if I could have finished those races I probably would have won the two easiest races in my career, that's how well the car was working. I just felt that I was in command, I was on pole in both of those years." One suspects that a certain Rufus Parnelli Jones might not agree with that assessment of the 1967 race, and Mario freely concedes that the Turbine car was the one exception, as he quickly discovered once the race began. Andretti explains, "I wouldn't have had that kind of advantage. We had to carry about 75 gallons of fuel, whereas Parnelli only had to carry about 35. Talk about an unfair advantage, it was ridiculous the advantage he had with the Turbine."

There were further revisions as the 1967 Brawner-Hawk was modified into a semi-monocoque, with the sheet metal riveted to the frame. Mario recalls, "That's all there was. Then we worked a little bit on the aerodynamics. With the '66 car, when you look at the bodywork wrapping around the gearbox, it was pretty tight and we used to run excessive gearbox temperatures. Brawner and McGee thought they would enlarge it and thought we'd just put on a ducktail." The resulting imbalance led to the wooden front nose-dive planes to balance the car discussed in the previous book, *Making it FASTER*. Andretti continues:

> I was looking at the chin spoilers of the Chaparral and so we put that on the Hawk. It was the perfect balance and in turn 1 at Trenton it was a tight radius, the car would just rotate so quickly with the throttle, and that was really fabulous, that was awesome how I could blow out of the turn. But I was so pissed, because the wooden nose was so heavy – but I set a new track record and led every lap. That was what was so much fun about those days, was finding those things, you just kept searching. The rule book didn't know any of those things could happen, so they were not prohibited. I think us in Indycar and Colin in Formula 1 were the first to do things like that with open-wheel cars.

The semi-monocoque Brawner Hawk of 1967 at the Rex Mays 300, Mario Andretti at the wheel (Jonsey Morris)

Mario took pole at Indianapolis by 1.5 mph over Dan Gurney's latest Eagle-Ford, running a similar 4-cam engine to Andretti at Indianapolis. Mario says, "In 1967 we dropped out of the race when the right front wheel came off, early on in the race. The reason for that was the tire companies owned all the wheels. They would bring everything mounted and the wheel they put on the right front was never registered where it goes against the hub, it was never machined. The wheel came loose and I lost the wheel, parked in Turn One, end of story."

Mario Andretti and Clint Brawner had to feel like they had copied the right stuff with two straight USAC Championship Car titles, and seeing how the Brabham-Repco cars were also winning the World Formula 1 Driver's and Manufacturer's Championships with their space frame cars in 1966-67. There have been some comments making the rounds about how the water pipes running along the bottom outside edges of the chassis tended to help channel air on the underside of the car and how that flat bottom of the raked chassis could create a lot of downforce from the underbody. Mario disagrees, "I don't know about that, because the only way you can kind of speed up the flow is with some kind of diffuser in the back to encourage the air to speed up. I think where we gained downforce was with the ducktail on the rear bodywork, the back end was planted."

Andretti was to take the championship down to the finale at Riverside Raceway in one of the greatest races in Indycar history, the 1967 Rex Mays 300. The race was decided when first AJ Foyt had to jump into Roger McCluskey's Eagle midrace and was able to score enough additional points to win the title after race leader Andretti had to stop for a splash of fuel with a handful of laps remaining. The race came right down to the end, with race winner and hometown hero Gurney sweeping around Bobby Unser in turn 9 headed for the white flag. Between the musical chairs by Foyt, a hometown hero winning with his Eagle (in the same year Dan won Le Mans and the Belgian Grand Prix), and the championship battle between Foyt and Andretti coming down to the wire, it may have been the greatest Indycar race ever held. Despite eight wins during the season, Mario had narrowly lost a USAC Championship Car title for the first time, to Foyt with only five wins. It was an echo to 1965's title chase as again the driver with fewer wins was to take the title.

The full monocoque Hawk arrived in 1968. Early in the year the car was slender and conventional in appearance. Mario retells, "We bought a Lotus chassis to be at our disposal, which I didn't really like. It was like what Clark had driven from '65 to '67 [Lotus 38], so we tried to cannibalize something else. If you look at what I drove at Indy, it is slim and it looks like a Lotus." That season it was musical drivetrains for first trying the 4-cam Ford, then the Turbo-Ford, then went to the Turbo-Offy, then back to the Ford. All of this undoubtedly created more than a bit of chaos as the team struggled to contend with all of the changes in the engine bay. He continues, "I had a lot of second-place finishes. I didn't win a lot of races, but I finished second almost everywhere, but came on strong at the end. The first race with the Offy engine was at Trenton again, and I led flag to flag. Mid-season engine changes really don't work."

By Riverside, Mario's Hawk had a distinctively different look with sloping sidepods and the overall appearance was closer to a '74 trapezoidal monocoque Brabham BT44 or 1970 BRM P153, than anything resembling a Lotus 38. The Hawk was back to running the 4-cam Ford and the engine blew up by midrace. Mario says, "Here was the thing: To try and protect their own drivers, in 1968 Goodyear was Bobby Unser and so he had plans that if he blew up, he had cars he could take over. The deal we made was that if we blew up, I could take over the Firestone car that was farthest up – closest to the front. When I

The final version of Andretti's 1968 Brawner Hawk at the Rex Mays 300, alongside Gurney in the Eagle (Gary Hartman)

blew up somehow Joe Leonard, through pit stops or whatever, was running third with the jet car."

That ended in disaster with Mario taking out Art Pollard in the sister turbine car in turn 9. Mario shares, "In retrospect if I had taken over Ruby's car at that time instead of Leonard's, I would have been the champion. I ultimately finished third in Ruby's car, and we had to share the points based upon the percentage of laps." In 1967, if Foyt hadn't been able to jump into McClusky's car, he wouldn't have won the championship then. Car hopping never worked in Mario's favor.

What was Mario's impression of the Mongoose? "It was very close to what we had in so many ways, but I didn't fit worth a damn... it was the longest race of my life as Lloyd was quite a bit taller than me, I was rattling around in there. I thought that race lasted three days." Joe Leonard was another tall driver. Mario says, "It was the same thing with the Turbine car."

Small wonder that Mario would find the Lloyd Ruby driven Mongoose to be a familiar proposition, as it was based upon the BT-11 Formula 1 Brabham, much as the original BT-12 Indycar was copied by Brawner. The Mongoose was another Brabham clone that was making its mark in Champ Car. Chief mechanic Dave Laycock built his BT-11 clone for 1967 and arriving at the first race in Phoenix, Lloyd Ruby put the new car on pole position and then won the race, leading every lap. He qualified second to Mario's Brawner Hawk at Trenton, and was clearly a threat to the established order. Midseason, Ruby was to take pole at IRP on the road circuit with the Mongoose. Lloyd was eventually to finish sixth in points.

Motorcycle racing legend Everett Brashear was a lifelong friend of Lloyd Ruby as they shared similar roots. Everett reminisces, "Lloyd Ruby and I started out racing the same year in Texas. He was a fantastic racing car driver, and he should have won Indy five times. Lloyd asked my son if he was racing motorcycles, and I laughed. 'Lloyd, you know we were the stupidest idiots in the world to do what we did.'" 1966 could have been that first Ruby win, as he led the most laps in the new Eagle, and in 1968 he nearly won again.

For 1968, an improved Mongoose was built and Lloyd was to turn fastest lap during the Indianapolis 500. He could well have won the three-way battle between himself, Bobby Unser in the Eagle, and Joe Leonard in the Lotus 56 Turbine car, had the car not run into coil problems when leading on lap 174, misfiring its way to a fifth place finish. Ruby followed that performance with wins in both races held that season at the Milwaukee Mile. Added to his five other podium performances, Lloyd was to finish out the 1968 season fourth in the championship. The points Lloyd gave up in turning over the Mongoose to Mario Andretti at Riverside may well have cost him third as he finished with 2,799 points to the 2,895 of Al Unser.

A pair of Mongoose at Indianapolis Raceway Park in 1967, Lloyd Ruby #25 and Arnie Knepper #20 (Pepper Bowe)

For 1969, the success of the Lotus 56 Turbine car wedge shape brought a new similarly shaped Mongoose to the team, which had performed well at the first two races of the year, getting the podium at both events. However, Lloyd preferred his old space frame 1968 BT-11 clone at the Speedway and persevered with it for the 500. Again, he was in contention for the win.

The factory Brabham team was back at Indianapolis for 1968-69 with the new factory BT-25s but they were normally off the pace, primarily due to various issues with the Repco powerplant. In 1969 Peter Revson was to win on the road course at IRP with his BT-25, but at Indianapolis the Hawk and Mongoose vanquished the latest factory Brabhams and everything else. The BT-25 was similar in concept to the 1968-69 Formula 1 BT-26, with its space frame chassis being reinforced with aluminum panels as was done in 1967 to the Brawner Hawk. The BT-26 was also the first Formula 1 car to go into full Biplane configuration, Jack Brabham showing up at Monza with high front and rear wings. Being as Monza was likely the worst track for such a layout, it wasn't raced until Canada. The Repco was the Achilles' heel of the Formula 1 car as well, and the cars only scored two finishes in 1968.

For 1969, the Brabham Biplanes switched to Cosworth DFV power and ran in that configuration until at the third Grand Prix of the season until they were banned midweek at Monaco. New team driver Jacky Ickx recalls being at the top of Formula 1 during the high-wing era of 1968-69. "I ran the first wing on Ferrari at Spa in 1968 and had it at Rouen in the French Grand Prix [which he won]. The idea was to get some top speed and to put it [down] before the corner. It was an interesting idea, but not so practical in reality. It was mounted on the roll bar in the middle of the car, before the engine behind the cockpit. The Brabham was mounted on the uprights and it was always broken. Everyone had a broken one at Spain, and that was the end of it for a while."

Grand Prix racing had reverted to low wings from that point on. The low wings were not always an improvement. Jacky Ickx says, "I remember I spun on a fast corner at Watkins Glen. I stopped and I didn't understand what had happened. It was the low wing had gone. I didn't see it immediately. I went around and thought I had lost a wheel and then realized there was something missing." Ickx had great success with the BT-26, taking wins at the Nurburgring and at Canada, to finish second in the World Championship. For the following season

Jack Brabham in his BT-25 at the 1968 Rex Mays 300.
It featured a semi-monocoque chassis (Gary Hartman)

Brabham switched to a full monocoque chassis, with Sir Jack winning the season-opening race at South Africa for his final win. It was the end of an era, the end of Ron Tauranac's space frame Brabhams.

Back in the USA, 1969 was to bring huge success for the STP Hawk team, now owned by and a part of the Granatelli effort, running Andy's trademark bright orange/red cars labeled with STP logos. This latest Brawner Hawk did not figure into the equation as the team headed to Indianapolis. Mario says, "The whole practice there at Indy, I was practicing with the Lotus four-wheel-drive car. When that thing came apart they decided to withdraw it, not only for us, but also the other two Lotus cars as well. When they did that I was relegated to the Brawner car, which hadn't even been cleaned up since winning at Hanford, California."

The Lotus 64s that were withdrawn were incredibly quick. Mario remembers, "I loved the car, but I knew there was no way that thing was going to be reliable. Number one, the gearbox was right behind me, the engine was in there reversed. It was just burning up within five, six, or ten laps on the track. The gearbox temperatures were up to 270 - 280 degrees. We didn't know what the heck to do, and it was a blessing that the hub broke and we got rid of it. We would have started from pole 100% for sure, but it would have lasted ten laps." Mario had suffered painful facial burns in the crash to where in the official Indianapolis qualifying photo of the front row, Mario had his twin brother, Aldo, stand in for him.

Mario had been far quicker than Graham Hill or Jochen Rindt in the Lotus 64s, despite it being Mario's only experience in a four-wheel-drive car other than a couple of laps in a Lotus Turbine car the previous year. Mario says, "Yeah, anything that feels good to the driver. It felt good to me, it would respond to the changes we were making in the aerodynamic balance, as we had moveable front wings." From this experience at Indianapolis, Mario was willing to race the Formula 1 version later in the 1969 season, when Rindt and Hill refused to have anything more to do with it. Mario explains, "Yeah, the Lotus, when you look at the body shape, we had some downforce. They were timing turn 1 and nobody had gone faster through 1 [at Indianapolis] than I had during the practice days. It would have been small satisfaction as it would not have lasted in the race." Mario's run with the Formula 1 version (Lotus 63) at the Nurburgring didn't last long either, the car landing hard and grounding out coming off a jump during the race while running with full tanks. The ensuing crash meant that Mario was unable to complete a single lap, which was better than how the accident ended for Vic Elford, upside down in the forest with a broken shoulder.

Vic Elford had finished seventh at Monaco with an uncompetitive Cooper Maserati entered by Colin Crabbe. As Elford says, "After that Colin said, 'That's not good enough, I'm going to buy you a real car.' So he went to McLaren and bought the M7B. It was a very good car and I was getting results with it, finishing in the points and then went to the Nurburgring and qualified sixth."

Mario put the four-wheel drive to full use at the start, rocketing from fifteenth on the grid up into the top ten. As Elford retells, "I made a balls-up of the start, Mario got by me and Beltoise got by me. I spent the whole first lap stuck behind them. There is one corner just before Bruchen, a long right hand 100mph corner over the top of a crest. Mario was driving the four-wheel-drive Lotus, and he'd never done a lap on full tanks. Over that crest he bottomed out and slid off the road on the left. Of course there were no guardrails then, he took off the two left hand wheels on the fence posts. One of those wheels bounced across the road, Beltoise managed to squeeze by, but by the time I got there it had done one more bounce and I hit it with my left front wheel and went over the top of it. I was upside down and into the trees like 'that.' I couldn't see much, and you are disoriented upside down. I could see the marshals standing around doing nothing. I knew where the fire extinguisher button was, and I got my finger out from behind the steering wheel, but I couldn't reach the button."

It was a dire situation, and thankfully it wasn't a similar situation for Mario Andretti as he climbed from the wrecked model 63. Elford continues, "Mario got his helmet off, walked back up the road and found me upside down. Then he started kicking ass to get the marshals to do something to help me get out of the car. They tipped me up and got me out, but they were frightened because they thought it was going to blow up. They had every reason to be as it was the first lap, and the car was full of gas. When we looked at the back, the extinguisher had been ripped off and the wires were dangling together. Had I pushed the button, there would have been a nice flame-up." What had saved Elford was the combination of Mario Andretti and that Vic couldn't quite reach the fire extinguisher button.

It is hard to say that Mario took much of anything positive away from his experiences in the Types 63 and 64, but bits from the Lotus 64 were to provide dividends for Mario and the quickly modified Brawner Hawk. Mario admits, "I stole one of the moveable front wings from the Lotus and put it on the left side of my Hawk. I only needed one, the right side would have been too much, and I won the race with only one wing on the left side. It was on purpose, I had qualified and started the race with the one wing on the left side." So, is this the aerodynamic version of transferring fuel from one side of the car to the other? "[Laughing] Yeah. This happened on the Thursday and I only had Friday to practice with the Brawner and put it in the middle of the front row. I knew I was competitive and had something for anybody out there."

This was the last year of the Brabham-influenced Brawner Hawk. Mario explains, "What we did with the Brawner Hawk is that we copied the Lotus chassis, and the Brabham suspension points. It was a whole new and different monocoque with some sidepods on there for radiators. That car worked quite well, but we had some overheating problems that really plagued us." The car now sported angled sidepods that raked up towards the rear to create downforce, versus the late-1968 trapezoidal-shaped car with sides that sloped up to the cockpit from the side.

From the earliest parts of the race, Mario kept ducking out of the draft to cool the Hawk, which suffered with chronic overheating. Thinking back to the live race coverage on closed circuit, the commentator was saying how perhaps Andretti just wanted some refreshing cool air, when it was obvious what was really happening. Mario recalls, "Here we go again, the thing is going to come apart. Some of the other guys dropped out, like Foyt or Ruby and so forth, but if push came to shove I could handle both of them. But it was one race where I drove really smart, trying to see if I could nurse this thing to some degree." Much of the 1969 Indianapolis 500 saw Andretti and Ruby running 1-2 between the Hawk and Mongoose. Lloyd was to lead eleven laps near the middle of the race with the old BT-11 based Mongoose when he pitted for fuel from the lead on lap 105. Starting to leave the pit early with the fuel hose still attached, it tore the fuel fitting out of the side of the car and its race was run. The Mongoose was the last car to seriously challenge Andretti for the lead that year at Indianapolis, the highlight in what was probably Mario's greatest season. Revson qualified his factory Brabham BT-25 in thirty-third but finished fifth, the highest-finishing rookie, albeit three laps behind.

Mario looks back on his years in the Hawks fondly. "When you look at the amount of years that we had that we were a factor, I look at it as the best part of my Indycar career for sure with three Championships and two seconds in five years. How in my first five years from 1965-69, how close I came to having five championships. The two years that I finished second were 1967 and 1968, without scoring a single point from Indy either of those years, and Indy was two points per mile, 1,000 points if you won. If you won Indy you were almost a shoe-in for the championship." It certainly worked that way for AJ Foyt in 1967, and for Bobby Unser in 1968. Mario says, "They beat me by the closest margins in the history of that points system."

In 1969, Andretti was to score eight wins, including the Indianapolis 500, and scored nearly double the points of second place Al Unser's Lola by seasons' end. In the end tally, following on the overwhelming success of the Watson-Offenhauser (and the Watson clones) up to and through 1964, USAC Championship Car domination then shifted over to the rear-engine designs. Despite the stellar performances by Team Lotus and their many innovative racers from 1963-69, the most successful Indianapolis cars of the mid-to-late 1960s were the Brawner Hawks.

CHAPTER 6

GURNEY'S EAGLES

Gurney's 1967 Belgian Grand Prix winning Eagle-Westlake.
A beautiful and timeless design (Norm DeWitt)

Dan Gurney was driving in Formula 1 for the Brabham team from 1963-65, getting the first Grand Prix wins for the manufacturer in 1964 with their stellar BT-7. Closing out his Brabham years with five straight podium finishes at the end of 1965 with the BT-11, there was nobody in a better position to understand the strengths and weaknesses of the Brabham and be in a position to make an improved Indycar version than Gurney. No doubt following with interest the success of the BT-12-based Brawner Hawk in 1965, Dan instead chose to follow the Team Lotus monocoque design path of the 1965 Indianapolis 500-winning Lotus 38 by designer Len Terry. Dan explains his logic in pursuing the Lotus path for his new Eagle:

> Well, a lot of people were trying to understand what constitutes a better car. Ron Tauranac, the designer of the Brabham, set a very high standard and the cars were user friendly to the drivers. But we were not in a position to do that, we worked with Len Terry who was at Lotus. Ours had a bit of Colin Chapman thinking in it. Naturally it's more complicated than that, but I got to know Colin Chapman and Jimmy [Clark] of course when we went to Indianapolis in 1963. I was familiar with all the things that they were trying to do and all the problems that we tried to solve. We tried to include as many of the things that we thought were important, but the car was not very good the first time we ran it. We had one flaw and we fixed it.

There were issues with the initial suspension geometry. Gurney confirms, "Yeah. When we came out with the first one at the beginning of 1966, I ran it at Riverside up the straightaway on the rise under the bridge, and when it came down it wanted to take off one way or the other. It took us a while to get that fixed, but once we got that done, it was darn good."

There was no lack of ambition at All American Racers and the 1966 Indianapolis 500 saw five Eagles on the starting grid. The AAR team had qualified two of their new Eagles together on the seventh row, Dan Gurney on the inside and Joe Leonard in the middle with his "Yamaha Eagle-Ford." Roger McCluskey was next in the Lindsey Hopkins owned version, having completed the all-Eagle seventh row. Jerry Grant had qualified his AAR Eagle-Ford in tenth place, and Lloyd Ruby was best of all with a fifth-place grid slot for the AAR entered Bardahl Eagle. It was the fiftieth running of the Indianapolis 500 and there was no bigger stage upon which the Eagle could bring its great challenge to the established teams.

An Indy Eagle that raced in the 1966 Indianapolis 500, powered by the 4 cam Ford V-8 (Norm DeWitt)

Gurney's Eagle was eliminated in the start melee, along with ten other cars. The strongest drive was that of Lloyd Ruby's, who had led almost seventy laps of the race when the car suffered a cam stud failure that ended Lloyd's race at lap 166. Clearly the car had shown speed enough to contend for victory with the right setup and the right driver. Joe Leonard wasn't among those who was convinced as he struggled with the setup:

I drove for Dan in '65 and '66, I drove the Yamaha Eagle in 1966. It was a bad car, Len Terry did the engineering and he decided to raise it up. At Indianapolis, if you can lower your car 1/16th of an inch, you gain 2 miles per hour. Now they scrape the cars with a full load of fuel, and they only carry about 40 gallons [18.5 gallons in 2017]; we used to carry about 75. With about 75 gallons of fuel, the car handled like crap until you got about twenty laps or so. Then it would get pretty good, but when it got light it would float around on you. It wasn't no trip to Paris, that's for sure.

As a result, Leonard confirmed that he would set his car to work best on half-tanks for the race and then just hang on to it before and after. In all fairness it should be said that Joe's '66 Eagle was "a bad car" when it had a bad setup, as obviously Ruby's similar Eagle had performed at a race-leading pace.

Entered by "Anglo American Racers" (given the Westlake-Aubrey Woods V-12 under development) the Eagle had its Formula 1 debut at the Belgian Grand Prix on June 12 with an interim version powered by a 2.7 litre Climax 4-cylinder. Gurney put the car into the points at its second race, with a fifth place finish at the French Grand Prix at Reims. The debut for the V-12 Eagle-Westlake car was at Monza, Gurney bringing in his former Ferrari Formula 1 teammate, Phil Hill, to try and qualify the Climax-powered car. It was a forgettable debut with Gurney qualifying nineteenth on the last row of the grid, and Hill failing to qualify in his final Grand Prix appearance. Dan's V-12 car only lasted seven laps before retirement due to engine problems. During that Monza weekend, Bob Bondurant, having had his fill of a privateer BRM ride, discussed a ride for the USGP with Gurney. Bondurant explains, "I asked him who was driving his second car at the US Grand Prix and he said, "Well, no one." I asked if I could drive it. I drove the Climax-engine car at the USGP, and didn't finish." Bondurant had found himself disqualified for an off-course transgression and Gurney had lost the clutch after thirteen laps.

At the season finale, the roles between Gurney and Bondurant were reversed, with Bob assigned the newer V-12 Eagle-Westlake car. Bob remembers, "Dan told me that he was going to take the Climax car, because it was, 'quicker than the V-12 car right now.' The V-12 had overheating problems. He asked me, 'If you'd like to run it, back off a little bit when it gets too hot, and then pull it on in.' So, I ran it and man, it was good." After twenty-four laps the V-12 car retired with fuel system issues, but Gurney was able to achieve a fifth place finish with the Climax-powered car.

Bondurant says, "Before the race, Dan, Richie Ginther and I had breakfast together and Richie told us that he was leaving Honda after the race. Richie was always a great chassis guy, if you wanted a car sorted, put Richie in it and he will sort it out. So, Dan was thinking it might be better to have Richie because the car still needed sorting out. So, I didn't get the ride in the Eagle." At the end of its debut season in Formula 1, the Eagle had achieved three finishes (all Gurney, and all with the

Climax car) from ten starts and one DNQ (Hill). With a winter to work on the teething problems, Anglo American Racers would make their mark in racing history the following year.

Bobby Unser, who was later to have great success in Eagles, including two Indianapolis 500 wins and two USAC Championships, had been struggling with an impossible situation at Indianapolis, driving the unloved Huffaker car in 1966. Bobby recalls:

We started with that car with a Roots Blower and broke... it didn't work, so we put a turbo in it at the last minute. It was almost the very first of it, the very start for turbocharged Offy and nobody knew much about them. None of us knew anything about turbochargers in those days, which means your fuel curves and so many things like that." The team was obviously just throwing stuff at the car. "We had one engine and two or three guys worked day and night to get it together so we could try to qualify it and we got it in the race. But it didn't even get the hood painted. I mean it was a tough job. The car was not fast, the chassis was just a s**tbox. It was a terrible car.

Needless to say, Bobby was delighted to find a new Eagle under his Christmas tree for the 1967 season, and he didn't experience the same setup problems that bothered Leonard with his '66 Eagle. Bobby explains:

Things got a lot better. [laughs] It was a good race car, especially for its day, and '66 and '67 Eagles were the same car. Ride height is determined by a number of things. Number one, the center of gravity needs to be low, and how do you get it low? You get as close to the ground as you can, to get the thing where it touches the pavement as it goes around the track, and that's as low as you can get it. You use bump rubbers to stop it from getting too low, and soft springs to let it go down that far. They might have thought it was [a problem], but they probably didn't have enough to drink, you know? They needed some better Scotch, that's all.

Asking Gurney about the difference in spec between the '66 and '67 Indianapolis Eagles, he confirmed that the differences were minor. Dan says, "A different version, yes. It had a little bit different suspension." Gurney again was not rewarded with a decent finish despite qualifying second and running strong in the race. "It dropped an inlet valve and was still firing every time and blowing out of the inlet like a shotgun. My left ear has never been very good since then." Don't recommend a V-7? "That's right. [laughs]"

1967 and mid-70's Eagles returned to the Speedway for the 100th Running of the Indianapolis 500 in 2016 (Norm DeWitt)

After a DNF with the Climax-powered car at South Africa, Gurney's Formula 1 team was going from success to success in early 1967, with Dan winning at Brands Hatch in the March non-championship Race of Champions, sweeping both heats and the final with the improved V-12, with Richie taking second and third in the heat races with his. At Monaco, Richie Ginther decided to retire from Formula 1 after failing to qualify the second V-12 car. Gurney replaced him with Bruce McLaren in the second Eagle for three races in the mid-season, and a one-off for Scarfiotti ('66 Italian Grand Prix winner) at Monza, but otherwise the year was a single-car effort. At Monaco, Dan qualified seventh and was out almost immediately with fuel pump failure. It was pretty much the same tale in Zandvoort, with the car dropping out after eight laps, except Gurney had put the car second on the grid, behind only Graham Hill in the new Lotus 49.

Dan had a special titanium and magnesium car built for himself to race in 1967 and the speed was certainly there, even when put up against the fast and fragile Lotus. The first Grand Prix finish for his new V-12 car was surprisingly at the blindingly fast Spa circuit in Belgium, where Gurney was to make American history getting past Stewart's H-16 BRM to win. It was no fluke, as Gurney had also set the race's fastest lap while chasing down the BRM. Not only was this the highlight for both the Eagle and H-16 BRM programs, it was the first time that an

American had won a Grand Prix in an American car since Los Angeles' Jimmy Murphy won the French Grand Prix with his Duesenberg in 1921. Now, nearly one hundred years later, those are still the only two examples of Americans winning a Grand Prix in an American car, both from Southern California.

At France, on the unloved Le Mans Bugatti circuit, Dan qualified the Eagle in third, and was running second when a fuel line broke at mid-race distance. Dan qualified fifth at Silverstone, and fourth at the Nurburgring, where he not only set the fastest lap of the race, but was leading when a halfshaft failed near the end. Gurney was to finish third at the soaking wet Canadian Grand Prix. At Monza, although the cars were fast, Gurney and Scarfiotti were both out by lap 5 with engine problems. Despite being a factor in nearly every Grand Prix, Spa and Mosport were the only two finishes for the Formula 1 Eagle in 1967. Coming to Watkins Glen for the USGP, Dan put the Eagle third on the grid and was battling with Jim Clark for second when the suspension failed. Third on the grid again in Mexico led to another early DNF. The Eagle had made American Grand Prix history, but like the Lotus 49, it was as fragile as it was fast. As a result, Hulme won the World Driver's Championship in his durable Brabham, whose cars were again the Manufacturer's Champion.

Dan's son Alex Gurney takes the Formula 1 Eagle-Westlake
V-12 out for a few laps (Norm DeWitt)

The Eagles were finding success in the USAC Championship Car series that year as well. Bobby Unser was to win two races at Mosport with his 4-cam Ford-powered Eagle, and could have won the season finale, the Rex Mays 300, qualifying third on the grid behind Dan Gurney in the factory stock block Eagle, and Jim Clark in a one-off appearance with Rolla Vollstedt's 4-cam Ford-powered racer (Clark's last race in America). Clark led early, but soon after started to smoke. There are many versions of the tale, but the most likely, after hearing the tale first hand from numerous sources, is that Clark ran the nitro mix of fuel normally only used for qualifying, in the race. One needed to pull the pin on the grenade to stay with Gurney in the latest normally aspirated Gurney-Westlake-Ford powered Eagle, complete with chin spoiler and "tail feathers" (exhaust mounted aluminum wing/spoiler) around Riverside raceway.

Dan Gurney was recently reunited with his 1967 Rex Mays 300 winning car. "I had no idea that the boys had managed to find that and get it all restored by John Miller and his understudy... terrific job." It was the dominant swan song for that 1966-67 Eagle design as the following year's Eagle was a whole different car, a lower, flatter design. Given Gurney's dominant performance at the season ending Rex Mays 300 ahead of Bobby Unser in the similar Leader Card Eagle, one could understand if they had stayed with and further developed that original Eagle chassis design. Did Dan think he had reached the end of the development for the '67 car and it was time for the new Eagle in 1968? Gurney explains, "Yes, but the word you use is 'think,' and you are right. We didn't know... very few people knew. When I was with Chapman, or Ron Tauranac at Brabham, nobody really knew. They were all probing and looking for something and that was true with the '68 Eagle, and it was a good car. That was one of the nice things about that era was because nobody knew very much and we were all hungry to come up with something that was better than the other guys, and there was a lot of enjoyment in that."

A young English designer from Lola named Tony Southgate had been brought in to create this new second generation Eagle. His resume included the Lola T-90/92 Indycars that morphed into the famous "Hondola" mix of a Lola Indycar chassis and the Honda Formula 1 V-12 engine which John Surtees had taken to victory at the 1967 Italian Grand Prix. Bobby Unser in the Jud Phillips car was magic in the new Eagle it from the word "go" in 1968. Gurney recalls, "It was a good car, but it was also the beginning of the Turbo-Offy, and it was particularly good on the straight parts." Dan Gurney could be forgiven for having regrets about committing to the stock block Eagle Westlake Ford in a turbo world, but that is not the case. He says, "No use having any regrets, we won some races with it. [Mosport twice and Riverside]. Later on we ran the Offy, you've got to have a pretty good stock block to run that kind of power."

Bobby Unser reflects on that time:

We started off really good with Gurney and the Eagles. Plus we had a good team, Jud Phillips was top of the line mechanic, and 'Little Red' Tom Herman could have been anybody's chief mechanic, so I had an extremely good team. And in those days you didn't have a lot of people. I had two mechanics and a clean-up guy and we didn't get him until later. He could sweep the floor, put the tools away, clean the s**t up a little bit and get the car clean. It was a two-man team with a gofer, period.

It was the polar opposite of Unser's experiences with Granatelli's outfit from 1963-65, with a dozen people standing around the cars.

The first Indianapolis 500 winning Eagle. Bobby Unser's Eagle-Turbo Offy also won the Championship (Norm DeWitt)

Bobby laughs, "But my cars got fixed properly. The guys that did the work, they did absolutely precise good work. I just did not have mechanical problems with those guys. I may have things that break, but nothing was going to fall off or be out of adjustment. I always helped with the chassis setups as that was my department. If something doesn't handle right, part of it is me. We had three engineers: Jud, Little Red, and Bobby."

Bobby continues, "We only had one 1968 car, the '67 was the backup, my spare. The '68 car just worked, it wasn't a good road race car, but it was good at Indianapolis, in fact that was the only place where it was super-good. There it was untouchable." Looking back upon the '68 season, Bobby had been on pole for the opening race, and then won the next three races headed to Indianapolis. That is overwhelming dominance when you add in the Indianapolis victory, winning four of the first five races in the new Eagle. He tells more about that season:

Most of that was the team operating. But, there is another thing that we aren't covering there and that is the Goodyear tire and rubber company. I was becoming one of Goodyear's prized possessions. If you didn't get paid by Goodyear, you got paid by Firestone, as that was where the money was in those days, and I still think that's how it should be today. I also think mechanics should have to do something other than exchanging parts. These guys today don't know much about what's going on. Hell, they don't see the inside of an engine, they don't see the spark plug guy anymore. There just isn't much to do, but we did all of that and we became smart.

Jud, Little Red, and myself, we learned how to do bump steer, almost none of the American mechanics understood that in those days. We had a Pikes Peak car – a Lotus rear-engine car – and I had to learn what bump steer was in a hurry. When I made the steering faster, I ended up with bump steer, and you lost all your geometry. I owned the thing, and Chapman built it as a Formula 1 car, just like an Indycar with really slow steering. At Pikes Peak you need fast steering, so we had to figure out how to get started with it. How did we learn it? We went down to the library and rented a goddamn book if you can believe that, and took it to the race shop and we just started doing it. Before you knew it, it came to us, why this one had to be higher, why this one had to be lower, why this had to be shorter, and this had to be longer. That's where I learned bump steer was working with that car, via that book. Once we got that, we started going to town once we understood… and nobody at Indianapolis knew it yet. So we started getting down to the thousandths with dial indicator gauges and I used to do some of the strangest things with bump steer that you'd have ever believed. It [the '68 Eagle] was a good race car, but was basically a store-bought car that we didn't know much about.

There was also the switch from the 4-cam Ford to the new Turbo-Offy to contend with. Bobby explains:

It pulled out so much more power that it was just scary. Now, how to perfect all this and get it working to where it doesn't break and getting the carburation to work, meaning to get the fuel metering right. That's really tough to do. We had eight nozzles, two per cylinder, with one set of low-speed nozzles and one set of high-speed nozzles. If you want to get your head f***ed up, pardon my French, that's how to get it f***ed up, that's the way to do it. The low-speed nozzles were basically for getting in and out of the pits. Then once the turbo lights, it automatically has to kick into the other four nozzles, so you have eight nozzles squirting every stroke. You've got to remember that this was Hilborn injection or bypass injection, it was not direct injection, so that is doubly hard to do. But it goes back to Pikes Peak, as we were able to work this out faster. If you could make something run at Pikes Peak with the same parts or concepts, then it's going to make it a lot easier for running on a flat track somewhere.

The Rubik's cube of figuring out the proper fuel metering still wasn't resolved as there were tracks that required a lift mid-straight, to say nothing of the complex variety of throttle openings that result from a lap around a circuit such as Mont Tremblant or Riverside. Bobby continues:

You take Trenton, you've got the back straight and going into Turn 3 and you had to back off, and then going back down into

Turn 1 it was the same thing. But Trenton had that reverse dogleg in the middle, so you'd back off there at really high rpm and then get back on it again. That's a whole different deal entirely and with bypass injection I still don't know how the hell we ever we did it. Now today the computer does all the thinking and you just change it with a typewriter, you don't change it with brains anymore. It was a whole different thing, and we learned faster than the other people did. Most of the guys would buy their cars from the one of the manufacturers. When you got a Gurney car, most of the people had to just assume they are going to get the car right when they buy it because they don't know how to do it [set it up]. And that was a giant thing.

At the end of the season it again came down to a game of musical chairs at the Riverside finale. Bobby says, "In those days you could drive as many cars in one race as you wanted, which was rather stupid, but nonetheless, those were the rules. Mario was in three cars [laughs] and as I go by this damn Turbine I thought, 'Goddamn that looked like Andretti,' and bigger than s**t it was. And then there was another car that he was in." By the end of the day, Unser had won the closest battle for the USAC Championship Car title in history, finishing second with Andretti in the Lloyd Ruby Mongoose third. It was the first USAC Championship for an Eagle.

Dan Gurney on his way to a second successive victory in the Rex Mays 300 with his stock block Eagle (Gary Hartman)

It was no cakewalk, as Bobby Unser had to struggle throughout the race with a frustrating and dangerous situation with the braking system in the Eagle. Unser recalls:

The brakes went bad, the balance bar came loose. There were two times in my career I can think of where that happened. The other one was the Fresno-built Gerhardt 7-inch offset car at Trenton, the first time I ever drove for Leader Card. The goddamn balance bar came loose. They had nothing but little circlips inside the tube that locked the balance bar. That circlip came loose and I wrecked that car bad. And then it came loose that day at Riverside, which prevented me from having much of a chance of winning it. At Trenton you used the brakes twice each lap – at the end of each straightaway. But at Riverside… Oh s**t, you'd better have your brakes done just right. It wasn't that you lost one or the other, you lost too much of one or the other. Even the brake balance bars, if they came loose were made so that if you went down so far then it would get into a bind and you still had a little bit of whichever end. It was like having an emergency brake that you just have to have, but you don't want to use. It didn't happen very often.

In Formula 1, Gurney had persevered with the 1967 chassis for 1968, and the newer lower Tony Southgate Eagle never made the switch over to Grand Prix racing like the 1966-67 design had. Despite the difficulties of running a two-year-old design, Gurney put on a stellar drive in abysmal conditions at the Nurburgring, running as high as fourth, showing what the Eagle could still do, the fastest car on Goodyear tires. This was a day forever associated with Jackie Stewart's disappearing act in the Matra-Ford, helped enormously by his superior Dunlop wet-weather tires. That was the last and only finish for the Formula 1 Eagle in 1968, and their final entry was at the following Grand Prix at Monza.

Bobby Unser and Leader Card had this amazingly well-sorted Eagle and so for 1969 they made an indecipherable move in switching to the four-wheel-drive Lola. Unser reminisces:

That was a Bobby/Jud/Little Red mistake, no question about that. I didn't like that car at all, it just wasn't a good car. We kept the '67 Eagle as I'd totaled out our '68 car at St. Jovite, Canada. That [the '68 car] is the car I won the Speedway with and it is currently in the Speedway museum. Watson took it all bent up, took it out to his shop and straightened that son-of-a-bitch out. That car, until just recently, had only been run once some years ago when I drove it. It was the only time it had ever been run since I'd wrecked it, the harnesses were the same, black tape was still on them, everything. It's really amazing, you know.

Bobby was stuck with the unloved four-wheel-drive Lola, or the three-season-old Eagle – a somewhat unbelievable situation given his status as the defending Champion.

Southgate introduced a significantly different Eagle design for 1969, Gurney again moving away from the car that dominated the season-ending race at Riverside, much as he had done the previous season. The 1969 version could be best described as a flat (not BLAT) Eagle, much flatter and angular in appearance, embracing the latest trend of the Lotus 56 turbine's doorstop-wedge shape overall, not unlike the new Mongoose of Lloyd Ruby. Maurice Philippe's latest Lotus (56 Turbine car) had started an Indycar design trend. Tony Southgate's latest car was a big departure from 1966-68 cars as the beak had been clipped completely, removing the distinctive and signature look that previous Eagles had carried. It was Southgate's last Eagle, as he departed in 1969 for the position as designer at BRM where he penned the Grand Prix-winning P153 and P160 BRMs as mentioned in Chapter 4.

This was a fascinating era in both Formula 1 and Indycar, with a

The 1969 Indianapolis Eagle, the last of the Southgate Eagles.
Gurney finished second in the 500 (Norm DeWitt)

technological smorgasbord of creativity, one that seems hard to comprehend in a current Indycar world that celebrates a single chassis with one approved aero kit. A far cry from today, it is the antithesis to what interests an innovator like Gurney. He says, "Now you have to kiss a bureaucrat to get permission to do anything. I don't want to come across as somebody who is negative on what is going on now, but if you start taking away freedom, I don't like it, and I don't think the fans like it either."

Gurney had finished second in the 1968 Indianapolis 500 with the stock block Eagle and for 1969 he returned with the new chassis, again powered by the Gurney-Westlake Ford V-8. Joe Leonard was back driving an Eagle again for 1969, only this time in a Smokey Yunick-entered car. Joe recalls, "I still hold the motorcycle record on the Daytona Beach course, about 99.9 miles per hour. Fireball Roberts only went 101 in Smokey Yunick's Pontiac convertible. When I drove Smokey's Eagle at Indianapolis in '69, something came up about that and he said, and you know Smokey didn't mince words, 'Hey Jose... Hell, I knew you was a good racer because I remember you ran that little chickens**t bike. You had what, about 40 horsepower? And you run within about a mile per hour of Fireball in that Pontiac with 400-plus horsepower?'" Leonard had qualified alongside Dan Gurney on the fourth row of the grid at Indianapolis with his City of Daytona Beach Special, and was running second until he was black-flagged for a coolant leak. Until that point, it appeared that the Eagle-Turbo Ford was as much or more a competitive proposition than the Dan's AAR Eagle with the Gurney-Westlake stock block.

In the end, Gurney again took second place at Indianapolis and showed the pace of the car throughout the season. Pole at Continental Divide raceway led to another second place finish. At IRP, Dan was to win race one from pole, and then won race two at Brainerd, the car at mid-season having sprouted a variety of winglets to the sides of the engine cover. Kent, Washington, held the Dan Gurney 200, and again showing his pace on road circuits, Dan started race one from pole and then closed out the year with pole at his beloved Riverside, finishing third behind the dominant Brawner Hawk of Andretti, and Al Unser's Lola. Dan Gurney finished fourth in the Championship despite only entering nine of the twenty-four races.

The 1970 Eagle was a bit less radical in its angularity of shape, and although the nose remained somewhat flat with the vertical fences atop the nose, its shape had as much in common with the 1968 car as the moderately successful flat Eagle of '69. Picking up where he left off with a strong run at Riverside, Gurney won the race at Sears Point Raceway.

If one had any doubt as to the basic goodness of the 1966-67 Eagle, they need look no further than the 1970 season for confirmation. In 1970 Bobby was still running the Leader Card '67 Eagle with nose wings and other mods. Johnny Rutherford recounts his Eagle experiences:

In '66 I was out for the year with two broken arms from a sprint car accident in Eldora. In '67 I came back and that Eagle was available and we were able to obtain it and got it together. I came back to racing at the 500 that year and those were good race cars. They handled well, and they didn't give you a lot of grief. In '70 Mike Devin was my crew chief on the car and we designed a body on the thing to give it downforce. It was a flat car all over, we struggled and struggled with it and the thing slowed way down. The Friday before the first day of qualifying, Mike came back and cut out a piece of aluminum and bolted it to the bottom of the nose... a little chin spoiler. It made the car and we did well, but had a broken header during the race and spent a lot of time in the pits – finished eighth. Year to year things change and people take what they've learned previously and decide what would make it a little better.

Rutherford qualified second at Indianapolis, only 1/100th of second behind Al Unser in the Colt. 1970 was also the year that Bobby had his audition at AAR. Bobby remembers:

We were at the Speedway, not during May, testing cars for Goodyear. In the meantime, Gurney was going to retire but we didn't know that, we'd heard rumors. Somebody called me from Goodyear... it was probably Truesdale or Ed Alexander, and asked me if I would mind trying out a Gurney car if they brought it out from California. They were our bosses, and I said, 'Yeah,' but I'm not that dumb, I'm starting to figure out what they are doing. They decided where Bobby was going to be, that's where the money came from. So they asked if I would try out Dan's car a little bit during the tire test.

Gurney had agreed to replace the irreplaceable Bruce McLaren in Formula 1 and Can-Am cars after Bruce was killed. It was to be Dan Gurney's final season and the Indianapolis test was the audition. Unser recalls:

They had to get a driver, this is Goodyear's possible number one team. They've got Foyt, they've got Gurney, but Gurney was kind of on top of the pecking order because he built and sold cars. I was confused, as I wasn't thinking of Wilke and Phillips, that wasn't even in my thinking process, but of course if Goodyear asks me to do something, it's where my paycheck comes from so I said, 'Yes.'

Dan comes back there and he's got the heavy hitters... both Wayne Leary and John Miller the magic engine man are there with him. Wayne was an old good friend of mine and we go way back to the sprint car days. I didn't know Dan that well, but he had brought the first team. I was running the tire test and we shut down while they were going to remount tires. I had like half hour or hour that we were going to be down, and in the meantime Dan and Wayne were getting their car ready and all heated up and ready to go for me. Well, this is show biz, and I am still a little bit confused. I get into the car and now I can see what is going on... this is a 'dog and pony show' to make Bobby happy because they are looking to hire me, it was just obvious by now. So, I got into the car and went around a couple of laps. That wasn't known to be a supercar by any means, but boy I'll tell you what, I went fast in it immediately. I did a few laps and the car was better than the '67 Eagle that I was running right then, my race car. On the phone that night at the hotel, Dan speaks so softly on stuff like that, sometimes it's a little hard to understand him, but he asked if I was interested in getting together with him. I knew then that Goodyear was orchestrating this and that Dan was going to quit and they had to find somebody to replace Dan as his was their big team. So I went to talk to Jud and Little Red and told them that it was part of the deal that Dan had to hire them and keep them as part of my team as I wouldn't part with those guys for nothin'. Dan didn't like it, but he agreed to it. As it turned out, Jud said 'no' as he just didn't want to work with a big team. Little Red came and stayed with me for a while until he decided he didn't like being with a big team either and he took off. No hard feelings, it was all really good. Wilke understood it as #1 he will not spend the money that's necessary as each year everything cost more than last year. I demanded fast cars and Goodyear was paying me the big bucks to drive fast.

Bobby somehow got second in the 1970 championship with a five-year-old Eagle design, while struggling with unusual tire issues. Bobby - "You know that '67 Eagle used to blister the left rear tire, not the right rear tire. I've never seen a race car do that before or after. Let's take Phoenix, Indianapolis, Trenton, any race track you want to take, it would blister the LR. The tires in those days… we had gumball tires, if that's what we wanted. I did most of the testing so sometimes I had different tires." Unser started at Langhorne from pole, winning the race.

Gurney had finally joined the turbo brigade for 1970, running a turbo-Offy at Indianapolis and finishing third. Despite having a third and a first from the first four races of the season, Gurney only ran one additional USAC race in 1970, the debut Indycar race at Ontario Motor Speedway. Although it was the end of the trail for The Big Eagle and his racing career, at the season finale at Phoenix, Swede Savage won the Bobby Ball 150 with his AAR Eagle-Ford. All American Racers had closed out their final season with Dan Gurney behind the wheel, victorious.

1971 brought Bobby Unser onto the team with a sleek new Eagle designed by Len Terry and Roman Slobodynsky where the beaked nose had returned. Fast does not begin to describe the new Eagle, as it shattered records everywhere it had raced. Unser started seven of the twelve races from pole, winning at Trenton and Milwaukee. He reflects:

On the left front I'd use a really soft tire and the rubber would rotate on the tire itself and the rubber would become egg-shaped. The front end was shaking so bad at Phoenix that my hands went numb and I couldn't hold the wheel. On the brakes it would shake so bad that you couldn't even see. The damn thing would go fast and a lot of times I could set it on the pole and a lot of times I would lead races, but it didn't make any difference as the car would be shaking so bad. There was something wrong with the design, it's the only car in my life that would ever do things like that, but the car was fast. We can look back today and tell you what some of the problems were, but we didn't know that then. John Miller knew how to make that car go fast, I was never short of horsepower when I had John Miller. I ran the last race of the year at Phoenix for Dan and then the car went to the home office in Santa Ana and got changed for the next year. We had the Gurney flap that I called the wicker bill. That was totally Dan's deal, but that worked better than anyone would believe. We were doing a lot of changes and were way ahead of everybody with the wings.

The "Gurney" or "Gurney Flap" became one of great contributions of this era to aeronautical development, regardless if the topic is racing cars or aviation. This small vertical piece added to the trailing edge of the aerofoil section, if done properly, gave additional downforce without a corresponding increase in drag. Unser reflects upon the culture of innovation:

In '72 Hughie Absalom and their group [Parnelli] couldn't get those speeds we were running and thought we were cheating. It was just that we had some really good secrets from our winter testing [that presumably didn't find their way onto the production Eagles]. The two guys in the world that I drove for an awful lot were Roger Penske and Dan Gurney, and neither one of them would ever cheat. They couldn't even sleep at night if they had to cheat to win a race, they just do not cheat. Did we have special stuff? Boy, you ain't kiddin' we did! You know that I wouldn't cheat too, but I'd be willing to get awfully close. I'd fess up to that.

A few years ago Al and Parnelli and I were doing a deal down in Florida and they were talking about how Bobby cheated all the time and had a different turbocharger than anybody else had during that timeframe. Well, of course I did, but number one, there was nothing illegal about that. And number two, they never bothered to check and see what I had at the time – they didn't know that I had it. Somebody, years later, told them about it. I could tell this Air Research turbocharger from 30 or 40 feet away, as there were only two of them ever made, and I'd see one laying on the bottom shelf in Rutherford's garage. And they never ran theirs, they never put it on the race car. How could it be illegal when there were no turbocharger rules of any manor, shape, or form? You could run a giant one, a little bitty one, you could do whatever you wanted to do.

There was some serious internal drama going on at AAR between the factory team side and the customer sales side of the endeavor. One is reminded of Dick Scammell's comment about how the monocoque Lotus 25 "...was a nice surprise except for the reaction of the customers who had just bought their cars from the other side of the business." As always there is the tension between the bits the "factory" car is running versus what the customer has available for purchase in comparison. At Eagle, there was apparently a quick pipeline between the factory car and the customer's access to the development bits. Bobby continues:

The 1972 car was a happy race car. That thing it had a big smile on its face when it was born. It didn't take much of a driver, that car just liked going fast. He sold so many of those things, but I'd get into big fights with Dan about the guy who did the selling for him, and all I wanted to do was to just smack that guy. I'd go out and develop stuff, usually doing all the testing at Ontario. I'd develop some new wings, real nice stuff and that guy would sell it to Bignotti before I would even get a chance to run it in a race. I had an agreement with Dan that I would have two to four races, whatever the agreement was, before anybody else got the new stuff. But that guy, he would always know about it and just was doing his dealing, especially with Bignotti. I showed up at Trenton for the spring race and looked at Johncock's car and he's got my goddamned new front wings that I'm not even going to let out yet. I was hiding them, saving them for Indianapolis. I just hit the freaking ceiling. Mine was the factory car, it should have gone fast. I talked with Dan a lot about that stuff. If you aren't careful, that's where your driver starts lying to you in testing. I made a deal with Dan that I'd never lie to him. If I found something good in testing, which I always did, I just wouldn't lie to him about it. I gave him a straight shot and ran it as fast and as hard as I could. When he started selling it to the other competitors before I'd was even going to run it, that was not good as hypothetically I could test a new front wing, decide that it's good, not go as fast as I could and not put together a full lap. Just set it aside and tell Wayne to hang onto it for me and not tell anybody about it, but that wouldn't be right. Everybody was getting smarter and I'd try two or three different wing designs at every test. The car was 18mph faster when they got to Indianapolis than the year before.

Unser's years with Gurney were feast or famine. A stellar pole at Indianapolis in 1972 was one of seven that season (from ten races) and Unser won at Phoenix twice, at Milwaukee, and Trenton. However that season was best described as "hero or zero," and once the other abysmal results were added up, Unser had an eighth-place finish in the points chase. 1973 was worse, with a single win at Milwaukee and a second at Trenton for twelfth in points.

Although the factory team wasn't having much success in 1973, their cars were selling in record numbers. Joe Leonard had won the Indycar title in 1971-72 with the VPJ cars, but their '73 racer was a disappointment and it was hardly surprising that he was eyeing his options. Joe recalls:

Art Pollard came up to me at Indianapolis in 1973, about fifteen or twenty minutes before his last practice run. He was smiling and said, 'Joe, you've got to get an Eagle, you've got to get a Gurney Eagle.' He was running about fourth quick, and he was so happy... he was like a kid. Art was older than the rest of us, he was well preserved, stayed in shape and all that. He was drafting right behind Bobby Unser and was in dirty air and when he went to turn it in turn one, it wouldn't turn... he hit the wall, barrel rolled and ended up over in turn two.

Art Pollard was killed in the accident. 1974 saw Joe Leonard heed his former teammate's advice and he joined the many racers driving Eagles. Sadly, Leonard suffered serious injuries that ended his long career in a grinding crash against the wall at Ontario soon after.

1973 was to bring Eagle their second Indianapolis 500 win, and it was the Patrick Racing, George Bignotti car of Gordon Johncock. Tragedy had struck during the race as former Gurney protégé Swede

Savage in the other Bignotti Eagle had an enormous fiery crash coming off turn four. A month later Swede succumbed to his injuries in the hospital. The rain-shortened race was one of those events at Indianapolis best forgotten.

1974 saw the AAR turnaround with a dominant Bobby Unser achieving ten of fourteen race finishes on the podium, four of them wins. Fifteen of the cars starting at Indianapolis that year were Eagles, an incredible success story for Dan Gurney, AAR, and as a manufacturer. By season's end, Bobby Unser had won the championship by over 1200 points from Indy winner Johnny Rutherford and McLaren. The following year was entirely different, Bobby finishing first when another downpour during the Indianapolis 500 ended the race early. It was his only win during the season, and the final win for an Eagle at Indianapolis.

Bobby Unser soon left the team, but returned to AAR in 1978, taking a single podium before leaving for Team Penske in 1979. The next great era for Gurney and his Eagles were the BLAT Eagle years beginning in 1981, analyzed with Trevor Harris and Geoff Brabham in *Making it FASTER*, and the 1983 Eagle with Al Unser Jr. and Trevor Harris later in this book.

Dan Gurney is at home on two wheels or four. He raced motorcycles and was the Montesa importer (Norm DeWitt)

CHAPTER 7

AJ FOYT AND THE COYOTES

*AJ Foyt leads the parade of winners at the 100th running of the
Indianapolis 500 in 2016 (Norm DeWitt)*

AJ Foyt's first Indianapolis 500 was in 1958. AJ Foyt says, "I was a rookie and was nervous as I didn't know nobody. We had that big wreck down at the end of the back straightaway in turn three and I spun through it. Somebody run over the top of Paul Goldsmith, I think that was Jerry Unser. I didn't hit nobody and nobody hit me, I was lucky that all I did was spin. I wasn't sure if I wanted to go back there and run again, because I didn't really know if I liked it or not." One can assume that a situation like that might make one reconsider the career path. "Well, they will tell you about how you get the draft and this and that. I wanted to call them back and say, 'What about this first lap wreck? You didn't say nothin' about that.'"

Foyt persevered and teamed with George Bignotti in a variety of Watson or Watson clones, he was to win the Indianapolis 500 in 1961 and 1964. Foyt was held in the highest esteem for his driving abilities, which was perhaps best summarized by Joe Leonard, who tells what happened when his USAC-stock-car teammate Foyt asked him to try out an Indycar. Joe recounts, "When I switched to stock cars, I was rookie of the year for the Chrysler Motor Company. I won at DuQuoin, AJ Foyt was my teammate and he asked me, 'Hey Joey, you run pretty good. Why don't you come and drive a real race car?' I asked him what he was talking about and he said, 'Come on down to Indy and I'll get you a ride.' Well, they [Chrysler] said no to my going to Indy. I said 'Why? Foyt drives the stock cars, Foyt drives the midgets, Foyt jumps into a sprint car every once in a while... Why?' They said, 'When you are AJ Foyt, come and see us.' It was a slap in the face after I drove my butt off for them."

AJ had hedged his bets for 1965, ordering one of the new Lola T-80s designed by Eric Broadley and the sorcerer's apprentice, Tony Southgate. Foyt also had the Lotus 34, which in the end he selected to race, Al Unser getting the T-80 for the 500. It was the right decision as AJ Foyt put the year-old Lotus 34 on the pole at Indianapolis, up against Jimmy Clark in the latest Lotus 38 factory car. AJ says, "In 1965 I'd just got hurt out there in California and kind of was just coming back. Then Jimmy Clark had gone out and set a new track record. I went out behind him in his old car and said, 'I want to bring the track record back to the United States.'"

Keep in mind that this was an era where nobody in a year-old Lotus was going to out-qualify Jim Clark in the latest factory Lotus. Foyt recalls, "You know, earlier that week I wrecked it. We broke a rear upright and I spun coming off of two. Then we worked real hard to get it ready. Theoretically, after I broke my back in California, I wasn't supposed to be driving. When I spun and was going backwards all the while I was thinking how the doctors said it could be serious and I could be paralyzed, so I laid my back as flat as I could against the back of that seat. I think we got it back together a couple of days before qualifying. Fortunately George Bignotti, my chief mechanic, had it set up pretty well and we got the pole back, which was great."

Were you surprised to out-qualify Clark in the latest Lotus 38? Foyt says, "Nah, I wasn't. We were handling pretty good even before I wrecked it... and I felt I knew a little bit more about Indianapolis than Jimmy Clark. Jimmy was a great race driver, probably one of the best Formula 1 drivers I've ever seen come over, but I felt like I had my car working a little bit better than him and Chapman had theirs working, even though it was the old car, you know?"

The first Coyote of 1966. Background – The Lotus 34 that Foyt put on the '65 Indianapolis pole (Norm DeWitt)

It was the rare exception when Lotus would show up with a new design and not instantly make the previous model obsolete. AJ agrees, "Yeah, it really was, you know. In '64 I run the roadster because I knew it would run all day. I knew it wasn't the quickest car at the start of the race, but I felt that at the end of the day that it would be just as fast."

1966 brought the first Foyt Coyote and it certainly wasn't the fastest car at the Speedway. Foyt also entered a modified Lotus 38 for George Snider, which started on the outside of the front row alongside Clark and Andretti. In comparison, AJ started on the outside of the sixth row after crashing on pole day and having to qualify on day two. The silver lining was that his lap had been the seventh fastest in the field of thirty-three. In the race, both Foyt cars were out from crashes by lap 22. AJ looks back on that first Coyote, "Well, we just didn't have a lot of suspension the way it should have been, it just didn't work very good."

That first year Coyote was heavily influenced by the Lotus 38, which had won the previous year. He continues:

It was kind of close to it, but we did other things. With the 1967 car with Quinn Epperly Limited and Bob Riley out in California, they did a lot of work for us on that too. Bob was working for the Ford Motor company and I was working for Ford at the same time. Bob and I got real close and he kind of helped us. They'd come by and we'd work on it in the wintertime, it was a clean sheet of paper, and that's how it came together. We'd seen a lot of stuff, and maybe some of it was like a copy, but we had come up with our stuff, our own molds and everything… it was our own design. Klaus Arning [head of Ford's department of advanced suspension design] was a very smart guy at Ford who figured out the suspension for us. Bob was involved with the body and suspension design. Those were two brilliant guys.

Foyt had been the thirteenth-place finisher in points for 1966 and decided instead to run #14 for 1967, his first year running the number that would become his trademark. Superstitious guy? AJ says, "Not too much. Bill Vukovich run #14, and you know, Wilbur Shaw run #14 too."

The 1967 car that won the Championship as well as the Indianapolis 500 had asymmetrical suspension for oval track racing with an offset chassis to the left. Foyt had the Coyote available with either asymmetrical or symmetrical suspension for the road races. He explains, "I think a lot of people went symmetrical because of the road racing and all that. I really liked my cars offset for an oval, but it just got to be so much money that everybody went symmetrical. Who really brought the first car that was really run fast and was symmetrical was Dan Gurney and the Eagle. That's when everybody just started going symmetrical. With the offset car you needed to change the suspension, this and that… it was a cheaper way to build a car." The Brawner Hawk was the other contending car that ran a symmetrical suspension, whereas the Lola and Lotus in 1966-67 were also running asymmetrical.

The view everyone had of AJ Foyt's Coyote in 1967. Note the offset in the asymmetric suspension (Norm DeWitt)

Or, you could just buy a symmetrical car for the road races. After the Indianapolis 500 win with his Coyote, AJ purchased or rented the Eagle that Dan Gurney had qualified second at Indianapolis, appearing at Mosport with a bright orange Eagle to race, taking seventh in both races held there. Switching to the Coyote, presumably as the symmetrical version was now available, AJ finished seventh at IRP, and then second to Mario Andretti in both races at the Circuit Mt. Tremblant road course in August.

At the 1967 season finale, Foyt ran the symmetrical Coyote at Riverside. Some drivers didn't have an option to run anything but an asymmetrical car. AJ had also brought along the ex-Gurney '67 Eagle-Ford to Riverside, unraced since Mosport, for Jim Hurtubise to race. Jack Starne has been AJ's right-hand man for fifty seasons. Jack says, "We just borrowed the Eagle from Gurney. Hurtubise drove it at Riverside and when AJ crashed out we wanted Hurtubise's car, but it blowed up or something, so AJ ended up getting into McCluskey's car. I think we kept that car for Donnie Allison [in 1970] who crashed it at Indy. I don't know what the arrangements were, but I do remember taking it back to Dan's shop."

Foyt recalls, "In the race, somebody spun, I went out across the dirt and I'll be dammed if they didn't come clear out there and hit me. That was the bad thing. All I had to do was finish to win the Championship. So, McCluskey pulled in and let me drive his car and I finished the race in his car." Those races in 1967 (Mosport and Riverside) were the only times Foyt raced an Eagle. "I would have to say, yeah… that was the only times I drove an Eagle. [At Riverside] I was just trying to win the Championship, and that was the biggest thing. It was a good car, don't get me wrong, but I was just trying to get it home." Race results show that Jim Hurtubise in Foyt's orange Sheraton-Thompson Eagle had retired at half-race distance with an oil leak.

1968's Coyote was a development of the 1967 car, and was quite severely outclassed in terms of speed by both the Turbine and turbocharged opposition (Bobby Unser, Lloyd Ruby, etc.) at Indianapolis, as Foyt had continued with the normally aspirated 4-cam Ford engine. Jack Starne says, "The way things first started out with the Turbocharged Ford, I would say that the problem was reliability, to be honest. We worked on that project for quite a while but finally got it under control." Foyt was to win at Continental Divide Raceway and Hanford (along with the Hoosier Hundred and Sacramento dirt races) to finish sixth in the points. 1969 was to bring a new car and the new Turbo-Ford powerplant to the team. Raw speed was no longer the problem.

AJ Foyt in his modified Coyote of 1968, with 4 cam Ford power,
at the Rex Mays 300 season finale (Gary Hartman)

The turbocharged Ford wasn't exactly the most cautious choice to make, noting the variety of engines playing musical chairs in and out of Andretti's Hawk throughout 1968. Given AJ's ties to Ford, there wasn't much choice to make for the Foyt team, and the new Coyote certainly proved its potential with the pole at Indianapolis, the only car to qualify at over 170mph that year. But this was also coming off DNFs at Phoenix and Hanford in much shorter races (150 and 200 miles respectively). The car sported dive plane wings with endplates on either side of the nose, and small wings on either side of the roll hoop, ahead of the main wing over the engine bay. The overall shape was flattened from the previous car, with the wide, flat nose similar to the Brawner Hawk, although the Coyote lacked the raked sidepods of the Hawk. The other cars that shared the front row at Indianapolis were Andretti qualifying second and Bobby Unser in the four-wheel-drive Bardahl Lola with a turbocharged Offenhauser on the outside, that car sporting a small, low wing hung off the rear of the car. The Lola was decidedly different in appearance from the other two cars, which could well be described as wing factories. It was aero downforce versus four-wheel drive on full display in the search for grip.

In the race, Foyt was on the pace as the race approached half distance (66 laps led) when he had to pit for a cracked manifold, which really kills the performance of a turbocharged engine. Eventually rejoining after repairs, Foyt was to finish eighth, nineteen laps behind winner Andretti. Teammate Jim McElreath fared better, turning a thirty-third qualifying performance into a fifth place finish.

Other than at Indianapolis, 1970 was a terrible year for Foyt as the season highlight was a pair of third place finishes, one in the Coyote at Colorado Divide, and one in the dirt. The season was a runaway for Al Unser, following his Indianapolis win along with victories in seven of the last nine races. However, of those two races that Al Unser didn't win, one was McElreath's victory in the inaugural Ontario 500, driving the Coyote-Ford. The season's final tally saw AJ finish ninth in points, with McElreath taking third.

1971 saw a bit better fortune, but the ever-more angular Coyote shape had finally reached the point where it resembled a wedge with a front radiator. The car had a distinctive asymmetrical floor or low sidepod extended out to the left between the wheels. Jack Starne explains, "I think it was for fuel, because we were short and that was an add-on part on that side." That low sidepod was also raked to provide an aerodynamic benefit as well. Qualifying was a McLaren benefit, but Foyt managed to take sixth for a decent starting position. In the race, AJ managed to finish third, although he never seriously threatened winner Al Unser or second place finisher and pole position starter Peter Revson in his McLaren. Winning the final race at Phoenix gave Foyt the nod over Bill Vukovich Jr. and Al Unser in the tight championship battle for second place behind champion Joe Leonard. In contrast, McElreath was unable to qualify for the Indianapolis 500, and in the four races he ran with the Coyote, his best was a sixth place in the season finale at Phoe-nix. It was at this point that Ford ceased involvement in their turbo-charged Indianapolis engine, and the program was taken over by AJ. The cam covers were changed from saying FORD to FOYT.

1972 was forgettable to the point of being beyond mention. The 1972 Coyote was a more traditionally designed sidepod car that looked similar to '72 Eagle. Jack Starne agrees, "Yeah, they all looked about the same, with that Eagle look. I think that's how all of it came along, you see what everybody else is doing. You kind of put your stuff together and go for it."

1973 saw Foyt's new racer by Bob Riley make its debut. It had a low monocoque with a wide wedge-shaped sports car nose akin to what was going on at teams like Tyrrell with their #005 for Jackie Stewart in Formula 1 that year, except the new Coyote had an adjustable wing section in the middle of the "sports car" nose, versus the conventional narrow nose with adjustable dive plane wings on each side. The new car was quickly to evolve into a contender, winning at Trenton and Pocono. Jack Starne recalls, "We didn't run that [center wing] too long. The thing worked alright, but you couldn't really make it strong enough… it wanted to fold up because it worked so hard."

The team would throw a little black cover over the wing as if to hide a setup from prying eyes, which seems pointless when absolutely nobody else was running anything remotely like it. Starne laughs, "Well, you have to keep everybody alert, keep them on their toes even if nothing is going on. We needed to make some changes and the next year we took some big radiators off of the rear and moved them all to the front. Just kind of made the package more aero-wise."

Bob Riley had built a new car for defending USAC Champion Roger McCluskey and team owner Lindsey Hopkins in 1974, and that car shared many features with the new Coyote. The Riley-Offenhauser was wide, low and flat with a narrow cockpit. This was an era of a wide, low and flat monocoque, which was pioneered by the Lotus 72 in Formula 1. That concept was really poor in torsion and longitudinal bending, but apparently the low CG and aerodynamics trumped everything else, as AJ confirms, "Right, I think it was center of gravity more than anything. I had the car in '75 and run off from everybody, but something happened. In '76 I was running second when they stopped it for rain. It seemed like every time I was there with this car something happened, in '77 I didn't know why I hadn't won this goddamn race three times in

The 1977 Coyote with the wide flat tub that put the upper wishbone connection mid-upright (Norm DeWitt)

a row... it seemed like every time something happened. In '77 we ran out of fuel and got into the pits... then I turned around and made up 32 seconds. It was my favorite car and my favorite win. It was great to have the first win and be lucky enough to win a race there... but if I had to pick through all of them, I'd say 1977 was probably the greatest year I could ever have."

Jack Starne adds, "AJ drove that car which was built in '73 with the high sides, then run through '76, but we built a brand new car for 1977.

It looked similar, but there were a lot of changes. The cockpit area was down lower. We ran one bar that went from the roll hoop forward and down to the side that took care of the torsional part of it." This bar also helped to keep the cockpit intact in the event of an accident. There was also high front wing that mounted above the sport car nose that was sometimes used. Starne continues, "We ran the wing on short tracks. The [sports car] nose thing just worked for us so we kept with it."

Have no doubt that there was a favorite Coyote looking back across the years and this one was it. Foyt reminisces, "I'd have to say the 1977 car because I was able to take the name 'Ford' off the motor. We developed the supercharged motor and it had 'Foyt' on it. We built the motors in Houston, Texas. We built the car and it was designed by Bob Riley. The owner, the driver, the car, the motor, was all built by us. It was all a Foyt deal. Of course I loved sprint car racing and I'd come out to Ascot and race. But as far as Indycars, I'd have to say the '77 car was my favorite." The last season for that design was in 1978.

Low and flat taken to extremes with the 1977 AJ Foyt winning Coyote, and a Turbo Foyt V-8 engine (Norm DeWitt)

In late 1978, the team shifted over to the Vel's Parnelli Jones (VPJ) car, which wasn't exactly the newest car in the paddock. Parnelli had pretty much closed up shop by that point. Starne recounts, "We were still struggling with engine problems a little bit – the Ford engine as it was awfully heavy and awfully big. You are pushing a lot of air and we did our best in that area. AJ and Parnelli struck a deal to drive the car in Ontario. We broke something on the gear selector in the transmission. We had some gearbox parts made that fixed that. It all kind of came together for the rest of the season. The VPJ car was a really nice race car, John Barnard and Gordon Kimball did a nice job on that car." The torsion bars were gone by this point and it was a reliable, quick, traditional car. Starne continues, "It had coils all the way around and I don't know what they had done to it before." The Foyt team raced the VPJ car through 1980.

The 1981 Coyote appeared to be influenced by the Chaparral Yellow Submarine. It was the first ground effects Coyote, and the last. Foyt's team had been pretty late to the party given that the Yellow Submarine had debuted at Indianapolis in 1979. Jack says, "Yes, we were a couple of years behind, but we had faith in the VPJ car, as it was still quite fast. There always comes to a time where you have to change, and that still applies today. We always used to build a new car to take to Indy every year. We never tested the car, just went straight to Indy, and this car first ran at Indianapolis. It [the '81 Coyote] was a nice car, and that was a Riley car too. It was a big learning curve for us and we learned an awful lot when we finally got a handle on it."

Going from a Parnelli to a full ground effects car was a big leap, and running a few races in 1981 meant that this had the shortest lifespan for any Coyote. Still, the car impressed with its speed and Foyt qualified on the outside of the front row for the Indianapolis 500. Foyt was also to

Bob Riley's first ground effect Coyote was the last. Foyt started on the front row at Indy in 1981 (Norm DeWitt)

win the 1981 June USAC race at Pocono with a March, leaving the new ground effects Coyote to George Snider, who finished fourth. Snider was to become the 1981-82 USAC Champion, oddly enough counting both Indianapolis 500s in the final tally. This was the era where drivers split their allegiance between CART and USAC, to the detriment of both.

At Michigan's CART race, Foyt had a big accident when he slid on some oil and suffered a heavy impact with the wall on the right side of the car. On that sad note, so ended the Coyote era. The new Ground Effect Coyote had a racing career of only three months. Jack Starne remembers, "It was a nice car and there was only one of them ever made. But, I don't know of any highlights for that car. That was the car that crashed at Michigan and AJ broke his arm. We went to the March right after that. The only one that had highlights for me was the 1977 car, it was pretty much my responsibility to change the drawings, different things here, different things there, for his fourth win."

In the final tally across sixteen seasons, the Coyotes had brought AJ Foyt two of his four Indianapolis 500 victories as a driver. In this day of single-sourced spec cars, it is hard to visualize a grid made up of unique Coyotes, Eagles, McLarens, Penskes, Chaparrals, and Parnellis, just to name a few. In their era, this blizzard of technical variety was normal, and captivating. As a result, much of this book is a celebration of what we have lost.

Chapter 8

The Lola

The first race winning Lola in USAC was Rodger Ward's T-90, leading the Championship going to the 500 (Norm DeWitt)

Lola was yet another of the British-based teams to race at both Formula 1 and Indianapolis, along with Cooper, Brabham, McLaren, and Lotus. They first entered Formula 1 in 1962 with Reg Parnell as team manager with promising rookie John Surtees and Roy Salvadori driving their V-8 Coventry Climax powered Mk-4 and Mk-4A spaceframe Lolas. Future Shadow designer Peter Bryant was one of the team mechanics at the time, and the team won two non-championship races in the early part of the year, Surtees winning at Mallory Park, and Salvadori winning at Crystal Palace.

That said, the true highlight had to be John Surtees qualifying second on the grid and then finishing second at the British Grand Prix to Jim Clark's revolutionary new monocoque Lotus 25 (the second Grand Prix win for both Clark and the Lotus 25). The Lola team followed that up with second at the German Grand Prix at the Nurburgring. In the end, Surtees had finished fourth in the World Championship and although the battle for the title was between Hill's BRM and Clark's Lotus, Lola was a legitimate contender in their debut season.

1963 saw the team struggle and with Surtees' departure to Ferrari, they brought to Formula 1 a number of new drivers, most notably Chris Amon and Mike Hailwood. The best results for the Lola (the team also raced a Lotus and a BRM) were a pair of seventh places for Chris Amon. Lola faded away from Formula 1 for 1964.

For 1965 a new Indianapolis car effort had arrived from the fertile minds of Lola's Eric Broadley and his assistant Tony Southgate. As with many others, the Ford 4-cam V-8 powerplant was used. The car was no record setter, but it was reasonably quick. In the Indianapolis 500, Bud Tinglestad had run Lindsay Hopkins' American Red Ball Special (Lola T-80) as high as fifth before dropping out. Tinglestad had the services of Jud Phillips as his chief mechanic. Jud would go on to great success with Bobby Unser at Leader Card Racing, so it was a highly competent team. The T-80 Lola was a conventional car with asymmetric suspension. Al Unser finished ninth with his Lola in his first Indianapolis 500, using an AJ Foyt-owned T-80. Having had reasonable success for a new team at the Brickyard, a new Indianapolis Lola was being designed for 1966 and driver John Surtees – the T-90. A savage wreck for Surtees in the Lola T-70 sports car at Mosport in 1965 ended those plans, and as Surtees fought back from his injuries, the driver lineup for Lola at the 1966 Indianapolis 500 became Walt Hansgen and Sir Jackie Stewart.

At the races before the Indianapolis 500, Rodger Ward had taken second at Phoenix and then took first place with the Mecom Lola at the Trenton 150 in April. That Offenhauser-powered T-90 was the first winner for the marque in USAC, and Ward then drove it in the 1966 500, his final race. Oddly enough, it was a supercharged Offenhauser, with a Paxton Roots-type blower, which was not a common sight in USAC as the turbo-Offy was similarly beginning to make its presence felt. As with the previous T-80, the car was similarly offset on its suspension to favor ovals. As a result of these early strong performances, Rodger Ward and the new T-90 Lola was leading the championship, headed to Indianapolis.

Ward's Lola ran a supercharged Offy, where the Hill and Stewart T-90's used 4 cam Ford V-8 engines (Norm DeWitt)

Just before the Trenton race, Walt Hansgen had been killed in the April trials at Le Mans when he skated off the circuit at high speed in the rain with his Ford Mk II. Joining Jackie Stewart on the team would now be his BRM Formula 1 teammate, 1962 World Champion Graham Hill.

The Indianapolis 500 was familiar territory for two-time winner Rodger Ward, as he qualified the Supercharged Lola in thirteenth on the grid, with the Ford-powered T-90s of Stewart and Hill in eleventh and fifteenth respectively. Somehow, all three Lolas avoided the first-lap melee triggered by twelfth-place starter Billy Foster touching Gordon Johncock, Foster starting directly in front of Graham Hill. In the car directly behind Hill was AJ Foyt, who along with eleven others, were eliminated in the crash.

It was not an Indianapolis 500 of domination, between Jim Clark (Lotus), Lloyd Ruby (Eagle), and Jackie Stewart (Lola), they had collectively led 174 laps of the race by lap 190. By the end of the 500, Stewart was leading Graham Hill by a substantial margin when the oil pressure plummeted for the Wee Scot. Stewart looks back on 1966, "It broke ten laps before the end – suddenly the scavenge pump let go, and therefore no oil was getting up into the top of the engine. I was two laps in the lead, 5 miles, but I finished sixth. Graham Hill won the race, Jimmy Clark finished second, and I won the Rookie of the Year, which I liked as you got a year's supply of butcher meat, but it didn't arrive too well in Scotland as in those days it arrived by sea." There are many who claim that Clark actually won the race (unsurprisingly, many of those on Clark's STP team such as Andy Granatelli), and it remains a topic of debate that will likely never achieve resolution, similar to the next generation's Tracy versus Castroneves fiasco of 2002. It should also be mentioned that Stewart's drive had been sufficiently impressive that fellow rookie and race winner Graham Hill did not win the Rookie of the Year. On the flip side, Ward had struggled with his Lola, until finally retiring the car before half distance. So far in 1966, team Lola had two wins from three races and was a strong contender for the title, but at the Indianapolis banquet following the race, Rodger Ward announced his retirement.

At the non-championship Fuji 200 in October 1966, Jackie rebounded to take pole position with the Bowes Seal-Fast Special over teammate Hill as the Lolas started 1-2 on the road course. In the end, Stewart won over Bobby Unser in his first race with an Eagle, while Hill finished fifth. Interestingly, Stewart's Lola had been converted over to a symmetrical suspension setup for the road-course race versus the offset setup run at Indianapolis, while many other cars ran Fuji with their standard offset suspension. Sir Jackie explains, "It seemed a good car, but I had trouble qualifying it [at Indy '66]. The Lola was a better road racing car for me, but I didn't know it at all well. They took away the offset, it was central. Most of the American drivers had done no road racing, and I enjoyed it. It was good for me!"

1967 was to bring the T-92, in many ways unchanged from the 1966 winner – an evolution of the T-90. Stewart was again in a position to get a stellar finish, but it wasn't to be. He reminisces:

In 1967 it was the same thing, the scavenge pump again. I had more trouble qualifying it and the car wasn't as good the second year, but I was still competitive. Eric Broadley was a lovely man and George Bignotti was the chief mechanic with a great reputation. I was laying second ahead of AJ, actually, and he went on to win the race. Everybody said, 'You must get bored at Indianapolis, always turning left all the time.' I loved Indianapolis. You had to be very clean and very smooth. Jim Clark was who I learned most of my lessons from, a fellow Scot who I shared an apartment with in London. When the Europeans went over to Indy, Jack Brabham started it as an Australian with a little Cooper, and then suddenly we all went there. It was Jochen Rindt, Jimmy, Graham, Denny Hulme, myself – a whole raft of Europeans went over because we just loved your American money. As a wee Scotsman, that was a very important thing. When I didn't win it, the Scottish economy really did never get right after that, it was very sad.

When the Lola broke in 1967, Jackie was running third behind only Parnelli's Turbine car and the AJ Foyt Coyote. Once again the Lola may not have been the fastest car there, but was consistently in a position to contend for the win when others hit trouble. Those two years in the Lola were the only years Stewart drove at Indianapolis.

Al Unser had returned to the Lola team after racing the Lotus 38 alongside teammate Jim Clark in 1966. Al says, "I had run the Foyt Lola in 1965, and in 1967 I drove the Lola. It was in my younger days, I didn't know much about what was going on, but I thought it [the 1967 T-92] was a great car, a real balanced car." At Riverside in 1967's season finale, Al Unser and John Surtees were teammates in the Lola T-92. Surtees qualified fourth in the Bowes Seal Fast Special (entered by John Mecom, the Lola importer) for his only Indianapolis car start, which raised many eyebrows around the paddock. However, there was another T-90-based Lola that Surtees had raced for the previous two months, and that car was successful beyond all expectations.

John Surtees recalls, "I only drove the Indianapolis car in one race, but I was supposed to drive it many other times. After I was injured, Graham Hill ended up winning the Indianapolis 500 in what was to have been MY CAR. It worked out as that Indianapolis Lola was the chassis that formed the basis of the car I won the Italian Grand Prix with."

The Honda Formula 1 RA273 racer of 1966 through mid-1967 was nothing short of a beast. Overweight and cumbersome in its packaging, it was another of the overly complex behemoths in the first two years of the 3 litre Formula 1 alongside Cooper-Maseratis and the H-16 BRMs and Lotus. Surtees, with his connections at both Honda as the team driver in Formula 1, and with Lola, brought the two teams together to essentially graft the Honda powerplant into the Indianapolis T-90 Lola chassis. In its debut at the Italian Grand Prix, Surtees raced with what would become known as "The Hondola," Honda RA300, or Lola T-130, depending upon who was asked. After a long battle with Jack Brabham for second place, Jim Clark's late-race issues (fuel delivery problems after perhaps his greatest drive from a lap down) meant that the Surtees/Brabham battle was now for first, and an overly aggressive move in the braking for Parabolica sealed the deal as Jack Brabham slid wide while Surtees repassed him on corner exit and won the drag race to the finish.

The concept of racing in Grand Prix and at Indianapolis harked back to 1913 with Jules Goux and his 4-valve/cylinder twin-cam Peugeot Grand Prix car that won the 1912 Sarthe Cup at Le Mans and the 1913 Indianapolis 500. This time the technology was moving in the opposite direction – an Indianapolis 500-winning design that was to win a Grand Prix. It was the Lola that Conquered the World.

Lola continued with the "Hondola" development and built the RA301/T180 for 1968, the sleekest of the Honda Grand Prix cars and, by all accounts, having enormous power. The car struggled to find reliability, starting the season with five straight DNFs after qualifying fourth at both Monaco and Spa. The French Grand Prix was on the schedule for July 7th at Rouen, and that was enough to send shudders up the spine as Jim Clark and Mike Spence had recently died on the 7th, supposedly a lucky number. Sadly that premonition was to prove accurate when Rouen saw the debut of the new air-cooled V-8 RA302, an all-Honda effort with a magnesium tub to try and get the weight issues resolved by any means available. After previously testing the car, Surtees refused to race it, staying with the fast and drivable RA301 V-12. The V-12 car also wasn't skinned in magnesium, and being made from aluminum, was the safer and more conventional proposition. France's Jo Schlesser was brought in to drive the RA302, with disastrous results. Starting sixteenth on the seventeen-car grid, the car cannoned into an embankment on the second lap, overturning and catching fire. Full of fuel, the magnesium car was a no-hope proposition for Schlesser and he died in the car. One can only speculate as to John Surtees' mindset as he circled in the rain for 60 laps, passing by the raging inferno, then smoldering hulk with his teammate trapped within for almost two hours to finish second behind Ferrari's Jacky Ickx. Surtees' warnings about the RA302 had been vindicated in the worst possible way.

John Surtees' RA301 took second in the 1968 French GP. Jo Schlesser perished in the RA302 (Bernard Cahier)

When a replacement RA302 was built, Surtees again refused to have anything to do with it. At Monza, new team recruit David Hobbs was brought in as the second driver. Hobbs recalls:

Well, it came about because Honda had contracts with Eric Broadley to build a new chassis for the V-12 engine. Honda had their own car, the magnesium chassis with the air-cooled V-8 engine, which Jo Schlesser had been killed in during July. John [Surtees] suggested to Honda that they put me in the car, as I would be doing F5000 for John. So they put me in the car and we did some testing at Silverstone and in fact with the V-12 car I broke the lap record, which at the time had been held by Chris Amon in the Ferrari. So, the car was getting to be promising. When we went to Monza they wanted me to practice both the V-8 air cooled and the V-12, so I never really had a decent crack at either of them. Finally on Friday night they asked me which one I wanted to drive and I said that I wanted the V-12.

It was essentially the same specification magnesium RA302 that had cost Schlesser his life in France. David was not impressed:

The whole thing was a magnesium car and once it starts burning you can't put it out. Of course, they all caught fire then in those days because of the lack of protected tanks. There was no safety in that way at all. Drivers in those days weren't safety conscious hardly at all and if it was lighter, it would be quicker and they would go for it every time. The V-8 was terribly peaky as it had no power until about 7 [thousand revs] and it blew up at about 7.6. The car was sort of flexible, it kept changing. It was just the same [as the French Grand Prix car] with that engine hung off a cantilever. You'd change the springs and couldn't tell the difference.

Surtees meanwhile put his V-12 car on the pole, averaging over 150mph. It was looking promising for Surtees in the race until Chris Amon had a failure of the rear-wing hydraulics, putting him over the barrier and into the forest at Lesmo (for Chris Amon's tale of that hydraulic wing failure, see *Making it FASTER*). Surtees got caught up in the accident and was also eliminated. Getting through all the carnage, the race was now looking promising for David Hobbs... for a while. "I started fifteenth, passed a lot of people and was up to about fourth when it dropped a valve."

Hobbs felt he would be offered the Honda for the following year, but it wasn't to be. "They wanted me to drive for them in 1969, but then during the winter they decided to pull out of Formula 1 and come on straight to build the Civic CVCC engine with no catalytic converter for the American market. Honda used all their engineers from Formula 1 and put them on that. They still put their engineers on the racing programs." It would be thirty years before Honda returned with plans in 1998 for another Formula 1 car, and then that was abandoned with the death of designer Harvey Postlethwaite. With Honda's similar withdrawal from motorcycle Grand Prix racing at the end of 1967, it was a shocking change of direction for one of racing's greatest manufacturers. It was also the end for Lola in Formula 1 until 1974, when Embassy Hill racing ran the Lola T-370 with what could charitably be called moderate success (Hill scoring a single point in Sweden being the highlight).

The new four-wheel-drive T-150 Lola arrived in USAC for 1968. It was a highly advanced car, taking the proven four-wheel-drive concepts from the Granatelli Novi/Turbine Car program and combining that trend with a flat and wide shape, which was the coming thing versus the cigar shape of the previous Lolas. Anyone who had watched the opening laps of the 1967 Indianapolis 500 understood the advantages of four-wheel drive, as Parnelli Jones passed cars at will around the outside. Al Unser confirms, "It was very good handling, I liked it once I drove it here at the Speedway. When I took it to the road courses, I definitely had an advantage because of the way it worked with the four-wheel drive. Bignotti was my crew chief on that car." Unser won five races in a row when healed from motorcycle crash injuries, taking victory at Nazareth, both road races at IRP, and both races at Langhorne. Starting from the pole at Trenton came later in the season. Al finished third in points, narrowly ahead of Lloyd Ruby and his Mongoose.

Al Unser in the four-wheel-drive Bignotti Lola T-150 with 4 cam
Ford at the 1968 Rex Mays 300 (Gary Hartman)

Al explains, "Eric Broadley from Lola is the one who designed it and it was in the early stages of four-wheel-drive race cars. It had some unusual steering habits to it that I had to get used to." As Brian Redman once said, "Eric is a notorious changer of things." Al laughs, "Yes, he was. I got along with him. We changed the ratio from front to rear, which Eric did in England and sent it over here. Bignotti was able to help me, me telling him what the car was doing and he could transmit it to Eric Broadley. So we had a good relationship working together, all three of us."

Eric Broadley, like many gifted designers, had his mad-professor side. David Hobbs worked with Eric for much of his long career, including the Lola GT at Le Mans in 1963 and a number of sports racing cars including the stellar T-70. David recounts, "Eric Broadley was a semi-genius. You'd go testing with him and you'd realize he's got two different shoes on. You'd say, 'You've got a red shoe on one foot and another over there,' and he'd say, 'Oh, yeah, so I have.'"

Bobby Unser had switched from his Championship-winning Eagle to drive the four-wheel-drive Lola in 1969. Al says, "They could see that we had an advantage, that wasn't hard to see. At each racetrack I was always running well. Whether I finished or not, because of whatever happens, it doesn't take long for a team to say, 'Hey, we need to go in that direction.' At that time you can't say he knew more than we did, nor if we knew more than they did." Did you ever discuss the car with your brother when you were both driving? "What we were doing versus what he was doing – we never discussed it, we wouldn't tell him and he wouldn't tell me."

Now in the Lola, and it didn't take long for Bobby Unser to have second thoughts. Bobby admits, "That was a Bobby/Jud/Little Red mistake, no question about that. I didn't like that car at all, it just wasn't a good car." Obviously Al Unser had driven the similar four-wheel-drive Lola in 1968 with a great amount of success, and was again a race winner in 1969. Bobby continues, "Bignotti, obviously in that particular time, was smarter than we were and he could get his car going better than we could get ours going. It did well with Al. He broke some things with it, and there were some faults with it, but with Bignotti he figured out how to make the car go fast and you really have to give him credit for that."

Although he didn't care for the car, Bobby certainly understood the advantages of the four-wheel-drive Lola. "Common sense, we've always been into engineering and all-wheel drive obviously has to be better. The [Gurney] Eagles became the backup of choice, so the team had to rely mostly on the '67 Eagle." Prior to the switch back to the Eagle, Bobby Unser had put his four-wheel-drive Bardahl Lola on the front row of the Indianapolis 500, behind only AJ Foyt and Mario Andretti, so it wasn't all bad. We will never know how Al might have fared in Indianapolis qualifying or the race, as a motorcycle crash in the Indianapolis infield left him with a broken leg and out of the 500. Al Unser won the

Milwaukee race and the late season race at Phoenix, heading into Riverside – previously Gurney territory. However this was 1969, and the year belonged to Andretti. Al took second to Mario in the race, lifting himself to second in points over his brother.

Bobby and Al Unser weren't the only drivers that season who threw in with the four-wheel-drive Lola, as Penske Racing joined in, having run a Sunoco Eagle Chevy at a couple of road races in 1968. It was the birth of Team Penske in Indianapolis car racing. The car had a bizarre mounting of the Offy's turbocharger to the left of the chassis, almost alongside the cockpit. Somehow it worked, with rookie Donohue qualifying fourth, one spot behind Bobby Unser in his Lola. Mark's stellar drive, running much of the race in third, resulted in his being named Rookie of the Year. However his best finish during the year was a fourth at Brainerd and the Lola experiment was no more of a great success for Team Penske than it was for Leader Card Racing. The difference being that Team Penske stayed with the Lola for 1970, switching to the turbocharged Ford. Donohue was to lead five laps of the 1970 Indianapolis 500 to finish second, but the day belonged to Al Unser in the VPJ Colt.

A memorable and strange turbocharger installation on
Mark Donohue's Penske Lola at the 1969 500 (Norm DeWitt)

There was a short period of time circa 1969 when Formula 1 also built a number of four-wheel-drive versions of their two-wheel-drive cars, but there it was an attempt to regain the lost traction from having the tall wings clipped after a practice session in Monaco 1969. The wings in USAC were modest at best, and so the four-wheel-drive concept literally gained traction on this side of the pond, where it didn't so much in comparison to the low-wing Formula 1 racers of 1969-70, whose wings were far more effective than what was allowed in USAC. For 1970 USAC outlawed four-wheel drive.

1969 had seen the birth of the Vel's Parnelli Jones team effort, and they were racing their factory Lola. The new for 1970 VPJ Colt certainly resembled the Lola, and appeared to be a two-wheel-drive copy of the '69 Lola. It was known as the Johnny Lightning Special. Al Unser remembers, "It was a copy of the Lola in a lot of ways, sure. Bignotti kept what was the very best of the car and what needed help, he tried to change. In those days we had that working relationship and it just came together. It was a copy, you can't deny that. The Lola only held 68 gallons of fuel, and it had to hold 75 gallons. It didn't have to, it wasn't a rule, but the rules allowed 75 gallons of fuel and it shifted where the weight was in the car." One can see where the left side of the monocoque had an extension to accommodate the fuel on the inside of the car right below the fueling connection. This was the car that dominated the 1970 Indianapolis 500, leading from pole position for 190 of the 200 laps. From the outside, the basic shape of the tub largely followed the contours of the previous Lola. It had won the season-opening race at Phoenix, won on the road course at IRP, Milwaukee, and at Trenton. At season's end, given Al's dominance on the dirt as well (wins at DuQuoin, Indianapolis, Sedalia, and Sacramento), it was a complete runaway. Headed into the final race, Al had won eight of the previous nine races and in the end, Al had more than double the championship points of his brother, Bobby Unser.

The 1970 Indianapolis 500 winning VPJ Colt of Al Unser. Lola based, it was the dominant car of 1970 (Norm DeWitt)

Al Unser says, "When the '71 car was made, we copied the '70 car and we'd learned different things about the suspension and it was reinforced." The sidepod wings that ran up to the engine cover had also been extended forward. The nose had grown small winglets on either side with endplates. The tub was an overall sleeker and more refined proposition versus the 1970 car, with the fuel fillers now on the sides of the car near the rear of the tub versus atop the tub ahead of the screen.

For 1971, Al Unser was to win five of the first six races, including another win at the Indianapolis 500. From that point forward the season went straight into the tank while his consistent teammate Joe Leonard was to win the California 500 and, combined with four other podium finishes, ended up bringing another USAC Championship to Vel's Parnelli Jones Racing.

The evolving VPJ Colt was the end of the line for Lola in USAC Championship Car racing, as for 1972 the team embraced the new Maurice Philippe-designed VPJ-1, complete with its dihedral wings set at an angle between the wheels, a failed concept that was soon removed from the cars.

Lola was to briefly return in 1978 with their new T-500 that was fielded by Jim Hall and his Chaparral team. The Lola was another from

The 1978 Lola T500 was another low, flat monocoque car of the late 1970s, Al Unser winning at Indy (Norm DeWitt)

the low, wide, and flat school of monocoque design, with side radia-tors and a narrow protruding cockpit surround. Perhaps its signature feature was the fully exposed front rocker arm suspension, an unusual sight on a 225mph car, even in 1978. You'd have thought that the last place the car would work well would have been on superspeedways. Hughie Absalom laughs, "I know, we're still trying to figure it out now. I can't figure that one out either. Everywhere else we broke down or something happened."

Al Unser won the Triple Crown of 500-mile races, Indianapolis, Pocono, and Ontario, with the car, finishing second in the Championship to Tom Sneva. Sneva had no wins, but achieved his second straight Championship by way of ten podium finishes, which cost him his ride at Team Penske. 1979 was the first "split" with the emergence of CART and with the exception of AJ Foyt, most teams moved across to contest the CART series. The Lola was repainted in yellow for new sponsor Pennzoil, and ran the first races of 1979 until Indianapolis saw Chaparral introduce the John Barnard-designed "Yellow Submarine," which was a game changer. Lola was again gone from the scene until their 1983 season's return with the Newman-Haas T-700 and driver Mario Andretti.

CHAPTER 9

BRUCE McLAREN MOTOR RACING

*The first Grand Prix winning McLaren was the M7A of 1968 with
Bruce McLaren at the Belgian Grand Prix (Norm DeWitt)*

Bruce McLaren was part of the 1960's trend by the most mechanically inclined Formula 1 drivers, to create their own teams with the goal of manufacturing their own cars. Preceded by the creation of the Brabham Racing Organization by Jack Brabham in 1962, Bruce created Bruce McLaren Motor Racing the following year, although he continued to drive the factory Cooper Formula 1 car through the end of 1965. These initial efforts were with McLaren's modified Cooper T70s, being prepared to run in the Tasman Series. The Tasman series was triumph and tragedy as Bruce was to win the New Zealand Grand Prix, along with two other races to win the championship, but teammate Timmy Mayer was killed in a savage crash during practice for the season finale at Longford, Tasmania.

Howden Ganley recounts, "I joined in June 1964, and I was employee number three. The car with the stacked exhaust was the Xerex Special modified with the Olds engine, and it was renamed the Cooper-Oldsmobile. During the winter the team went off to the Tasman series with the Cooper, and I stayed behind to build the M1A."

Back in England the team focus was primarily on the sports car. Howden continues, "That M1A became the 1965 car and Trojan made them as McLaren-Elvas. There was only the works car that were built by us, and all the rest of the customer cars were built by Trojan. The car that Elvis Presley drove [in *Spinout*] was a McLaren-Elva, not a McLaren-McLaren. They had slightly different numbering systems. McLaren get their own like M1A, M1B, M1C, Trojan used Mk. 2 or Mk. 3, and it was very confusing. Similarly, there was never a works M12 [Can-Am customer car of 1969], when the factory was running the M8s. They've all been made into M8s now, probably."

The team started work on their first proper McLaren Formula 1 car, the M2A, designed by Robin Herd. The car was advanced for its time, not unlike future honeycomb tubs of the 1970s, only using Mallite, an aerospace material which used balsa between the aluminum sheets versus the later aluminum honeycomb. After testing, the car became more conventional in its construction and was known as the M2B when it was raced in 1966.

Howden says, "I went to Monaco for McLaren's first ever Grand Prix." It was a disaster as the engines were underpowered. Since the four-cam Ford was unacceptable, the team reverted to the Serenissima V-8, which proved little better. The season highlight was a fifth at Watkins Glen when they had reverted to a still underpowered but improved Ford. 1967 was somewhat better as Bruce started with a McLaren Formula 2 car, installed a BRM V-8, modified the car to Formula 1 specs, and took fourth place at Monaco. Bruce was supposed to be running the new BRM V-12 that season. Howden explains, "The engine was very late like all BRMs, so they built an [McLaren] M4 and made it into a 4B with a V-8 BRM. Bruce then went to drive for Gurney in the second Eagle."

Late in the season, the new car with the V-12 BRM engine arrived. For the debut in the Canadian Grand Prix, the V-12 was installed in Robin Herd's new M5A. Howden recalls, "They nearly won Canada with the V-12. The customer BRM V-12 engine had two-valve heads [as McLaren used in 1967], but then Aubrey Woods at BRM converted them into four-valve heads later." McLaren out-qualified the entire factory BRM team (running H-16s) around Mosport, taking sixth on the grid and running as high as second in the race, Bruce having one of the great races of his life in the drenched conditions. As it started to pour again, the conditions looked to be playing into Bruce's hands when the car began to miss, as the electrics failed. It was a stellar debut for the M5A.

The new M5A showed even more potential at the Italian Grand Prix, qualifying third and starting alongside Jim Clark and Jack Brabham. The engine failed in the race, but the car had shown remarkable speed.

*The Bruce and Denny show with pannier tanked M7As at the
1968 Spanish Grand Prix (Bernard Cahier)*

Its last race was at the South African Grand Prix that kicked off the 1968 season. The stage was set for a car/engine combination that would put McLaren on the Formula 1 map.

Robin Herd had designed the new M7A, a most successful car that immediately was a contender for Grand Prix wins with its Cosworth V-8 power. They tried pannier tanks, but that was abandoned after the first few races, not showing up again until 1969. Hulme was to finish second at Jarama, which was improved when the team achieved their first Grand Prix victory with Bruce McLaren behind the wheel at Spa. The win wasn't based upon sheer speed, as Hulme and McLaren had qualified fifth and sixth on the grid. Oddly enough it was the second year in a row that a Grand Prix car was built and raced by a team founder/constructor to take their team's first victory at Spa. Dan Gurney had won with his Eagle in 1967, and now Bruce McLaren had won with his McLaren. To win the race, both had finished ahead of BRM's number one driver, previously Jackie Stewart, and in 1968 ahead of Pedro Rodriguez. At this point, Hulme and McLaren were second and third in the championship.

From that point on, the title contender on the team was defending World Champion, New Zealander Denny Hulme. At Zandvoort, the cars qualified seventh and eighth, a tenth apart, but both cars were soon out of the race. Mediocre results appeared to be the pattern for most of the year until Hulme broke through to win at Monza, and although McLaren qualified second for the race, he was an early retirement.

The next round in Canada was the first anniversary of their debut the previous year in the V-12 BRM-powered car, and Team McLaren took their first 1-2 finish in Grand Prix racing, around Circuit Mont-Tremblant, also known as St. Jovite. Winner Hulme was now in contention for the world championship, and he was tied with Graham Hill for the points lead at 33. Watkins Glen did not see much success for the team, as McLaren limped home in sixth, Hulme's accident near the end of the US Grand Prix locked him in as the underdog in a three-way duel with Jackie Stewart in the Matra MS-10 and Graham Hill in the Lotus 49. In the Mexican Grand Prix, Hulme's retirement sealed his third-place championship finish, but Bruce's second-place finish elevated the team to second in the manufacturer's championship. This was a huge step for the young team in defeating Matra, Ferrari, and BRM in the battle for the constructor's prize.

The M7 series was to have a three-year lifespan in Formula 1, continuously being developed as was the not unusual at the time (Lotus 49 1967-70). Hughie Absalom remembers:

The M7... there were some minor changes with the shape of the tub, and Bruce had a couple of ideas for changes like the fuel tanks between the wheels. In 1969 we did the four-wheel-drive car [the M9A] as well. They made a big mistake. I think the power ratio was something like 40-60 and since then it's been discovered that you only need to be 10-90 or 15-85 [front to rear split] or something for a ratio. Weight was one of the reasons it didn't work, but I think there must have been some power loss somewhere in the system, putting all that drive to the front where it didn't need it. There was Lotus, McLaren, and Matra, three of us all on the same wavelength. We turned the engine around, the transmission was behind the seat. We [the four-wheel-drive Formula 1 constructors] all did the same.

The four-wheel-drive Formula 1 cars were running down a dead end as the concern over having sufficient traction after the high wings were outlawed had been much of the reasoning behind the concept. However, between tire development and the new lower wings allowed under the rules, there was sufficient traction without the additional weight and complication of four-wheel drive. It had been an expensive dead end for the three teams. As Hughie says, "Pioneering is like that."

McLaren decided to enter the 1970 Indianapolis 500, so Gordon Coppuck designed the turbocharged Offenhauser-powered M15A with a rectangular and wide tub/body shape. The car had a front radiator and twin-nostril exit similar to Herd and Coppuck's M7A Grand Prix car. Unfortunately, Denny Hulme was burned in practice and was no longer able to drive. Peter Revson turned the best qualifying time for sixteenth on the grid with teammate Carl Williams in nineteenth. Revson was a mid-race DNF, but Williams got his best Indianapolis finish with ninth.

The loss of Bruce McLaren in testing at Goodwood in June 1970 shook the team to its core, but they recovered quickly with Dan Gurney taking Bruce's place in what was to be Gurney's last season of racing. After three races, Peter Gethin took over Gurney's car with the best result for both being a sixth place. The M14A was obviously just an update on the earlier M7A, which was now looking a bit long in the tooth, compared to competition such as the Lotus 72. However, Hulme's

The first Indianapolis McLaren was the M15A of 1970, a wide, rectangular design by Gordon Coppuck (Norm DeWitt)

four podiums through the year had allowed him to take fourth place in the final championship tally, in this saddest of years. Piers Courage had been killed at Zandvoort, and at season's end, Jochen Rindt was declared the posthumous World Champion, having died in practice at Monza.

Coppuck had designed a new car for the 1971 Indianapolis 500, the M16A. Again the car used the turbocharged Offenhauser engine, now placed into a wedge-shaped low/flat monocoque with dive plane front wings, not unlike the Lotus 72. McLaren had obviously put some serious priority on the Indianapolis car program, given that they needed to obtain the services of Ralph Bellamy to design 1971's new Formula 1 McLaren, the M19A. The overall plan view of the M19A was another of the "pregnant guppy" outline, with the bulging sides along the cockpit, not unlike the Tyrrell or BRM P153. The car was not successful, having only a single podium in 1971, oddly enough achieved by Mark Donohue in the Sunoco Penske M19A at the Canadian Grand Prix, one of only two Grand Prix that Penske contested.

In contrast with the floundering Formula 1 program, at Indianapolis the new M16A turned in a shocking performance, with Peter Revson on pole position. Alongside in another M16A was Mark Donohue in the Penske M16A, and the other team car of Denny Hulme gridded fourth. By the end of the race, Al Unser in the Colt had won his second consecutive 500, with Revson coming home second, ahead of AJ Foyt. McLaren hadn't won the 500, but with their cars qualifying first, second, and fourth, it was obvious that nothing in Indianapolis had the sheer speed of the M16A.

The following year, Indianapolis saw massive increases in speed given tire advances and new rules allowing rear wings separate from the bodywork. In the case of the updated McLaren 16B, the car now carried a massive rear wing. It was going to be Eagle (Bobby Unser/Jerry Grant) versus McLaren (Peter Revson/Gordon Johncock for Team McLaren, and Mark Donohue/Gary Bettenhausen at Penske Racing). Gary Bettenhausen led most of the race until running into ignition issues with 25 laps to go. Jerry Grant then stayed out front with his Eagle until a fumbled pit stop. With twelve laps to go, Bettenhausen teammate Mark Donohue moved into the lead, and Roger Penske had achieved his first Indianapolis 500 victory.

In Formula 1, an M19A driven by Hulme was to win the South African Grand Prix in 1972, with teammate Revson also on the podium in third. Later in the season, McLaren had both drivers on the podium at Austria and Canada in the updated M19C, with Revson in pole position at Mosport. Despite sometimes being on, or near to, the pace, the 19C wasn't really a championship contender and Hulme did well to finish third in the championship with Fittipaldi and Stewart sharing nine wins from a twelve-race schedule.

Better things were to come on the Formula 1 side with the introduction of Coppuck's new M23, entered for Denny Hulme at the South African Grand Prix. The M23 was obviously much along the lines of the Indianapolis M16 series of cars, not altogether different from how the Lotus 56 Turbine car was the predecessor to the Type 72. Hulme put the new car in pole position, shockingly the only pole of his long and distinguished career. In the race, Revson took second with the M19C, showing there was still some pace in the old car, but the M23 was obviously the future, with its wedge shape, wide flat monocoque, and radiators on either side just ahead of the engine. Peter Revson was to win two races that year (Britain and Canada), and Scheckter seemed destined for a career of "win it or bin it" as he battled for the lead of the French Grand Prix with Fittipaldi, and then caused the largest accident in Formula 1 history at Silverstone where eleven of twenty-nine cars were eliminated due to Jody's opening-lap crash. Hulme took a win in Sweden, where he went from fourth to first due to mechanical issues for those ahead of him, though he had also set the fastest lap of the race.

At Indianapolis, May 1973 was to be a nightmarish month of injury and fatalities, with the loss of Art Pollard and Swede Savage. Johnny Rutherford was now driving for McLaren and set a new fastest lap, at over 199 mph to take the pole. Johnny explains, "I always told Betty that if I found a team that wanted to win as badly as I did, we would win." He had found that combination in Team McLaren. Rutherford had quickly become McLaren's primary threat in Indianapolis car racing, a combination that was to last for seven seasons.

Johnny Rutherford in his 1974 Indianapolis 500 winning M16C/D,
the most successful of their cars (Norm DeWitt)

Revson touched the wall on lap 3 and had to retire the car. When shortened by rain, the Indianapolis 500 had gone to Gordon Johncock in Pat Patrick's Eagle. The M16Bs were absolutely on the pace, but this was one race where everyone just wanted to put it behind them. Team owner Lindsey Hopkins had been in the Indianapolis car game for a long time and had seen both the horror and the ecstasy of the event, having lost both Vukovich and O'Connor in the 1950s. By season's end,

Hopkins was the car owner that finally brought McLaren their first championship in big-league open-wheel racing, when Roger McCluskey in an M-16B with a turbo Offenhauser won the 1973 USAC Championship. Team McLaren had won Indianapolis, and now a McLaren car had won the USAC Championship Car series, one year before Emerson Fittipaldi won the Formula 1 World Driver's Championship in his M23.

Fittipaldi had moved from Lotus to McLaren for the 1974 season. These early M23s had a rising rate front suspension similar to the M19C, which in later years was replaced with a more conventional design. Emerson Fittipaldi thought that although he preferred the Type 72 Lotus over the M23, there wasn't much comparison between the teams. Emerson recalls, "McLaren was a better team than Colin's. Alistair Caldwell was the team manager and he was fantastic. And there was Teddy Mayer, an American, and Tyler Alexander, so that was two Americans, but everyone else was from New Zealand."

It was one of those rare seasons without a dominant car or driver. Emerson had three wins including his home race in Brazil. Carlos Reutemann had three wins in the Brabham BT-44. Jody Scheckter had two wins in the Tyrrell 007 at Sweden and Brands Hatch. Ronnie Peterson had three wins in the Lotus 72, and Niki Lauda had two wins in the Ferrari. In the end it came down to a three-way fight at Watkins Glen, between Ferrari's Clay Regazzoni, Tyrrell's Scheckter, and McLaren's Fittipaldi, who won the title after the Ferrari challenge imploded. It was Team McLaren's first Formula 1 World Driver's Championship and the team had also won its first Manufacturer's Championship. McLaren had

Mika Hakkinen approves of Emerson Fittipaldi's 1974 World Championship winning M23 McLaren (Norm DeWitt)

now won both premier series in open-wheel racing worldwide. They were not to win another Manufacturer's Championship until 1985.

Team McLaren had a tremendous 1974 in American racing, taking victory in the Indianapolis 500 with the M16C/D, leading most of the race after starting on the ninth row. Rutherford also was to win Milwaukee and Pocono on his way to second in the championship behind Bobby Unser's Eagle. The following year he started out with a win at Phoenix and second at Trenton headed to Indianapolis. In the rain-shortened 500, Rutherford finished second. There were no other wins, and Rutherford again finished second in the championship, this time to AJ Foyt.

The Grand Prix team was to see changes in the driver's lineup. 1974 was Denny Hulme's last season in Grand Prix, scoring a single win in Argentina. McLaren's driving link to the "Bruce and Denny Show" of the late 1960s had retired from the sport. Jochen Mass drove for McLaren in the last two races of the '74 season (after Hailwood was injured at the Nurburgring), and was brought onto the team to drive alongside champion Fittipaldi for 1975. Jochen Mass on the M23 McLaren:

When you get in a good car from the start it is easier. It was a wonderfully designed car. The Lotus 72 was probably quicker, but it was unreliable. The McLaren was a versatile car, you could really make it good in all kinds of conditions. You could feel the changes made quite quickly. It was very sensitive, or sensitive enough to give you all the information needed. This was the days of the aluminum chassis and no one was like the other. They sometimes were different batches of alloys, or stainless. We didn't build our own monocoques, we had them built by a company, and there were inconsistencies. The stiffness of the car was very dependent upon those factors. Sometimes the softer ones were better than the stiff ones. There were minor differences, but you could feel them and that was good.

Emerson left McLaren at the end of the '75 season to create an all-Brazilian Formula 1 team and car, the Copersucar. 1975 had been a Ferrari year, Lauda showing complete mastery. With the withdrawal from Formula 1 by Lord Hesketh's operation at the end of 1975, James Hunt was looking for a decent ride, and ended up at McLaren. Winning the Spanish Grand Prix, Hunt was disqualified when the car exceeded the allowable maximum width. The irony was that the M23 was the widest car in Formula 1, and the FIA had used that to determine the maximum, however the new Goodyear tire had increased overhang on the rims which made the essentially unchanged car now illegal. Eventually that decision was overturned and the victory restored to Hunt and McLaren.

Much has been made of the drivability (or lack of) on the 1976 car after the protest over the width on the M23 that had led to disqualification at the Spanish Grand Prix. Jochen weighs in, "Aerodynamically it was slimmer and on the straightaway we gained in top speed. We didn't have any particular headaches with that car. There was fiddling around with it, but the information we had was not solid enough to go one way or another." Hunt was to win the World Championship in 1976 after his late season form coincided with a horrendous injury to Niki Lauda at the Nurburgring. It was a battle for the ages, and although what was presented was as much Hollywood as fact, their rivalry became the subject of a subsequent movie by Ron Howard called *Rush*. Insofar as how the personalities of the two rivals were depicted, the movie was uncanny in its portrayal. At the end of 1976 the McLaren M23 had completed its fourth full season of competition, and the team was hard at work on a replacement for 1977.

Hughie Absalom remembers, "I came back to England [for 1977] and went back to work at McLaren. I worked on the prototype M26. I think we won three races. When they first came out with the M26 at the end of 1976, in testing with it James didn't like it. By the time I came back in '77, we had a development test team to try and push it forward, but it wasn't until Spain that we ran it in a race."

Jochen Mass recounts his experience with the new M26 McLaren and the issues faced by drivers who were clearly the number two drivers on Formula 1 teams of that era:

> The M26 was not so nice. With the radiator in front, the weight distribution was not as good. It won a race or two, but only because it had 70 horsepower more than the others. Cosworth built some development evolutionary engines. Nobody knew it, not even the drivers, but the teams had the spec sheets from the tests and they said they had 14 or 15 more horsepower. In the 1990's Keith Duckworth told me, 'They had 70 more horsepower, you guys never had a chance.' They would just drive

away from you. The number one drivers got them, on occasion one would blow up so you had a chance at a podium. It was frustrating and a bit unfair. I don't know exactly when it [the program] stopped.

By the end of the season, Hunt in the M26 had won the British Grand Prix, and two of the last three Grand Prix of the season (Watkins Glen and Fuji). Finishing the season on a high note, McLaren's chances were looking better for 1978.

Teddy Mayer and James Hunt at Long Beach in 1978. The M26, a Grand Prix winner, was a challenge (Norm DeWitt)

Dick Scammell was working for Cosworth in that era, and was the project engineer on the automatic transmission for Formula 1:

We were also trying to develop and get some more power out of the DFV, and we did a deal with Williams, Tyrrell, Lotus, and McLaren to do special engines for them in a limited number. Of course they put them in the cars that had the best chance of using it. It caused absolute havoc. I would have thought 25 horsepower, enough to feel it. All engines aren't exactly the same, where you get some really good standard ones, and some not-so-good development ones. Seventy horsepower was a bit much. For sure, that way of running a team is very difficult. It's still the same today. The drivers hide everything and I don't blame them. That's their job in life, to be competitive and take every advantage you can. We had a situation at one stage where people were putting an ordinary engine in for practice, a special one in for qualifying, then an ordinary engine for practice on the second day, and then another development one for qualifying again. People try hard and it's a competitive world out there. If you don't try your hardest, you'd better look around and find another job.

Ron Pellatt had been a mechanic in Formula 1 for a number of years and was at McLaren during the transition from M23 to M26. He reminisces:

I got poached by McLaren in late '77 to work on the M26. I did the whole season in 1978. The 26 wasn't doing what they wanted, they were hoping to get great strides forward. The 23 was a good car and it stayed good to the end, the 26 was a development beyond it. The 26 was a lot narrower and the aerodynamics… it just wasn't the car that we wanted. We were overweight and struggling to keep the weight down. Then the Lotus 78 came along, and that just blew everybody away really, being right down to the weight limit. They got it right, you know. I think James was losing his edge, getting a bit disappointed with it. He just felt he wasn't getting the best of everything. He tried his hardest. We had a final party with him and that was something else! I don't think you could ever break the Hesketh level, as that's another level, but he was really good to be around. It wasn't that way every minute of the day, but we had fun with him in the evenings… good memories. He enjoyed being with his mechanics and he was absolutely genuine with us. Maybe the press got a little bit tired of him. They would ask him questions when he would get out of a car that is less than competitive, they would put a mic under his mouth when he was getting out of the car and expect him to give them a good interview when he was absolutely knackered. He'd be pissed off because he couldn't get what he wanted out of the car and wasn't competitive, so you don't always get the answers you want, do you?

Mika Hakkinen was reunited with mechanic Ron Pellatt at Laguna Seca in 2016 (Norm DeWitt)

1976 had seen the final year of the McLaren M16 at Indianapolis, the M16E. Rutherford put the McLaren on pole position, and was leading when the race was stopped for rain just after half-distance. Johnny Rutherford had led far more laps than anyone else, and his second 500 victory in the orange Team McLaren car was the third win in five years for McLaren the manufacturer. Yet again, Rutherford finished second in the USAC championship, just nipped in the final tally by former

McLaren driver Gordon Johncock, despite also winning at Trenton and in Texas. 1977 saw McLaren launch a new car, the M24. Despite being an improvement, the car never did achieve the stellar record of the M16.

At Indianapolis, the new McLaren M24 was the fastest car on the track. Tom Sneva took pole position in 1977 in the M24, finishing second in the race behind AJ Foyt. The following year Penske Racing brought out their PC5 Indycar. Sneva says, "Actually my favorite car would have been the old McLaren. In 1977 it was a McLaren. In 1978 it was what they called the first Penske [the first Indianapolis Penske, the PC5], but it was just a McLaren copy." For 1978, Sneva repeated what he had done at Indianapolis in 1977, taking pole, finishing second, and winning the series championship. Team McLaren continued to race in American Indy car racing (USAC/CART) for 1978-79, but that was the end of the saga. Johnny Rutherford stayed with McLaren until the end, taking two wins at Atlanta in 1979, the final season for the factory team at Indianapolis.

Mario Andretti with the McLaren M24 he raced for Penske Racing at Indianapolis in 1977 (Norm DeWitt)

In Grand Prix racing, McLaren lost James Hunt at the end of the 1978 season, and the team went straight down the charts. John Watson moved to McLaren from the Brabham-Alfa team, having been replaced by Nelson Piquet. The new M28 was a disaster, and Watson's only podium was at the first race, in Argentina. Teammate Tambay fared even worse, not scoring a single championship point. The replacement M29 provided little improvement and although 1980 saw Alain Prost brought in to partner Watson, both had to settle with best scores of fourth and fifth. McLaren was in disarray.

1981 saw the arrival of John Barnard's new carbon tub MP4/1 (discussed at length with team engineer Gordon Kimball in the previous book, *Making it FASTER*). The turbo era was picking up steam and the team brought out the MP4/2 with the TAG engine, designed by Porsche. The interim MP4/1E with the TAG engine had run successfully in 1983, Lauda leading in the South Africa finale. With new teammate Alain Prost on the team, it was a formidable combination and the team finished 1-2 on points, Lauda narrowly ahead of Prost (a half-point difference).

For 1985 Ferrari was in resurgence and at the midpoint of the season, it was a tie in points between Alboreto and Prost. As it turned out, changes to the Ferrari turbocharger manufacturer ended the challenge and with Alboreto ending the season with four straight retirements, Prost was champion.

1986 marked the end of the John Barnard era at McLaren and an unlikely championship for Prost when the Williams team had split their points between their two drivers (Mansell and Piquet). Stefan Johansson was brought onto the team for 1987. He remembers:

The McLaren MP4/3 was the first post-Barnard car. It wasn't an easy car to drive. I came into it relatively late, I only had about 25 or 30 laps of testing before the first race in Rio. It was quite a nervous car, and quite difficult to find the window in the car, where the sweet spot was. I think Prost would confirm that as I think it was one of the most difficult cars that he drove for McLarens, which the results would tell as he finished fourth in the championship. It was also at the end of the TAG engine development, the last year of the program. We had a lot of silly issues with the engine turbos, mapping, this and that. All in all, it wasn't an easy car to drive.

One often sees references to the Gordon Murray BT-55 Brabham "roller skate car" as having an influence upon the MP4/3 and the following Honda-powered MP4/4. Stefan disagrees:

No, not really, the McLaren wasn't anything like that. The main guy on the design side was really Steve Nichols. Gordon was the technical director, but I think the car was already quite far along by that point [when Murray arrived]. It was different times in those days, in terms of aerodynamics, and everything wasn't nearly as sophisticated in comparison to what it is today. Now it's like an ocean liner where if you change the philosophy of the design it takes a massive amount of time before you can make the ship turn.

The issue with the MP4/3 was a combination of the TAG motor and the mechanical side of the car. The mechanical was the tricky part, trying to get that right. Damping, geometry, just the feel of the car, it wasn't a very harmonic car to drive. It was very twitchy and nervous all the time, and it was difficult to commit because you didn't quite know where the limit was. It easy to go over the limit and then the back end wouldn't support it. So then you lose more time, you wouldn't necessarily go off, but would lose more speed. We had lots of turbo issues and there were lots of races where we didn't finish that we should have scored a lot of points in. Prost was fourth and I was sixth in the championship that year. Imagine if we'd had the car from the following year which was completely dominant.

For 1988, Ayrton Senna was brought in to partner Prost, Honda engines had replaced the TAG turbo, and Stefan was out at McLaren. "It is what it is, and I think it was pretty much a done deal before I even joined, but I felt it was the best option available at the time."

Stefan had just come from two years racing for Ferrari and one assumes that the 1987 McLaren MP4/3 would have been an improvement on the drama and underachievement from Ferrari during his tenure there. He recounts, "The 1985 car was good, until we started having lots of turbo problems later in the year, which caused most of the problems." Ferrari had changed the turbocharger supplier for the Formula 1 car mid-year, and the reliability of the engine, along with their championship challenge, went up in smoke. Literally, in the case of Michele Alboreto during the European Grand Prix at Brands Hatch, where he continued to drive the burning Ferrari around the circuit. Alboreto drove onto the pit lane, and delivered the burning Ferrari back to the Ferrari pit. It was fully on fire, as he stepped out and walked away in disgust. There are accounts where Enzo Ferrari, facing the end of his life, had apologized to Alboreto for orchestrating the turbocharger supplier substitution that cost him the championship. Stefan doubts those stories:

> I don't think Enzo did it, to be honest. I think that is hearsay more than anything. It was something that would have obviously been done by the technical department. I don't think that was anything Enzo would have directed. We had all sorts of problems to where it culminated where we barely qualified at South Africa for the last race, it was a disaster. I think we were something like twenty-second and twenty-third on the grid because of the high altitude where you needed the turbos more than anything. We went from winning races at the beginning of the year to really struggling by the end.

1986 brought a new Postlethwaite Ferrari, and the drivers were shocked. Stefan recalls, "That car was a whole different story. I

Super Swedes – Penske chief mechanic Matt Jonsson, with
Stefan Johansson who drove the MP4/3 (Norm DeWitt)

remember Michele and I looked at each other when we saw the car for the first time and we were going like, 'Oh f**k, this is going to be a long year.' It was a big old car compared to anything else out there at the time."

So much for the era of superstar designers. Stefan says, "It was never one guy anymore, not even in those days. You'd like to think that, and there is one guy driving the direction of it, but there is always a big team behind it and a big collaboration."

The 1988-89 seasons will always be notable for the utter domination by McLaren-Honda in Grand Prix racing. Prost left for Ferrari after an impossible situation was created within the team following the famous Suzuka Chicane incident, when Prost turned in on Senna, who was alongside. The Ferrari was the only other car that could come to grips with the McLaren for 1990, and Ferrari had signed Alain Prost. Again it came down to Senna versus Prost, and this time was again decided at Suzuka, when Senna turned in on Prost heading into turn 1. It was a much riskier proposition at far higher speeds than what Prost pulled on Senna the year before, and Senna admitted years later that he had done it intentionally. They say revenge is a dish best served cold, but this was madness. It gives one a clear view of Senna's ruthless mindset when another driver stood between him and his goal.

1991 brought the McLaren MP4/6 with a V-12 Honda onto the scene, and Williams-Renault had upped their game to become a title contender in the wake of Ferrari's decline. Senna's four straight victories combined with Williams' early season unreliability meant that Senna was to win his second World Championship. By 1992 the die was cast and the McLaren hadn't a prayer against the new Williams FW14B, which won eight of the first ten races, including the first five on the trot.

1993 had some success, but the McLaren active car was again light years behind the Williams FW15C of Prost (see The Active Cars Chapter 17). Senna obviously didn't get the memo that his title challenge was hopeless given his customer Ford V-8 powerplant, as he managed two wins from the first three races, including masterful drives in the wet conditions of Brazil and Donington Park. Winning again at Monaco, Senna in the MP4/8 was still leading the championship. Having won the other three races, Alain Prost was then to reel off victories in the next four races, and was well on his way to the championship. Damon Hill's abysmal luck in the other Williams finally turned around to see him win three straight later in the season, nearly nipping Senna for second in the title chase.

For 1994 McLaren teamed with Peugeot, and their engine program was essentially hopeless. That lost season led to the partnership between McLaren and Mercedes starting with 1995. Nigel Mansell made a short-lived attempt to drive for the team, with the rest of the year having Hakkinen and Blundell as the drivers. Other than Alesi's victory with the V-12 Ferrari in Canada, every winning car had Renault power. Mika had two podiums in '95 but was plagued with nine DNFs. A huge accident by Hakkinen at Adelaide ended the season on a low note.

1996 was little better, with Mika scoring a best finish of third on four occasions. For 1997 the team started to turn the situation around, with David Coulthard winning twice, and Hakkinen winning once, the drivers ending up fourth and sixth in the championship, each suffering seven DNFs. The MP4/12 was the first of the non-Marlboro McLarens in a generation, now sponsored by West cigarettes and painted silver/black. This was the year of the dual brake pedal, allowing the driver to brake a rear wheel independently. It was a performance enhancer, and as with most such things, banned for the following season.

1998 brought Adrian Newey's first McLaren, the MP4/13. The car was simply dominant, Mika winning eight races on the way to his first World Championship. Schumacher was the only one close, as he won seven races in the Ferrari. Yet, the season was not without its drama, as there were counterfeit bearings found in the gearboxes that year. You wouldn't expect issues like this to hit a Formula 1 team, but they did. Chief engineer Pat Fry says, "There were some. They tracked it down to some bocce company in Brazil who were making them and selling them as being from well-known bearing manufacturer. You were buying them from a reputable bearing supplier in the UK. To some degree a lot of the bearings in the McLaren gearbox were custom made bearings and you were working with their special bearing supply people to get everything and you knew you were getting a good product. It was some of the simple off-the-shelf ones like out of a standard bearing catalog that were the problem."

1999 was to bring another stiff challenge from Ferrari, until Schumacher broke his leg at the British round and the challenge shifted to teammate Eddie Irvine. It was high pressure as the McLaren dominance of 1998 was over and it was a tight championship race as the teams came to Monza. Somewhat surprisingly the biggest challenges to Hakkinen at Monza were coming from the two Williams cars of Zanardi

The West McLarens of the 1990s all in a row
(Norm DeWitt)

and Ralf Schumacher, and Heinz-Harald Frentzen in the Jordan, who had earlier won the French Grand Prix. Mika took his eleventh pole of the season at Monza ahead of the Williams of Zanardi. In the race Mika cleared off while Zanardi struggled with an underbody that was dragging on the track, having come loose from the car. Frentzen was putting pressure on Hakkinen and as Mika came charging into the Monza turn 1 chicane (the last year for the old layout), it all went horribly wrong.

Turn 1 at Monza in 1999 will forever be an example of when you get the clutch engagement sequence wrong. Mika Hakkinen explains:

> For qualification and for the race we have different ratios. For qualifications we wanted to use first gear for going through the chicanes, so then we had one extra gear to maximize the performance. But in the race, because of the start, we needed a low first gear. So with that short first gear we needed to adjust the throttle blip really high and to keep the clutch open for a really long time until the car speed reaches it, that way you can put the first gear on, otherwise you will lock the rear tires. What I did was that I came to the braking point and selected first, and it selected first. It should have kept the clutch open until the car speed would have reached the right speed. So it was my mistake for putting it into first gear, but the team should have put in the new numbers for that big gap.

At that point there was little that Hakkinen could do but retire the car, with it well and truly beached in the turn 1 kitty litter. It had instantly become an unsalvageable situation, despite Mika's claim that

"In my time the drivers were better." As I was on driver's right side at the entrance to the corner, I had a ringside view of the entire episode. Hakkinen had a meltdown, throwing the wheel, gloves, and everything else that wasn't bolted down, eventually coming to tears as the Tifosi all around danced with joy at seeing Ferrari's greatest rival out of the race. Mika had to feel that it could have been the moment that he would look back upon as what cost him the championship, but across the following three races, Mika was to score two more points than Eddie Irvine, and won the championship by exactly that margin: 76 to 74.

2000 brought the MP4/15. Pneumatic valve system failure in the first race forced the retirement of both cars, however as it turned out, it was a Ferrari year and Schumacher won nine of the races. Hakkinen's four victories and Coulthard's three were enough to secure them "best of the rest." Alex Wurz was being considered as a test driver for McLaren headed into 2001, but there was one minor problem. Alex tells the tale, "Shoes! I wear different ones on each foot. Ron Dennis obviously has a little bit of OCD and wants everything to be in perfect order. Obviously they knew I had one red and blue shoe, and Ron sent one of his Team Managers to ask me before he started the contract negotiations if I will insist on the red and blue shoe. I already knew with that question, that if I said, 'Yes,' there would be no contract negotiation, if I say, 'No, I'm okay with whatever color shoe,' then there will be contract negotiations. So I said I'm not superstitious, which I'm not, and I said, 'Whatever shoe that you want me to wear.' I fully respected him wanting to keep everything in perfect order with his color scheme, so I didn't think a minute."

Where things got interesting was when McLaren wanted to mess with Alex's helmet design. Wurz recalls, "They had much more difficulty with my helmet design because they wanted to put a halo on, and obviously that doesn't work with my graphics and my design so I just said, 'No.' The halo is the band around the helmet for the sponsors. By then I had a contract, I knew it would be a little bit difficult for them to fire me. My helmet is a globally registered trademark, and I was much more stubborn and I said, 'Nobody messes around with my helmet design.' I had the same argument two years ago when the FIA said that drivers cannot change their helmet design. I said, 'This is the last thing where we can portray our personality. If someone wants to have an entire career in a yellow helmet, that's his marketing, his brand, and his personality. If you have somebody who changes his clothes and designs every weekend, don't restrict it. We have our own trademark opportunity, and that is his brand and personality, and I will always fight for it.'" One can only imagine how Alex Wurz's position would have been received in 2017 if he were driving the bright pink Force India cars, with the drivers wearing matching pink helmets.

The 2001 MP4/16 was certainly a quick car, but the stars did not align for the car. Hakkinen had the Spanish Grand Prix won when the car quit on the last lap. Although Mika and David each scored two wins during the season, it wasn't enough as Schumacher was to win nine races, and most often finished second when he didn't win (five times).

2002 was even more discouraging for McLaren-Mercedes as Schumacher was to win eleven races, and never once missed a podium finish, a remarkable achievement for Ferrari. However there were many reliability issues with the MP4/17 as Adrian Newey began to push the limits in his aerodynamic efforts to close the gap to Ferrari. Alex Wurz remembers,

The problems started with the MP4-17. In the beginning there were issues with engine reliability, which wasn't Mercedes' fault. It had to be mounted really low for Adrian Newey's aero package, so the cooling of the car had to have a specific airflow. The engine ran too hot, had some issues with the oil, and kept exploding. They raised the engine by a few millimeters and all the problems went away. They had to make the sump a few millimeters deeper because the oil foamed up. The engine guys said, 'We can either keep exploding, but if we raise the engine up a few millimeters higher, the oil will be perfect and you can run smaller radiators.' At the end of the year the car became a regular winner. Adrian was rigorously only for the aero at this point, and he said, 'I want the engine and everything to be really low so we could have the best aero efficiency.' Mercedes' engine designers said the engine will be a problem, but Adrian won. Then after a few races where the engines kept exploding, then we raced with 40 horsepower more, a deeper sump, and the car was much faster and more reliable. Back then we could run the wings we wanted, because we had the power to carry it.

Then there was the 17D, which had a different rear end. It was an integral part of the whole development and that was a sensational car. The changes were mainly suspension geometries and it allowed the aero to be a bit different. The reason that the 17D didn't win the championship was because Adrian and McLaren told Mercedes that, 'We will race the 18 for the second half of the season, no matter what,' and the 17D had a different engine spec. And then they ran out of pistons in the final races, and they couldn't make them in time. When Raikkonen blew up at the Nurburgring, he lost the World Championship over it [by two points] because they only had nine correct pistons in the engine and they had to take one piston from where they knew the production line wasn't so great, and it blew up. A decision which was made by the chassis people in Woking, then made the engine blow up. When Mercedes was told that McLaren weren't going to race the 17D anymore, of course they stopped making engines for it. At this time the pistons cost an endless amount of money because of special materials, so no wonder they stopped it. It is that simple and sad that McLaren and Mercedes lost the World Championship for this reason.

Looking back, absolutely my favorite car was the 17D. I was the only guy that developed it, although Pedro de la Rosa did some testing as well. The idea of the different rear geometries was my idea as in tire testing we were not having the right amount of squat and lift, we worked a lot on that. Nobody really cared about the 17D because everyone focused upon the 18. That was the magic car being praised so highly during the development because it was revolutionary. You can still see some aspects in the past few years of what Adrian had in mind being used on the Red Bulls. It was just too far ahead of its time in the design as the structure of the car just didn't work. Adrian is still now 99% aero. Some of the Red Bull cars had very narrow designs and when it was really hot, they had to turn the engine power down. Everything is a risk in motor racing. As far as design, it was the name of the game.

A narrow window of operations is the optimal solution if you can hit the target. Or it becomes a disaster if you don't. Wurz continues:

I had a few crashes so I passed the crash tests, but the 18 never passed the FIA crash test. The car kept breaking. You'd fix one part and then the next part kept braking. The amount of technical off-site work on this car was more than any other experience I'd had. At Jerez the floor collapsed in turn 4, and that means it stalled and poof, you were in the wall with fire. One day at Silverstone, I had a few times with the brake pedal going to the floor. The engine blew up and the conrod tended to go out through the bottom and it cut the brake wires. Those brake wires were next to each other for both brake circuits, so you'd end up with no brakes. When you felt the engine blowing up, you knew you had no brakes. So you were hoping that the engine won't blow up just before the braking where there was no runoff.

When you typically have an engine failure is at the end of the straight where the engine is under its greatest stress. Wurz adds:

Exactly, and we had a few of those. In one day at Silverstone, three times we had lost the left front wheel in Becketts, flying off the car. The uprights were so weak and the wheel nut wasn't strong enough and came off, and one time it was very close to a shunt. Then there was one brake failure, then an engine change over lunch break. Then in Bridge corner there was a rear suspension failure, into the wall with the chassis here and engine there. That's pretty much a test day in the MP4/18. The race drivers didn't want to drive it anymore. That was the last day for the car. Actually, it was the first time in my life, after the shunt I saw the car. I went back to the pits and called Martin Whitmarsh and told him, 'Martin, I don't want to drive this car anymore.' He said, 'Oh, I thought you would call me a few tests ago with this phone call.' He had been waiting. But, I give them credit for trying all these new technologies and new ways of making a car, because they lowered the center of gravity by an enormous amount, but it was just structurally not possible yet. Everything was just far too extreme.

Pat Fry was one of the engineers frantically trying to keep up with the cascade of failures. He weighs in on the MP4/18:

I'm sure Alex has fond memories of that car like I have. It was problematic with everything to be honest. They kept on asking to push the crank height down to the point where they couldn't scavenge the oil out of the engine. But that was also the one with the twin clutch gearbox planned for it. There was a standard engine with a single clutch, and there was a split gearbox with twin clutches and twin shafts that was also done at the same time. There were lots of fun design projects being done around that time, twin clutches, torque steering gearboxes with twin final drive and you could break each final drive that could use torque steer to turn the back of the car. There were lots of interesting things, but we didn't get the basic concept right. The MP4/18 I think came out of the 17D which almost won the championship that year. It was one engine failure away from winning the championship with a car that didn't get any development until Monza. That was a sound solid car where we changed the rear suspension on it over the winter, but otherwise was a real standard thing. It was reasonable, sensible, reliable car where the drivers could push. A lot of people talked about 60-kilo ballast and I think we only had 5 – 7 kilos of ballast. So in the drive to save a massive amount of weight and lower the center of gravity, we came to the MP4/18, which never raced at all as it never passed the crash test.

For the 2005 season, the new MP4/19 made its debut. Wurz says, "The MP4/19 had some aspects of the 18, but it went back to the 17D geometries and gearbox design. It was a hybrid." One presumes that meant it included the parts that didn't fall off the MP4/18. Wurz concurs: "More or less." The MP4/19 had horrendous reliability issues as Kimi retired from five of the first seven races, Coulthard retiring from three of the first seven. Chief Engineer, Pat Fry recalls the MP4-19, "2005, challenges… I think a lot of the engine problems had been solved by then and it had a standard conventional single clutch gearbox and everything. There was a period where we should have concentrated more on reliability than we did and held ourselves down because of that." Engineering Director Paddy Lowe adds, "It was a terrible year as we should have walked away with the championship that year, it just wasn't reliable enough."

Alex Wurz on his way to third place in the 2005 San Marino
Grand Prix at Imola (Norm DeWitt)

In his only race during his tenure at McLaren, Wurz was to get a third-place finish with the MP4/19 at the San Marino Grand Prix in 2005, after Button's BAR Honda was disqualified for having a hidden fuel tank in their car. Alex reflects, "For average I have the highest points per start as a McLaren driver. If you look at the Prosts and Sennas, of course they had some races where they didn't finish. The McLaren guys checked it, but I'm not very proud of that."

The MP4/19B was the new car brought out for the French Grand Prix, and yet again McLaren had a mid-season renaissance Wurz recounts:

> I can't really remember all the details, but 19 was a bit temperamental and at some tracks the tires worked well, at some tracks not. The 'B' car was a big improvement and scored more points than any other car from the mid-season on. The 'B' was very good mainly because the reliability was fixed. With the 'A' we didn't test so much as we couldn't make the tires suit our chassis or visa-versa, because simply when at every test you have a reliability issue, you just don't complete the tests. In this period it was very important that you did the most mileage with your tire manufacturer, which was Michelin. They designed the tires for your chassis and the tires could be changed faster than you could change the aero or mechanical design.

Getting sufficient temperature in the tires was a problem faced by some of the McLarens from this era. Paddy Lowe explains:

> In 2005 you had to do the entire race on one set of tires by regulation. I think the knowledge back then wasn't as good as it is now and how to deal with those things. Pirelli is more sensitive than any of those previous tires to temperature and we've learned to control it better whether it is with wheel heating devices to take brake heat into the wheels. For example running in the wet this morning, we will have been playing with the amount of wheel heating we do from the brakes. When you study the cars with the wheels off, you will see all sorts of different configurations of brake ducts in terms of venting the brake ducts of how to blow onto the wheels.

Pat Fry fondly recalls the prevailing atmosphere of innovation at McLaren during that era of remarkable design freedom:

> I think the good thing at McLaren in the early 2000s if you like: A) There were a lot of good, young, and talented engineers working together coming up with on different concepts. B) We had the freedom to generate lots of good concepts. But we concentrated more on these kinds of things rather than actually going racing. Through that period we had a reasonable simulator development program and in 2006 we started actually using all those tools and skills in the simulations that people had developed to try and sort out what we wanted for a car. The 2007 and 2008 cars were the products of all that research that had been done earlier.

McLaren came within one point of winning the World Championship in 2007, Lewis Hamilton winning four Grand Prix. In 2008, the result was reversed as Lewis nipped Ferrari's Felipe Massa by a single point in a cliffhanger finish at the season finale in Brazil. There were other technical highlights to come, such as the development of the 'F duct' of 2010, which can be found in the Aerodynamics Chapter 19. However, in the end, Hamilton was fourth in the championship. Button then took second in the Vettel runaway of 2011. In 2012, there was relative parity, with seven different winners from the first seven races. Despite winning seven races between Hamilton and Button, they were only to finish fourth and fifth in the championship, Lewis suffering through five DNFs. For 2013, Hamilton left to find great success at Mercedes and eventually McLaren declined into the forgettable years of the McLaren-Honda partnership of 2015-17. Perhaps the highlight for 2017 in all of motorsport was the return of Team McLaren to Indianapolis, in cooperation with Michael Andretti's armada of Dallara-Hondas. Fernando Alonso had skipped the Monaco Grand Prix to compete in the Indianapolis 500 and was in contention for the race win, until the Honda engine failed him yet again near the end of a race. Fernando Alonso was "Rookie of the Year," a misnomer if there ever was one for a two-time World Champion. McLaren's long history shows endless patterns of decline and resurgence. Don't bet against them returning to the top of the sport in the near future.

The last Indianapolis McLaren was Johnny Rutherford's 1979 M24, winning both heats at Atlanta (Norm DeWitt)

*Fernando Alonso after climbing from his broken car in the
2017 Indianapolis 500, the story of the year (Norm DeWitt)*

CHAPTER 10

VEL'S PARNELLI JONES RACING

*Parnelli Jones in the Lotus 56 Turbine car during pre-race festivities
at the 2014 Indianapolis 500 (Norm DeWitt)*

Coming off two highly successful years with the Colts in 1970-71 (see Chapter 8), the Vel's Parnelli Jones team brought in Maurice Philippe to create a radical new car, the VPJ-1. It was a super-team with Indianapolis winners Al Unser and Mario Andretti teamed with current USAC champion Joe Leonard. Bobby Unser says, "The trouble with Parnelli's deal was that he had three high-class drivers, and how do you keep three of them happy all the time? Can you imagine what it must have been like in that garage? You couldn't give one a secret part and it brings the whole team to a slowdown. At Gurney, if Dan ran a second car like he did with Jerry Grant, he kept it completely away from me as I don't like having a teammate. In their world they have to run two cars at least, but Parnelli running three? Give me a break."

The car had been built around Mario Andretti's build and as told in *Making it FASTER*, six-foot-tall Joe Leonard showed up at the shop and couldn't fit his feet into the available space in the footbox, which required subsequent modification for his size-twelve feet. Considering that Joe was the defending USAC Champion, that episode had to be a wake-up call for Leonard that this wasn't "his" team.

The 1971 VPJ Colt was the final development of the Lola T-150, winning Indy and the championship (Norm DeWitt)

There was real concern at All American Racers over the latest Maurice Philippe-designed Parnelli with the dihedral wings. Considering the revolutionary recent designs by Philippe such as the Lotus 56 Turbine Car and Lotus 72, those concerns were well founded. Bobby Unser recounts:

There was that asshole Englishman that was supposed to be the greatest designer in the world, he was Parnelli's secret. I looked at that dihedral-wing car when they were testing with it at Ontario. I told Dan Gurney, 'Boy, if that son-of-a-bitch car works, we are really in a lot of trouble. When they'd test we had spies and that was a really big part of it. I had pictures, and they'd have spies at our tests too. Hell, I used to steal tires from them. But I told Dan that the thing won't work as they'd put it in the middle of the chassis. It was a moot subject as you can't run wings in the middle of a car. On that '70 car, Dan was so up on having standard wings in the middle of the race car where Parnelli's car had them, inside of the front and rear tires. I'd tested that with Dan and told him it's an unworkable thing, the car has no idea where it wants to go. We spent a lot of money on trying it. Can you imagine putting feathers on an arrow halfway up? I've had an airplane with that dihedral thing on it and it's a s**tbox, it won't go straight. It wasn't that we were smarter, we were willing to try, and until you try it then you won't know it's a failure. But see, none of us knew that then.

Mario Andretti's early 1972 VPJ-1 with the dihedral wings,
and the wind tunnel model (Norm DeWitt)

It wasn't long before the dihedral wings were ditched in favor of a more conventional setup, the car keeping its trademark shovel nose between the tall suspension perches on each side. At the season opening race in Phoenix, the new Eagle of Unser won, with Mario Andretti taking second in the Parnelli. At Indianapolis, the Parnellis of Al Unser and Joe Leonard took second and third behind Mark Donohue's McLaren. The first win for the new car came when Joe Leonard won at Michigan, followed by his wins at Pocono and Milwaukee. Although Bobby Unser was to win two of the remaining three races, it was too little too late and Leonard was again the USAC National Champion. Joe had been the only driver to win with the VPJ-1, which is saying something when your teammates are Andretti and Al Unser. Joe had more than 150% of the points scored by runner-up Bill Vukovich Jr.

The 1972 Championship winning car of Joe Leonard,
with the conventional wing arrangement (Norm DeWitt)

Somehow in the midst of all this success, the other racers began to rib Joe Leonard about his eyesight. Joe explains, "It was usually 20/30. Mario Andretti used to kid me saying 'Guiseppe, are you okay? Can you see?' Dan Gurney would come up to me and say, 'Hey Jose, can you see?' I said, 'Yeah, I can see.' I have no idea how that rumor got started. I'd tell them, 'Listen fellas, I don't know if my eyes are as good as yours, but as long as your asses are behind me there are no problemos!'"

1973 was to see the team struggle with the new VPJ-2. Leonard's best finish was a fifth place in two of the first three races of the season. It was all downhill from there to an eventual fifteenth place finish in the championship. Andretti and Unser fared better, but not by much, Al Unser taking a season opening win followed by finishing thirteenth in points. Mario had by far the best results with a win at Trenton, and a pair of second-place finishes at Ontario and Michigan, dropping from fourth to fifth in points at the final race. Given the strength of the driver lineup at VPJ, it was a disaster.

Joe Leonard in the problematic 1973 VPJ-2. The Colts and VPJ-1 were a tough act to follow (Joe Leonard/VPJ)

Leonard returned for 1974, the team now running an Eagle-Offy without long-term sponsor Samsonite, but extensive feet and leg injuries received from a crash at Ontario ended his racing career. Mario ran the Viceroy VPJ-3 as well as the Viceroy Eagle at Indianapolis. The VPJ-3 had small winglets ahead of each rear tire, showing the team hadn't given up completely on the concept of mid-mounted wings, although the team ended up racing the Gurney Eagles. Perhaps the biggest blow to the team was the end-of-season departure of Firestone from Champ Car racing. It was certainly a low point for VPJ.

There was a new Indianapolis car coming for 1975, very similar to the new Formula 1 Parnelli by Maurice Philippe. Hughie Absalom says, "I came to Parnelli in 1975. They might have just run the car one time at Ontario when I joined. We were still running the Eagles in the off-season actually, and at the second or third test we weren't doing very good, either. The funny thing was our first time to Indy with the Cosworth was 1975 and the engine was okay, but the chassis wasn't. I think Al [Unser] drove it most of the time and he couldn't pinpoint the problem. Mario turned up after one of the Formula 1 races and said,

171

'Let me get in this thing.' At the end of the day he said, 'Well… maybe it's not quite ready yet.' So they both stepped back into the Eagles."

On the Grand Prix side of the operation, Dick Scammell was to join the Parnelli team as part of the effort, another connection along with Maurice Philippe from the glory days at Team Lotus. Dick remembers:

> I knew Keith Duckworth and Mike Costin from the 49 days with the DFV. I was at a race meeting sort of wandering around unemployed and Keith said, 'Oh, come and talk to me on Monday.' I went to see Keith and after he interviewed me for a day and when I left I actually didn't know if I had a job or not, as that was Keith. So, I worked at Cosworth for a while. Then of course my old buddies Andrew Ferguson and Maurice Philippe came by and said, 'We are off to Parnelli, we've got sponsorship for four years, why don't you come with us?' I was running development at Cosworth that time around and I in actual fact built the engines which were going to Parnelli and ran them on the test bed at Northampton, put them in boxes and sent them over to Parnelli. So off to Parnelli's I went, and then the engines arrived from Cosworth that I had built. I was brought over to run the Formula 1 team.

Scammell continues, "It was a nice car, the biggest compliment I ever had for that car was when we took it to the first race. Colin came along and looked all around the car because let's face it, I only worked for Colin for eleven years, and he said, 'That's the car that I should have designed.'"

Hughie Absalom agrees, "That's true, because all the designers had like one train of thought, and Maurice Philippe's train of thought was to make the Parnelli a generation two Lotus 72, which is what he did really."

Mario Andretti raced the small-budget Parnelli Formula 1 car for 1975, and the VPJ-4 was shockingly off the pace initially. Mario recalls:

> We had Ferrari, McLaren, Brabham and such, all the giants that we had to contend with. Even though we had some degree of proven talent in design, we still were definitely struggling. I was still driving a Lotus 72, as the Parnelli was a take-off on the last car that Maurice Philippe had designed in Formula 1 for Colin

The Parnelli VPJ-4 of Mario Andretti was Maurice Philippe's improved version of the Lotus 72 (Norm DeWitt)

Chapman. The car was a big disappointment as when we first tested it at Riverside, it was a second and a half slower than my F5000 car, just before going to South Africa. They used all torsion bar suspension instead of coil over dampers. It was so soft by comparison that we used the rear torsion bars in the front, and the back we had to go back to coil over dampers and change the thing all around to get it more back to standard. But, I'm going to South Africa with a car that's one and a half seconds slower than a F5000 car? Give me a break. That was what we were up against. We had the same issues that a lot of the smaller teams had.

Dick Scammell weighs in, "That's what happens I'm afraid in racing. As I had said with the 72, the original wasn't very good, but it got better. That was true with the 49 as well. Racing is racing and you should always do better and with a bit of hindsight you go, 'Oh, why did we do that?'" Dick also confirms Mario's feedback and impressions:

It was a very tidied up Lotus 72, it was a very neat little car. It needed stiffer suspension, you could stiffen the front up by using the thicker ones from the back and so on. The team just fabricated thicker stuff for the back. It's one of those things. I take my hat off to the people racing today as, boy, have they sorted out the reliability. It's just incredible these days the reliability of the cars and engines. Why couldn't we have done that then? Budget obviously helps and when you consider that the Lotus Formula 1 team was really six people, and now you look and see more than six people on each wheel and another 300 back at the ranch. You didn't have the money, the amount of people, and the amount of equipment to carry out the amount of testing.

The Parnelli, for being essentially a warmed-over 1970 design, had a few remarkable moments during 1975. Mario reminisces, "The only race that was potentially memorable was that I could have won at Barcelona at Montjuïc. I'd passed James Hunt for the lead and the car was unbelievable on that street circuit, as it was soft and very pliable. You had to jump at Montjuïc, and a freaking toe link in the rear broke. The car actually felt like a total winner. We also had a good race at Sweden where I finished fourth." That was as good as it got for Parnelli in Formula 1.

There were similar problems on the Indycar side of the operation in 1975 with the Formula 1-based VPJ-6. Hughie remembers:

I think it was after Indy when we went to Pocono with the Eagle and then Vel and Parnelli decided to it was best to go back to California, drop out of the championship, and then concentrate on re-working the Indycar. While all this was going on, I think Maurice left in April 1975 before we went to Indy, about two months after I'd arrived. Jim Chapman was the Team Manager at the time and we were kind of figuring out what to do. I had worked with John Barnard at McLaren before I'd left, and I put his name forward. They had a few telephone calls, he accepted the job and turned up.

By this time we were already reworking the Indycar, plus doing engine development work at the same time. Then we re-hashed the car around, took the torsion bars off the rear, and put the Eagle rear end on it with the Weismann transmission. And then we put the rear torsion bars on the front because what we finally concluded after Indy 1975 was that the front end was way too soft. As it went into the corner it would dip, and it would take a bite with the front and the thing would become loose. A loose Indycar… nobody wants to drive one of those for very long. Then John and myself and a few others hashed the thing around and then went back to Ontario. We used to run at Ontario about twice a week, trying to get A) the car, and B) the engine sorted. We didn't have any boost pressure restrictions, and we were running 80 inches or even more. The dyno couldn't sustain running that much power for more than a minute before it started overheating. The only way to do this was to go out to Ontario and do it on the track. So, we'd make some changes and eventually we got the thing to where it would live for 30 or 40 laps. By this time it was coming around to the Phoenix end-of-the-year race. So we went and I can't remember, but we qualified sixth or something. We developed the thing from then on. Al did most of the testing.

Al Unser finished fifth in the race as the team continued to prepare for 1976, which is when the new VPJ-6 with the Cosworth turbocharged powerplant vaulted to the front of the grid. In contrast, the 1976 season ended early for the VPJ Formula 1 program, despite finishing sixth in the South African Grand Prix, the second race of the season. The next and final race was their home event in the streets of Long Beach, the first of what was to be many years of Formula 1 races through the streets there. After finding out that his Formula 1 team was closing shop, a stunned Andretti moved on to drive for Team Lotus the rest of the Grand Prix season, a move which proved providential given what was

Mario Andretti in the 1976 Long Beach Grand Prix,
the last Formula 1 race for the team and their VPJ-4 (Gary Hartman)

to come from that combination in the following two seasons. (See *Making it FASTER* for Mario's stories of the Lotus models 77, 78, 79, 80, and 81.)

At the return to Phoenix for the opening USAC Championship race of 1976, Al put the Parnelli on the pole. Although Foyt was to take seven poles from thirteen races that year with his Coyote, the VPJ-6B was to create USAC Championship car history. Al Unser won at Pocono, at Milwaukee, and the season-ending Bobby Ball 150 at Phoenix with the Cosworth DFX powerplant, scoring the first wins for that turbocharged engine in the series. It wouldn't be long before Cosworth was to be the dominant engine in Championship Car racing. Yet although the VPJ-6B was finding success in 1976, the following car might have changed everything had the Parnelli team and John Barnard the resources with which to exploit it.

Al Unser recalls, "The first ground effects [Indianapolis] car was with Parnelli and John Barnard. I left that team and went with Jim Hall, where John Barnard designed a ground effects car [The 1979 Yellow Submarine Chaparral 2K]." The Parnelli car that Al describes was the transverse gearbox VPJ-6C. Gordon Kimball had gone to work for Vel's Parnelli Jones during the end of the F5000 era. Gordon recounts, "I graduated in 1976 from Stanford, and I worked that first summer as a mechanic on Al Unser's F5000 car. Maurice Phillippe did those cars, and actually they were pretty awful cars, technically. They then hired John Barnard to replace Maurice. I don't know if it was because the Viceroy money went away or something, but that's when John started to do the Indycar side, and that's when I started as well. I wanted to work a couple of years as a mechanic, to see the other side of it."

Gordon was in the right place at the right time. As a result of his engineering background, Gordon got the opportunity to become an early associate of John Barnard, a relationship which continued off and on through the following two decades, through a number of landmark designs. Kimball says, "At the end of the year John was the only one in the drawing office, so we did the VPJ6C over the winter of 1976, the transverse gearbox car at Parnelli's."

The car was not the success story for VPJ Racing that it could have been. Gordon Kimball continues, "VPJ was winding down, literally. We came to Indy, but somebody spit a turbo compressor onto the track and we ran over it, cut a tire, and destroyed the car in turn three. So, it [the transverse gearbox car] didn't race at Indy, but we built up another one to race the rest of the year." Al Unser persevered with the previous season's VPJ6B for Indianapolis and qualified on the outside of the front row, eventually finishing third. Unser also was to win the California 500 at Ontario, but by all accounts he had again used his primary car from 1976, the VPJ-6B. Al could see that the team was winding down and left to join Jim Hall's Chaparral Racing team for 1978.

The team continued into 1978, but by the late season race at Ontario, AJ Foyt was driving in the familiar blue and white #21 American Wheels colors that Al Unser had driven to the win the previous year, except that Foyt was racing the later VPJ-6C. Foyt was to campaign the Parnelli VPJ-6C for the following two seasons (1979-80), with the car initially entered as a joint venture between Gilmore Racing and VPJ, followed by Foyt eventually dropping VPJ as an entrant by 1980.

Gordon explains, "AJ Foyt bought or leased a couple of them, and ran a number of races. Vel had just lost interest, we weren't getting sponsors, and they shut the race team down. When I left in October of 1977, I was the only one in the drawing office, there were maybe two or three guys on the shop floor, and that was it."

Interscope Racing was to fund racing their VPJ-6B from 1977-80, Danny Ongais scoring a number of wins with the car. Ongais first won the 1977 400-mile race at Michigan, followed by five wins in 1978, all reportedly being in the older and more successful VPJ-6B. Between these final efforts by Interscope and Foyt Enterprises, it was the end of the Vel's Parnelli Jones Racing team's saga at Indianapolis. Following upon those early successes with the Indianapolis 500 victories for the Colt in 1970-71 (also 1970 and 1971 USAC Champions), their other great success as a manufacturer came when the VPJ-1 of Joe Leonard won the USAC National Championship in 1972.

Joe Leonard at home with the Colt of 1971 painting on his mantle.
It was the greatest year for VPJ (Norm DeWitt)

CHAPTER 11

PENSKE CARS

Roger Penske at Indianapolis in 2008. His team is the most successful in Indianapolis racing history (Norm DeWitt)

Team Penske first raced in USAC Championship Car racing with a 1968 (Tony Southgate-designed) Gurney Eagle powered by a Chevrolet stock-block engine. The pairing of Mark Donohue, owner Roger Penske, and primary sponsor Sunoco had a successful debut, scoring a sixth and fourth place finish at Mosport in the two races that made up the weekend. No doubt the additional experience with the car and the development of the Ford powerplant with Gurney Westlake heads showed as Dan Gurney won both races. The team returned for the final race of the season at Riverside, the Rex Mays 300, yet another race dominated by Gurney. Although Donohue showed competitive speed throughout the weekend, he knocked a wheel askew on the half-buried tires that lined the esses (as I watched from turn six), and his day was run.

Mark Donohue in the Sunoco Eagle during the 1968 Rex Mays 300, Penske's second USAC race (Gary Hartman)

Penske Racing returned in 1969 with the Lola T152, Donohue joining Bobby Unser in running the four-wheel-drive turbo Offy-powered car. There were distinct differences in the turbocharger mounting between the two, with the Donohue car having its turbo mounted forward of the engine to the left side of the cockpit in what has to get the award for most awkward-appearing turbocharger installation ever. However, it did free up the airflow to the ducktail wing at the rear of the engine cover. Unser's layout was far more conventional, the turbo mounted to the rear of the engine directly atop the wing, which undoubtedly had a detrimental effect upon airflow over the low rear wing of the car. In terms of speed there didn't appear to be a significant difference between the two concepts of aero versus turbocharger plumbing, Unser qualifying third, with Donohue fourth. Donohue was to run much of the race in third place until hitting trouble near the end due to problems with – the turbocharger – and finished seventh, taking Rookie of the Year honors. By most standards, that would be considered a successful debut at the Brickyard, but Penske standards and expectations are far higher than most. The team was also to take a seventh and fourth at Brainerd, and those were the highlights of Penske's five-race season.

The Lola T-152 that Mark Donohue took to Rookie of the Year honors at Indianapolis in 1969 (Norm DeWitt)

For 1970, the Lola returned with turbo-Ford power, the car now sporting vertical fences atop each side of the nose. Taking second place finishes at both Indianapolis Motor Speedway and Indianapolis Raceway Park (IRP) were the highlights of a four-event USAC season for the team, in their last year as a part-time entrant. For the first time, a Team Penske car led the Indianapolis 500, Donohue leading five laps of the race. The stage was set for a new emphasis upon USAC Championship car racing for 1971, and a switch to McLaren.

The McLaren was a huge improvement, and Donohue scored his first USAC victory from pole at the Pocono 500, followed by another victory in the Michigan 200. No doubt these successes with the M16 and the Team McLaren connections helped Penske to secure a Grand Prix M19A, Ralph Bellamy's successor to three years of M7A/B/C Formula 1 cars. Reports at the time were that the M19A was the first car that Penske struggled to get a setup sorted from solely running on a skidpad, but obviously the car worked given Donohue's third-place finish in his debut Formula 1 race at the Canadian Grand Prix. Yet again, Mosport had seen Penske Racing's strong debut in the top levels of open-wheel racing.

In 1972, Donohue and Penske were to achieve the greatest achievement in their long and successful partnership with victory in the Indianapolis 500. Donohue was still not running the full schedule, but when he raced the newly revised McLaren M16B-Offenhauser, it was on the pace. 1973 brought an early DNF at Indianapolis, in what was one of those years at the 500 best forgotten.

Penske Performance President Tim Cidric and family with their
1972 Indianapolis 500 winning McLaren (Norm DeWitt)

1974 was when Penske's next champion came onto the team. Tom Sneva says, "I was a rookie in 1974 and Bettenhausen got hurt. I'd had a pretty good rookie year for a low-budget team. Roger wanted me to go drive for him, and I'll never forget when at the Speedway he said he wanted me to drive for him, and it would be 30,000 dollars. I told him, 'I'd love to drive for you, but I don't have 30,000 dollars.' Roger said, 'No, no, no... we're going to pay YOU.'"

Sneva signed with the Penske team for 1975 and qualified fourth for the 500. In the race his Norton Spirit Penske-McLaren tripped over the lapped Eldon Rasmussen on lap 125 and had a horrific accident in turn two, the car shedding nearly everything but the monocoque when it was launched up into the catch fence atop the wall. What was left of the car finally landed right-side-up with Sneva inside struggling to get away from the burning methanol. It was a shocking testament to the strength of the McLaren chassis, as Sneva had received burns to his face and hands, but was otherwise intact. Later in the season he won at Michigan, and with three other podiums in the last five races, took sixth in points.

Jerry Breon had started at Team Penske in June 1974, brought in from Trans-Am racing to work on the IROC cars. Jerry retells, "My brother-in-law had a large Chevrolet dealership in Rochester, New York, sponsoring Warren Agor who had ties to Penske Racing in the 60s [mechanic for Mark Donohue]. We were racing Camaros in the Trans-Am and then went to work on the IROC 'Easter Egg Cars.' I think we

*The Penske PC1 Formula 1 was the first Penske Cars design.
It led to the successful Penske PC4 (Norm DeWitt)*

The logo for Penske Cars in Poole, England (Norm DeWitt)

did Riverside and Michigan, but then right after that they advanced me pretty quick and transferred me over to the Matador NASCAR program as the lead fabricator from mid-1974 through 1976. They were doing the Formula 1 car too. I went to see the car race at Mosport. They had the Formula 1 car [PC1] in the Reading shops and that's where he did the unveiling, but then after that everything was moved to Penske Cars in Poole, England. They were hinting at wanting me to go onto the Formula 1 team at the end of '74, but I opted to stay in the country and do the NASCAR program full-time."

Geoff Ferris' Penske PC1 was unsuccessful and was replaced by a March, then followed by a March-based Penske PC3. Jerry Breon explains, "It had that rounded front end like the March. I went to Indycars in 1977. We did a similar thing on the Indy side as the PC5 was a McLaren copy before we moved on to the PC6."

Both the Penske run McLaren M24 of 1977 and the Penske PC5 of 1978 were to take Tom Sneva to Indianapolis pole positions as well as two National Championships, the first two for Penske Racing. However, in each of those championship years, Sneva had fallen just short of victory at Indianapolis, finishing second in both 500s. When he failed to win a single race in his 1978 championship campaign, he was replaced at Penske Racing.

The PC6 was similar in appearance to the Austrian Grand Prix winning PC4 Formula 1 Penske of 1976. That success was somewhat overshadowed by the drama surrounding Lauda's injuries at the Nurburgring, six-wheeled Tyrrells, and James Hunt's late title challenge, which combined to dominate much of the coverage that year. However, John Watson's win in Austria had been at the same track that had claimed Mark Donohue in the previous years' race, and was the best possible result for team morale. Surprisingly, at season's end, Penske pulled the plug on the Grand Prix program. In hindsight, with the Lotus 78 ground effects car debut in 1977 at race one in Argentina, any development of the PC4 would likely have had minimal success, although it may have accelerated the ground effects program at Penske Cars to the benefit of the Indycar program.

The PC4 Penske Formula 1 car won the 1976 Austrian Grand Prix, with John Watson driving (Norm DeWitt)

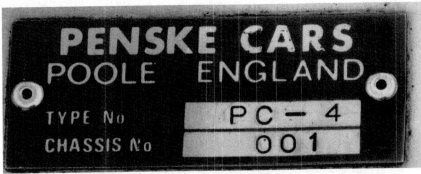

Chassis tag for the first Penske PC4 (Norm DeWitt)

As would be shown time and time again, Formula 1 cars do not make effective Indianapolis cars without significant development. Jerry Breon describes the complexities involved:

> By that time the engineering was quite a bit different between an Indycar and a Formula 1 car. When you set about building a car from the ground up for yourself, there are a lot of techniques that you have to develop in-house. There are fabrication techniques, material techniques, establishing trusted vendors… it's a pretty big task. Being able to take a proven design and go from there, rather than starting from scratch was an easier starting point. I would spend about six to eight weeks each year over in England at Penske Cars building the prototype car with those guys. There were several strong teams back in those days. There was us and Gurney, McLaren, Jim Hall, along with Pat Patrick and his cars.

185

The PC6 was the first Penske design to win Indianapolis, Rick Mears claiming victory in the 1979 Indianapolis 500. Also in 1979, Bobby Unser had run the PC7, the first ground effect Penske. Breon says, "We were just starting to dabble in ground effects in 1979 with the PC7 about the time when Bobby Unser came on-board with us. Of course he had his own ideas of how to do things."

Bobby Unser had a flow bench converted into a wind tunnel in his garage to test downforce (see *Making it FASTER* for more on that). Jerry reminisces, "I spent many hours in front of that thing in Albuquerque. I'd be stuck in the garage working with him, watching his brother Al across the street playing with motorcycles and dune buggies and stuff, wishing I was over there."

*The relatively modest underbody tunnel of the Penske PC7,
a first attempt at ground effects cars (Norm DeWitt)*

After all that effort, it had to be disheartening for Jerry when Al Unser showed up with the latest John Barnard wizardry, the Chaparral 2J "Yellow Submarine," at Indianapolis. Breon recounts:

> That was kind of a coup when he got that ride. Fundamentally the PC-7 was a pretty good car, but that wasn't a full ground effects car, it was a partial ground effects car. We were just starting to get into skirts to seal the car to the racetrack and developing all these different underbody shapes to move the pressure points around under the car to see what worked best. Back then it was pretty much seat-of-the pants engineering. When the driver and the engineer would come up with an idea, it was up to me to build it overnight at the racetrack so we could try it the

next day. It was a tough time for a fabricator like myself, but I managed to do it. We used to have some marathon test sessions, we'd bounce back and forth between Phoenix and Ontario in the winter over about a six-week period… at one track for a couple of days and then at the other track for a couple of days.

Obscure aerodynamic adjustments that offer improvements at 200+mph can be easily overlooked in the garage. Bobby Unser tells how incredibly small things can make a huge difference:

At the Speedway during practice days we had problems. You know when they put stickers on the wing for the sponsors, those letters were causing an air tumbling thing at the leading edge as we needed good laminar flow. So, I took those damn stickers off the Penske car that I was driving and the thing got better. I went back and told Roger and he said, 'We've got to have those on there.' I took sandpaper and took the little sharp edges on the vinyl and tapered them down to the aluminum. Now this didn't work that way with all wings, but it worked for that particular one, just the leading edge of the decal was wrecking me.

As was clear from the Eagle chapter, Bobby Unser did not like to share information about developments that he had been able to achieve independently. Bobby's teammate was Rick Mears, who had an entirely different mindset:

There were things done, obviously. Bobby was old school and that's the way it was, and he was one of your competitors. He was before learning the benefits of the team concept, and Penske has always been teamwork since day one, which is why I fit in so well right off the bat. My brother and I were practicing the team concept unknowingly as we grew up. We'd race against each other all the time and then go home at night asking what are you doing over here, and what's that over there? We knew if we worked together and raised our deal to another level, to where we'd have an advantage over everyone else, all we had to do was to race each other and that was a lot of fun. That was the Penske team concept. It could be difficult, depending upon your teammate at the time, and sometimes you come up with all the ideas and maybe your teammate doesn't get into the workings of the car too much, so it was a little one-sided at times. What goes around, comes around, and eventually they see the benefit of it.

But Bobby Unser wasn't buying what Rick Mears and Penske Racing were selling. It even reached the point where Unser had the markings on the springs changed to hide his true setup from Mears. Jerry Breon recounts:

That was the second time Unser just about got me fired. There were two instances at Indianapolis where I feared for my life. One of them was the 'Amsoil incident.' Penske had a long-time sponsor and supporter in Sunoco products, and we were advertising their products religiously. Bobby had this side deal on his own before he ever came to Penske with Amsoil gear lubricant, and he insisted on having that in his transmission for qualifying at Indianapolis. So that was one of those times we had to come in at three in the morning before qualifying and swap out the lubricant in his car the night before qualifying with this stuff. Fortunately it never caused an issue or problem, and I'm sure Roger knew... he let it go for a few days before he said anything about it.

The second time, Bobby didn't want to let Rick know what spring rate he was running in the car. So we would perform spring changes, we would perform fake spring changes, putting the same stuff back in. That kind of upset the apple cart a little bit with Roger and particularly with Derrick at the time. To do an end run around that, Roger came into the garage and insisted that we pull the springs out so he could look at them and see what numbers were on the springs. At one point in time Bobby had somebody on the crew – and it wasn't me – grind the numbers off the springs and put numbers on the springs that didn't accurately reflect what they were. That all came to a pretty severe head one night at Indy with Mr. Penske involved.

Bobby and Derrick Walker didn't exactly see eye to eye on much. Jerry admits, "Well, yeah I suppose there is an element of truth to that.

But Bobby was an old school guy and he was racing against everybody else. It was him against the world, so he would keep things pretty close to the vest and wasn't too anxious to share speed secrets with anybody, including his teammates. He was pretty intense about his own successes, but you know he's got the record to show for it. He gave Rick [Mears] a hard time but Rick just sort of sat back and smiled and understood that's the way it was and waited until he had his opportunity." Rick Mears explains:

Bobby was Bobby and he was a racer. I had my own pride, and I didn't ask him. That way when I beat him I felt like I beat him. I figured it out on my own, and that way I didn't beat him with what he did, I beat him on my own. I figured it out and that sounded a lot better to me. I always acted to Bobby like I didn't care and that drove him nuts, but that doesn't mean I didn't keep my eyes and ears open. I'd just mosey around doing my own thing, acting like I could care less what he did, and it drove him crazy that I wasn't trying to find out what he was doing. We laugh about it, and we got along great. I understood where he was coming from. We've been at functions together before where Bobby says, 'I taught that Mears kid everything he knows.' I'd get up afterwards and say 'You know, Bobby's right. He did teach me a lot, and he probably taught me more than he knows. He taught me to surf through the BS, and read between the lines. He taught me to keep my eyes and ears open, taught me all sorts of things that he didn't even know. Roger didn't like it because that's not the way his team works.

Jerry Breon recalls:

At that time, Bobby was Roger's guy, and he'd brought Bobby in to win races. Publically there was no number one and number two drivers, but Bobby was there to be the lead guy and win races for us, which he did. As mad as Roger would get about this, he would always give us that little wink of the eye when he left the garage, kind of making us feel, 'Keep doing what you need to do.' Looking back upon it, that wasn't the right thing to do, it was a pretty serious infraction and you know it wasn't fair to Rick. I suppose to this day, Rick holds a little bit of a grudge towards me and the others who were involved with Bobby back then. But, there were plenty of things I did for Rick too, over the years, and fortunately I didn't get fired. The end result was always that a Penske car won the race, and so there was no guilt in trying to accomplish that goal.

The Team Penske mantra of 'All for one and one for all' philosophy of raising everyone's performance on the team obviously wasn't always the equation in the 70s and early 80s.

The Penske PC6 and PC9B were both Indianapolis 500 winning designs by Geoff Ferris (Norm DeWitt)

1980 brought a new ground effects Penske PC9 [1981 was the PC9B]. The 1980 season was fought out between Johnny Rutherford in the Chaparral 2J and Unser in the Penske PC9. Johnny took the season opener at Ontario, followed by the Indianapolis 500. Bobby countered with the next two at Milwaukee and Pocono, and that was countered by Rutherford winning Mid-Ohio and Michigan, followed by Unser at Ontario. In the end, Johnny Rutherford in the Chaparral was CART Champion over Unser, although he had a disastrous flip in the Chaparral coming onto the front straight during the Phoenix finale.

There was probably as much drama going on behind the scenes at Penske racing that weekend as at Chaparral. Chuck Sprague was a new mechanic at Penske Racing for 1980 with the PC9. He recounts:

Bobby crashed his car in practice at Phoenix because he had insisted upon a special treatment for his skirts that nobody else had, so Bobby had no car for the race and he did not race in the 1980 season finale. Bobby had tricked out his car and his trick backfired on him, so he watched from the sidelines the next day. We only had three cars, and Mario was in the third one. This can happen when the guys you are hiding stuff from are not as stupid as you'd like for them to be. He had a theory, it caused a skirt to hang up and that was the end of the story. At Phoenix when something goes wrong, you are a string and a rock. When the string breaks, the rock takes off. There are no small crashes at Phoenix. That was the first place that we ever registered a composite 5G loading on a car. Even before that, every time I went to Phoenix I was always nervous, if we brought everything home without it being wadded up, I was a happy camper.

Meanwhile, Sprague had found the magical combination for a superior moving skirt on for the PC9. Chuck continues:

The PC9 had fully articulated skirts, so you are working on pressure versus wear, spring pressure, skirt weight, so on and so forth. In that '80 season Mario ran for us four times. At Indy we didn't run the articulated skirts, but we were constantly having skirts being shipped over to us from England. Bob Stroud and I, who were working on Mario's car, came across the one and only set that was made out of a smooth uni-directional carbon, and was therefore a little stiffer than the others. Basically the other crews weren't interested in them, but we found that because they were more rigid they didn't require as much support, and therefore we could run softer springs, and with softer springs we could run less ceramic rubbing blocks. That made the skirts even lighter so we could run the springs softer yet. That set of skirts worked particularly well and they always stayed on that one particular chassis and I think that was '01', which Mario won with at Michigan, the third race he did. If you look at the last race, Mario would have won that race except that we dropped a cylinder. Skirts were everything, and if you sealed the car to the track, you were golden. At the end of the 1980 series they were banned, and at the infield of Phoenix we burned them.

191

Their fight continued into 1981 as Unser started the season finishing second to Johnny Rutherford's Chaparral at Phoenix. At the following race, the PC9B was to bring Penske Racing another Indianapolis 500 victory in a contentious finish, with protests that gave victory to Andretti in the Patrick entry, then later awarded again to Bobby Unser. It was an episode best left undiscussed as it left everybody involved, spectators alike, disillusioned with the sport. After Indianapolis (USAC) the second CART race of the season was held in Milwaukee, and it was here that Unser's season began to unwind. Bobby reminisces:

As smart as I tell you that I am, how come I was so dumb all the time? Derrick Walker was the team manager and Derrick and I didn't get along even a little bit, but we went back to Michigan because I had some new tunnels I'd designed and Jerry Breon had built to try. With Penske's outfit I can just do things NOW, I don't have to ask permission. If I wanted to run Michigan for a test, I'd just say, 'I want to go do it,' and if it's humanly possible, it would happen. Roger would never say no. I had two cars, both 1981s. One of them I had at Milwaukee and we put air pressure into the wrong line, and it went into the fuel tank, swelled the tank up and broke the glue loose which holds all that s**t together. We ran it, but it was a terrible race that day. So, we took that same car, sent it to Michigan with new ground effects and I should have been running way into the 190s easy. Well, I'm down in the low 180s and I'm driving my butt off and that thing is going to bite me, it's just not working. I'd always done my own chassis setups and so I'm back in the garage area where I laid down my pad and I'm working the weight jacker checking the corner weights. I don't hear too good, too many race cars, too many years. Here, Derrick is walking by the car and looking at me going up and down with the corner weight. So he stops and puts his hand over his ear and says, 'Do that some more.' So I did it and he keeps going around holding his ear listening and pretty soon he says, 'The tub's broken.' I said, 'You're s**ting me.' Now understand that at that time I didn't know if a tub should be extremely stiff or if moderately stiff was okay. When it was being pumped up and down you could see that the glue was broken, the whole thing was floppy like a dishrag. So, Laurie [Gerrish] took the car back to Reading and I told them, 'I don't give a f**k what you have to do. Spill the glue all over, put twice as many rivets, three times as many rivets, I don't care what it looks like, make that freaking thing strong. I don't care what the guys in England say, I don't care what anybody says.'

They did everything that they could at Reading to make that tub like a piece of steel. We took that same damn car back to Michigan and the thing went so freaking fast you couldn't believe it, we were nearly at 200mph. You know how dumb we were? We did all these fancy things but nobody had brains enough to know that the tubs were flexing. They will all talk about it today like they knew it, but if you knew it, how come you didn't fix it? I saw the s**t going on and time went by before something hit me on the head and went 'ding-dong', and I realized the tub was bending. I learned it from making a mistake, and I had that tub for the second half of the year.

It wasn't just tub strength that Unser was fighting. Chuck Sprague explains, "The PC9 was a molded fiberglass tunnel, which was rudimentary at best. The 9Bs were fabricated aluminum, but had a whole pile of stiffeners on them. There was a test there somewhere along the way when we put a filler piece into the tunnel and pulled it right out during a Michigan 500 mile test, which incensed Roger Penske to no end, but it taught us a lesson." Bobby Unser confirms the struggles with underbody failure in the tunnels. "Some of the ground effects that we would design were pulling the rivets through the aluminum. I couldn't even believe it myself. Get bigger goddamn rivets, put more of them in there. Put bolts in it, do something!"

At the season finale in Phoenix the race went back and forth between Tom Sneva's March and Unser in the Penske, Sneva eventually coming out on top. As it turned out a third at Milwaukee and second at the season-ending race in Phoenix were only enough for Unser to score seventh in points (CART). Despite his injuries at Indianapolis from a pit fire, and missing the first Milwaukee race, Rick Mears had a dominating CART season in the sister car (which one presumes avoided any broken tub drama) with six wins, one second, one third, one fourth, and an eighth from the ten races entered. Bobby Unser recalls:

Geoff Ferris wasn't involved as he was working on the 82 car, and I didn't get involved in the 82 car until the very last. The PC10 was the best of them, by far. For 1982 I was going to run only five races with Roger, with Indianapolis being one of them, and that was our agreement. Roger called me up one day and said there was going to be a press thing for the new car, and asked if I wanted to run Phoenix, asking me if I also wanted to come test this thing since I was going to drive it. I told Roger that I had been reading Rick's [Mears] reports. I went back to Reading for the press conference and then Breon and I got drunk – I mean real drunk.

The next morning at seven o'clock, Breon and I were there at the shop with brown wrapping paper laid out all over and he is drawing up wings and we are going to town on things. I'm looking at the front sway bar, thinking we needed a bigger bar in there than that, or at least get me something need to try. Jerry Breon said, 'Can't do it, Ferris made it to where we could only have one bar and you've got to use little blades to tighten or loosen it.' I told him that isn't going to hack it, we need a bigger bar. That part of the tub was steel where the bar goes through, and he said, 'Okay, but I'd have to cut it and boy is everybody going to get pissed.' I had three sets of underwings laid out and at nine o'clock Derrick walks in and it came within a quarter-inch of a real fist fight. We were still hungover bad, in there designing race cars, that was really dumb. So I called Roger right in front of Derrick and told him I couldn't work with Derrick. Roger told me to, 'Cool it, go on home and I'll have everything taken care of.' Roger called me and said, 'Let's quit this fighting,' and I said, 'I'll try.'

Breon laughs, "Well, there again that was Bobby's efforts and desire to continually engineer the car. He had a pretty substantial history of what worked and what didn't work on Indycars, particularly coming from Dan Gurney's area, and he wanted to bring a lot of that information on board at Penske Racing. He had opportunities to do what he thought he needed to do to make his car better." Chuck Sprague says, "I wasn't privy to what went on there, but in all fairness to Bobby we did run the bigger sway bar." Bobby Unser continues:

So we went down to Phoenix, and Kevin 'Cookie' Cogan is the driver. I asked Jerry if all that stuff we had made was done. There was only one setup pad at Phoenix in those days, and I got that spot. We were working on it by ourselves and every time we went out, we broke the track record. All three sets of underwings that Jerry made were there. Put the next set on, set another track record with it. 'Pencil' [Geoff Ferris] came over with his clipboard and asked what I thought we should be doing. I told him that he should just put Cogan's PC-10 in the truck because it's not as good as the car with what we've been doing and he's just going to wreck it. You don't know what I've done and you just need to park the car and copy what I've been doing. They kept running Cogan's car just like they had designed it in England. He wrecked it, and knocked all four corners off it. That PC10 with the modifications was the best race car that I ever drove in my life.

The day was coming to the end, and we put the third set on it. It was now five o'clock. If I wanted to run until dark, I had the authority from Roger to tell the fireman to stay and they would be guaranteed pay to stay until dark. So we are under the car putting those wings underneath and it was Laurie and Jerry Breon and I. Laurie looked at me and said, 'You're leaving aren't you?' I didn't know it was that obvious and I told them, 'Yeah.' There were some tears and we put the car down, and I went out and blew the s**t out of the track record. For the first time, I went and took my helmet out of the Penske transporter, it was the end of the road. I hadn't told Roger yet, but my guys knew it. I flew home, called Roger, and told him about what we had done and how he had the greatest race car that I've ever seen in my life. I told him that you guys with Rick will win every race this year in the PC10.

Jerry admits:

Probably true. I was still pretty young back then so I wasn't privy to all the behind the scenes stuff, but I do recall that things came to an end there fairly abruptly. Certainly it makes you wonder, but my goals were a lot loftier than Bobby Unser. My goals were to advance at Team Penske and win as many races as I could with that group. I was one of a group of fabricators in the Indycar series that were fairly well known for what we were doing and the talents that we had. I had many opportunities to go elsewhere that probably would have turned out to be pretty short-term opportunities. If I had followed any one of them, I'd probably still be working today instead of comfortably retired. I was there for forty-two continuous years.

There was more to the 1982 success than just having a stellar new car. Chuck Sprague explains further:

The other thing was that we made some gains in terms of our approach to car preparation that helped us out in the '82 season. When we suffered various disasters, we were able to respond quicker. Up until 1982, each car came across [from England] as a basic roller, but they weren't even plumbed or wired. Each crew would then finish off the car, making decisions as they went, which meant that the cars weren't the same. It doesn't sound like much, but at Indy in 1981, when Rick's primary car suffered a problem, we had to qualify the backup car. It took us until one-thirty in the morning just to put the gearbox and rear suspension from the primary car onto that backup chassis. None of the clutch fittings were the same, the brake lines weren't the same, and the shift linkage wasn't the same. It dawned upon me that it was such an enormous waste of time. At the beginning of the '82 season, I spoke to Pete Parrott, my chief mechanic, and Clive Howell who was Bobby's, and said, 'Why don't we just get together and agree on the first car and then document the way we set the thing up so that when another car comes over, we can do them all the same.'

In 1982 there were more than a few crashes, both Kevin and Rick. We went from Elkhart Lake directly to Michigan, where we had crashes at both places. We were able to take the tub from the Michigan crash and dig the rear end from the Elkhart Lake crash out, they bolted right up and finished in sixth spot. When you are functioning like that, the efficiency allows you to spend the remaining time and effort making the car go faster rather than just trying to make the grid. Behind the scenes we were able to instantly respond and a lot of people don't appreciate it. Nowadays it's standard operating procedure everywhere. A lot of people like to decry the loss of individual ingenuity, but that's still encouraged. If you do something, you run it through an official test and if it works, it goes through all the cars. When I became Team Manager I was focused upon if there was one of anything, and if we were going to test it at an event, we'd better have enough of them for both cars. If you have a two-car team, and you don't run the same things on both cars, you don't learn the same things. It wasn't a big deal to run one part at a test, but by the next week at the race you make damn sure you are ready to make another one.

The PC10 of Kevin Cogan triggered the startline melee at the Indianapolis 500 that year. Bobby Unser was watching from the pits. Bobby recounts, "At Indy on the start when Cogan put the thing in first gear and got his turbo pressure going, and when he took his foot off the brake all hell broke loose. What got me so mad was that the tow truck comes over and picks up his PC10 by the roll bar and now everybody is taking pictures of the underneath of my car – all my secret s**t. I'm running Joesle Garza now, but that's my car, that PC10. There's Bignotti over there taking pictures of my goddamn tunnels. I ran out on the race track to the car and threw a blanket over it so nobody could take any more pictures."

What's good for the goose is good for the gander – as when Rutherford had flipped the Chaparral two years before in Phoenix. They were all doing the same, running for cameras to photograph the underbody of the 2J. Bobby continues, "Well, I'm the guy that was the benefactor of those pictures, and two or three days later I had a model made up and was already running it on my flow bench." So the moral of the story is, you either grab the camera, or you grab the blanket.

The Penske PC10 was an incredible design, by far and away the dominant car of 1982. Chuck Sprague explains:

I think there were a number of factors at play. The PC10 felt good right out of the box and we ran under the track record at Phoenix by a considerable amount, so we knew right away that we had a player. As for the season-long performance, most of the credit goes to Derrick for sorting out a flexible skirt system that basically responded to the vacuum under the car in such a way that it barely danced along the track surface, but it really didn't touch. We did a lot of testing with materials, backing strips, shapes, and angles, until we got a system where the skirts were molded where they faced in at a 45-degree angle. As the car got lower and the pressure dropped the skirt would flex accordingly so that it just danced just above the track surface and didn't wear. That was all the difference in the world as we got it wrong at the second Milwaukee race and I remember after the race talking to Rick, and there was a big 'wow-wee' worn in the skirt right at the lowest part of the tunnel near the radiators. Rick said, 'We got too greedy with the skirts, we got them just a little bit too low, so once they started wearing they got shorter, and when they got shorter they didn't flex as much so they started wearing even more.'

However, the PC10 was not to join the pantheon of Indianapolis-winning Penske cars. Rick Mears came close to winning the Indianapolis 500, battling to the wire in a memorable finish with Gordon Johncock. Jerry Breon says, "I wish we would have won Indianapolis that year, we did everything right except at the very end there, when we got caught out behind Herm Johnson in the last pit stop. Having to run down Johncock there at the end, it was close, but every time I've watched it we came in second. I think the PC10 and PC17 stand out as exceptional cars that we won a lot of races with." At season's end, Mears was a champion for the third time, having won at Phoenix, Atlanta, Pocono, and Riverside.

Chuck recalls, "The PC10 was not only mechanically a nice step forward, but we had a good skirt system and I don't know what resources Derrick was using to get these different materials, but I remember Derrick put a lot of effort into it. It was the first car that we used uni-directional carbon on both sides of the aluminum honeycomb for the tunnels. We would cut it to fit, and make shaped angles to go on either side to the tub and to the side panels. That was far lot more rigid than what we were using before in previous PC9B, which was just 063" aluminum, so you weren't maintaining a tunnel shape that was as smooth."

Mike Hull was at Arciero Racing for 1983, and they were to run year-old Penske PC10s. Mike tells about their relationship with Penske Racing:

I was really fortunate as we ran Roger Penske Jr. in Super Vee and in exchange for that we had an opportunity to run PC10s, when they went to the PC11s, which was a composite and aluminum honeycomb car. We were kind of like a sub-chapter of what Derrick Walker ran at the time for Penske Racing, and we ran Pete Halsmer. In that era of racing, Penske had set the standard. We had a customer car with a customer program and they treated us really well. Roger Penske himself was really good to us, and all those people who were there back then to this day are still great friends. I was impressed as I got to spend time that year in Reading, as they gave me a car bay and it really made a big difference for me in learning how to solve problems by being close to them and seeing how they did it. I have a great deal of respect for the consistency of their program. Mr. Penske decided that it was not only important to retain the quality and craftsmanship in his cars, but to hire the best drivers. The common denominator of any race team is the quarterback who is driving your race car.

Roger Jr. did a terrific job in Super Vee for us. We won a race with him that year, but I think we drove him into the business world. With the RT-5 Super Vee, Roger wanted his son to have the best car that there was, so we gave him every opportunity to find out what he should be doing with his life. We ran the same skirt system from the PC10 on the Super Vee. In those days we had just discovered bump rubbers. You had to figure out how to keep the skirt on the ground, and that was a big part of making those cars go.

Chuck Sprague says, "At the end of the season in 1982 was when Al Unser came out to test with us the first time. We took the skirts off the PC10 and we were appalled by how much worse the car was. It was a quantum leap backwards, it was gigantic. I'm sure part of it was the fact that the car was setup and sprung for all that downforce, and losing that downforce made the mechanical setup completely wrong, but it was an eye-opener to us just how much of the car's performance was strictly off of that aerodynamic package."

The PC11 of 1983 was a disappointment in comparison to the PC10. Mike Hull remembers, "By Saturday night at the races, Roger or

Derrick would come to us and ask us what we were using for a setup as we were out-qualifying their race team. They went backwards, and what they talked about was how the torsional stiffness of the PC11 wasn't what they thought it was going to be." When John Paul Jr. won with the VDS PC10 at Michigan, the scuttlebutt was that it's because the PC10 was superior to the PC11. Sprague explains, "VDS had done some development plus they stuck with one car all season long. When you are jumping back and forth you don't become an expert on anything." Jerry Breon shares the view, "Well, that's how it turned out. I think the PC11's problems were in the rear suspension, it had some strange configuration rocker arms at the back of the car and it was probably our first attempt at inboard shock absorbers. Looking back on it, it's hard to put my finger on any one thing, but I recall that car gave us a lot of fits from day one." Chuck Sprague explains:

In the case of the 1983 season, people talk about the PC10B, and I saw a chassis plate the other day that said that, but I don't remember ever building a car that said PC10B on it. I remember one time at Atlanta we tested a PC11 with a PC10 rear end on it, and we called that an 11B internally. We also took PC10 chassis with PC11 rear ends on them, back and forth. If it had been a clear-cut answer, we would have stuck with one or another. Part of the whole deal was that the PC10 and 11 were both very narrow chassis and were flexible. The 11 had high carbon side on it, and we may have made some pieces for the PC10 that did the same in aluminum. At the Michigan race [September '83] we had a PC10 with a PC11 rear end, and Rick had qualified something like eighteenth. We stayed up until two in the morning and made a new underwing for it, but not only that, we made a new set of headers for it that allowed us to run that underwing. Building an underwing is one thing, but building an underwing and a set of headers at a racetrack in one night? It was an uncanny accomplishment. We went out and won the race the next day, and that was a case of where we all looked at each other thinking, 'Great, now Roger can point to this the next time we had some massive project that was more than anybody wants to tackle.' None of us expected it to pay off, it was a shot in the dark.

There is something to be said for Engineer Scott telling Captain Kirk, "It will take at least forty-eight hours." Sprague continues, "Also, you run the risk of anytime you are building in the field, that it won't be of the same quality as you are building in the shop. The '81 red, white, and blue Gould Norton truck was the first one to have a generator, welder, and fabrication stuff in it. We showed it to Roger and he asked, 'Will it make us any faster?' I looked at him and said, 'Yes it does.' It was the first one and we could do things that we couldn't do with the other truck."

Despite the musical chairs between bits of PC10 and PC11 being moved around, Al Unser took the PC11 to the CART championship, while having only a single win, at Cleveland. In comparison, Teo Fabi had taken six poles and four wins with the Forsythe March, Sneva having two wins with his Bignotti-Cotter March, including at Indianapolis. The writing had to be on the wall that the March was the car to have.

1984's PC12 was the last of the Geoff Ferris cars. Breon says, "Geoff never left Penske Cars, he stayed on. When Roger brought in Alan Jenkins, Geoff concentrated on designing components, most notably our own design of gearboxes back then." After a couple of races with the PC12, the team committed to the March 84C chassis and Rick Mears won his second Indianapolis 500. Sadly, Rick also had a devastating crash at Sanair in the March that nearly cost him his feet and his career.

In 1985 and 1987, Penske was to win Indianapolis with a March. Danny Sullivan went on to win the 500 in 1995, the famous 'spin and win' year. Danny remembers, "I don't think there was much difference in 1995. There were certain years Lola made a stronger car, and certain years the March, and later the Reynard. My car was a little bit better in three and four, and that is why I made all my attempts to pass [Mario] in one and that was the entire reason. We were about equal in one and two, which is why I couldn't get a run on him in turn three. I had a Lola with Doug Shierson the year before and it was a great car. But in 1995 you could throw a blanket over them." In the end the title went to Al Unser in the Penske Racing March 85C, by a single point over his son Al Jr. in the Doug Shierson Racing Lola T900. As Al Jr. had noted before the race, "Dad taught me everything I know, but he didn't teach me everything HE knows."

It was one of the rare years where CART achieved relative parity between the chassis manufacturers. Danny Sullivan explains, "That's what you are after, but sometimes one gets a design that's a little bit better. And that's why Indycar was better in that respect than Formula 1, because if we had the money, we could go buy an Ilmor engine or a Cosworth. In Formula 1, you can't, and a lot of time those guys aren't even close. In certain years we were not competitive with the Penske, but in other years we were ahead of everybody. But with the rules, you couldn't get as far ahead of everyone as you could in Formula 1."

Where Penske Racing had their 'unfair advantage' was that they worked harder than the other teams. Bobby Unser says, "They worked harder, that's their unfair advantage. I was there, I know where his unfair advantage was, and it was that everybody worked hard." Jerry Breon agrees, "I can attest to that." Jerry continues:

> Every driver that we had come through there, and there were quite a few as you know, had their own personality and their own set of luggage that they brought with them. The goal was success for Team Penske and the best way to accomplish that was to pool the effort, share the resources, and to go as far down the road with the information as you can. There are lots of detractors out there like Paul Tracy, that say we race against ourselves, whereas Ganassi focuses upon Dixon and makes Kannan do whatever it takes to advance Dixon in the points. There are different approaches you can take I suppose, but there's no telling which one is right. This is obviously Mr. Penske's choice.

The PC15 of 1986 wasn't competitive and the PC16 wasn't much better. As Jerry tells it, "We were in a slump then... actually a giant slump. Phoenix and Long Beach were the first two races in 1987 running the PC16. The PC15 and PC16 were Alan Jenkins cars and those were beautiful cars with a lot of great craftsmanship, but they just were a little bit slow. At that time in conjunction with developing the car, we also were developing the Ilmor Engine. In fact the PC15 was the first car we were running the Ilmor engine in, so we had two major projects going there simultaneously." In 1986, Mears took pole at Indianapolis and led much of the race, narrowly beaten by the leading duo of Rahal and Cogan in their sprint to the finish. Third in points was the team's best result that year, Danny Sullivan leading the way with two wins: at the Meadowlands and Cleveland.

For 1987 Mario Andretti had Nigel Bennett's new Lola T-8700

which did an absolute horizon job on the field at Long Beach. Mario recalls:

That was the best chassis, the best Lola of the time, and the most unreliable engine at the time... the Ilmor engine, but I pushed to do it. I liked the feel of that engine in the chassis during testing at Phoenix. I was correct, the car was good and we were quick everywhere, but we couldn't finish the races. Most of the races that we finished we won. Partly it was that Adrian Newey was my race engineer. Nigel Bennett and Adrian Newey, that's the best of the best, and I began to really appreciate the value of Adrian Newey, as far as understanding the dynamics of how to set up a car, get the balance and every aspect of it. We were looking at the ambient temps, the degree of sun, all the little particulars and it taught me a lot along the way. We really clicked.

The victory at Long Beach was the first win for the Ilmor/Chevrolet in CART, the first of many. It was going to be a tall order for Team Penske to grapple with this combination. Jerry Breon concurs:

The Lola was good, in fact we bought one and we tested it. Roger would spare no effort or expense to see what the competition had. There were times where through a third party we would somehow get our hands on the competition's chassis and engines, and test them for ourselves to see what we were up against. We tested that car quite a bit at Road Atlanta one year and we knew what areas we had to work on. I think the fact that we never raced Lolas [or at least not until 1999], tells you that we had the confidence to take what we learned and incorporate that into our cars. In those days there was no spec car and you could build and run whatever you wanted within the rules. The incentive for developing your own cars was so you would have something no one else had. It was a double-edged sword. If you've got a good car, you are looking good. But if you don't have a good car, which we had several times, then you aren't looking so good and you have to catch up.

At Indianapolis, after a week of struggles with the PC16, they ditched the cars in favor of the previous years' March 86C, Rick Mears' best time in the PC16 being about 7mph off the pace set by Mario. Both Mears and Sullivan raced March 86C-Chevrolets (Ilmor) while Al Unser was brought back to replace the injured Danny Ongais, and was to race an older March 86C Cosworth show car, which tells you all you'd need to know about the situation. Breon says, "Things were getting down to the wire, we were already at Indianapolis trying to qualify and run the race. There is only so much you can do in a matter of ten days."

Mario Andretti led all but seven laps of the race until twenty-three laps to go when the car slowed with a short in the fuel injection system, and it was the end of a month of dominance. At first Guerrero led until problems with his car emerged during a pit stop, then Al Unser inherited the lead and the old show car became the winner. Cummins was the primary sponsor and they finally got the Indianapolis win they had so richly deserved with their innovative diesel racers, such as the 1952 car (see *Making it FASTER*). The 1987 winner sits in the Cummins factory showroom alongside their '52 pole-winning Kurtis diesel.

For 1988, Penske Racing brought Nigel Bennett into the fold, as he had left Lola to form a consultancy firm. Roger Penske says, "We saw that obviously that carbon was coming into play, both in Formula 1 and other types of motorsports, and as far as we were concerned someone had that background. We were running aluminum chassis, Geoff Ferris was our designer, and to get someone like Bennett who had that experience and also, as close as he was to Nunn... Nunn was not only a team owner but he was an engineer, and I think that rubbed off on Nigel and became a real asset for us." Jerry Breon agrees, "Yeah, that [carbon] was another developing technology of the time." The result was one of the greatest cars in Penske team history, the PC17. Jerry recounts:

> We had lots of success with that car, right out of the box. By then the engine was very well developed. We went back and forth between practice and qualifying with those flush and open wheels, so Rick had pretty much committed to the flush wheels for the race. Before the first pit stop he came on the radio saying he wanted open front wheels on the car so we had to do some scrambling as everything was mounted on flush wheels. Back then in addition to doing fabricating, I was doing the tires for Rick too. A number of people have claimed to have the same sort of feel as Rick Mears, but it didn't necessarily bear out in their

results. It was bias ply tires back then and stagger was a huge thing. He was pretty particular about knowing what he wanted, but he would want to try a number of different things. I would go through piles and piles of tires trying to figure out some sort of correlation between tire codes and potential stagger. We kind of figured out how to control stagger to a certain extent. We'd run the tire on the first run when they were brand new, and Rick had a talent for being able to micro-manage stagger a little bit from the cockpit of the car depending upon how he drove the car and what kind of line he took on the race track. So we thought we had little bit of an advantage with the tires over someone else. From that point on, everybody caught on really quick as to the importance of stagger was and how important it was to manage it. But there was a time period there where we had a pretty good handle on achieving what we were trying to achieve. The PC17 was good enough then and it responded to minute changes.

Penske Racing took the entire front row of the Indianapolis 500, Mears on pole. Hiring Nigel Bennett's consultant firm was showing immediate and positive results. Nigel Bennett was also the race engineer for Danny Sullivan and the combination took four wins and the CART Championship for 1998. Sullivan recounts, "Nigel was absolutely fabulous, not only a great designer, but a great engineer. More importantly, he was somebody who listens, so when there was a problem he would think about it and would say, 'I get it,' and work to solve it. It was a great experience. He sat in on all the meetings, sometimes he would sit on your car when you pulled in asking, 'What's the car doing?' I ended up winning the championship with him and then almost ended up winning Indy a second time. Roger wants it as bad as anybody, so he was always trying to get the best people."

For the following year Roger sold the new Penske to Patrick/Ganassi for Emerson Fittipaldi. Jerry Breon says, "That was the PC18 and that was the first time that one of our competitors had the same car that we did and they went on to win the Indy 500 with it. I'm sure that was a move by Roger to help his friend Pat Patrick out, and it was certainly understandable. There was a lot going on behind the scenes that a guy like myself wasn't really privy to." It seems on the surface to be a case of giving somebody the biggest hammer you have and then getting hit over the head with it. However, there was more to the deal as it also secured the services of Fittipaldi for Penske Racing, along with his Marlboro sponsorship, for the following years. Chuck Sprague explains:

That was Roger's decision to make and the only time we did that [with a current spec Penske] was 1989. With the exception of that year, it was always just a previous years' thing. It was still a distraction, but we made a commitment to the customer and whatever time it took to do that to the best of our abilities, we would do. My only concern about a customer car program, is that you have to staff it and arrange it in such a way that it's a self-sufficient operation that doesn't pull resources from the team. We were never in a position to do that. I did the best I could to insulate our team from it, but sometimes we had to provide spares out of our own truck. That was always nerve wracking as you didn't want to tell a customer no, but you didn't want to wind up in a situation where you didn't have what you needed. I'd go to Roger and tell him, 'If I give him this gear, then we won't have a spare for practice tomorrow morning.'

Again, the front row of the Indianapolis 500 was all Penske cars, this time with Fittipaldi's customer PC18 on the outside of the front row, with Mears on pole. Emerson Fittipaldi won the race after a late duel with Al Unser Jr. Danny Sullivan was driving for Penske Racing and had started alongside Mears. Sullivan says, "We made the mistake... they [Fittipaldi's team] kept it really simple and we over-engineered it. We had too many things and it was a car that was really kind of much more on the knife-edge. You had to keep it simple and if you got it a little bit wrong on either side, it was hard to get it back. Mo Nunn was really good, they went in and found the place and didn't move very far off the center. And they won the championship. Keep it simple."

Al Unser Jr. wasn't particularly impressed with Fittipaldi's winning move, and decided to let Emerson know after climbing from his wrecked Lola. Al Jr. tells it:

To be honest, I walked out there intending to flip him off. When I got out of the car and got my helmet off, I started walking out to the track. The safety crewman stopped me and said, 'Where you goin?' I said, 'Well, I'm going out there.' 'What, you want to flip him off?' And I went, 'Yes, of course'. So he stepped aside and said, 'Go right ahead'. I'm out there waiting for Emerson to come and I saw where I was. I'm not at a Saturday night sprint car race. So, I thought my answer should be first to congratulate him. A lot of that is because of who Emerson is. Emerson would never want to hurt anyone, especially another driver. I mean, we are going for the Indy 500, we are going for the win. It's racing.

No doubt this substituted gesture proved wise, as by 1994 Al Jr. was driving alongside Emerson Fittipaldi at Penske Racing.

The 1989 championship was close between Fittipaldi and Mears, Emerson winning by ten points in the end. Nobody else was even close, as the two drivers scored eight wins between them. The PC18 was the second dominant Penske in the last two years. Jerry Breon explains, "Emerson was well established and his Formula 1 career was behind him. He wanted to finish out his career in Indycars and his last few years would be fun for a change. He didn't have to prove himself by then, to anybody." The 1990-91 pairing of Mears and Fittipaldi was one of the most synergistic in Penske team history. Rick Mears recalls, "Emerson was a great teammate, plus he and I had similar driving styles and feel, and like similar things from the car. I could really count on him, he was a really straight shooter. If I asked him what was going on with the car, and he said, 'This and this and this.' I knew exactly what he was saying and exactly what it would do to my car."

Chuck Sprague was with Penske Racing's race teams from 1980 to 1998, moving up from mechanic to Team Manager. He describes some of their processes:

We always tried to work that way. All our debriefs had a program where each driver had five to ten minutes with their engineer at the guardrail. Then everyone went to the engineering office in the truck. We had a list of items we would go down in very strict order. First would be if a car needs an engine change or a differential change, or do we need gears, as those are big projects. Then once we worked through that, we would begin the individual summaries with each engineer reading off his notes in front of everybody. I would be there along with Nigel Bennett, so we had the logistical and engineering heads of operation there. Basically once we got to the end, we would make a decision if we were going to spread one setup all the way across. We would divide and conquer, to where we would start with a baseline for the next practice session and this guy was going to do aero stuff, this guy was going to do shocks, and it was all data management. Even back then we could measure anything we wanted to, but it was managing the data and learning from it. We would go to tracks where Friday night Emerson would throw his entire setup in the trash can and put Tracy's setup on the car. Then he would spend the next day adapting to it and then tweaking it a little bit, and then out of nowhere would come this qualifying lap.

Paul would say, 'What happened there?' Well, you were aware of everything we were doing, but your focus was where you were on the time charts, not the car's performance capability. So Emerson took your mousetrap and improved on it. Every driver always had all the information the other drivers had. There was no tolerance for anything being held back... never on my watch. You can't fall into that trap, because it doesn't benefit you in the long run. For each time you give your setup to other guys, there will be a time or two when the other guys bail you out. When we moved down to Charlotte, the stock car guys were astounded that our team was setup how it was. Their teams were set up so that each car had a crew chief and the crew chief was God. We had one team with two or three entries, and the whole philosophy comes down from there.

Given the pairing of Rick Mears and Emerson Fittipaldi's recent history and success of the Penske Cars, 1990 was a disappointment as Al Jr. and Mike Andretti ran off into the points lead. Two wins for Sullivan and one win each for Mears and Fittipaldi were the best results for the PC19, with Emerson starting at Indianapolis from pole. Sullivan swept the Laguna Seca finale from pole, but his fate with the team had already been decided. For 1991 the team would be Mears and Fittipaldi in the new PC20. Paul Tracy was brought onto the team as test driver and for the occasional race.

Mears was the master of the Superspeedways, and won both the Indianapolis 500 and the 500-miler at Michigan from pole position. It was the seventy-fifth running of the Indianapolis 500 and featured the unforgettable front row of Rick Mears, AJ Foyt, and Mario Andretti, one for the ages. The late race battle between Mears and Mike Andretti featured passing around the outside of turn one on consecutive laps and the roar from the crowd was amongst the loudest I have experienced at any race, ever. Mears had his fourth win at Indianapolis, but it was a Mike Andretti year, taking eight wins along the way to his only CART championship. The PC20 had a new transverse gearbox, and when looking it over and chatting with a team engineer the morning after the race, he confided that it was a lot harder to work on than with the old longitudinal gearbox design. Not exactly service-friendly, but with a better mass concentration, the PC20 was now Penske's newest Indianapolis 500 winning design.

Chuck Sprague confirms, "Access to the transverse gearbox was a bit restricted, because you had to dive underneath it and go in from the side. There is always the aspect of something new and different which makes it harder to work on, even if it is actually easier to work on. There were so many different new parts in that, and the shift linkage in particular was problematic." Penske cars were also provided to the Bettenhausen team, which then returned the favor four years down the road.

In 1992, the team could hardly have started stronger, with a 1-2, Emerson-Rick sweep at Surfers' Paradise with the new PC21. It was the highlight, and although Fittipaldi was to win three more races that year, the championship battle came down to Rahal versus Mike Andretti, with Unser Jr. in third. Mears had a big accident at Indianapolis when a mechanical failure put him upside down and into the wall, breaking his wrist. Another crash during the race (a fate shared by much of the field in that frigid year), certainly played a part in his retirement from the sport later in the year. However, despite being without Rick Mears for the first season in nearly a generation, 1993 was to see a resurgence for Penske Cars with the arrival of the PC22.

The first fully carbon honeycomb Penske PC22 was a factor everywhere, but particularly on road racing and street courses, with Fitipaldi taking second at Surfers, Cleveland, Toronto, and Laguna, along with wins at Portland and Mid-Ohio. The one race that didn't follow the script was Fittipaldi's stunning win in the Indianapolis 500, where he ran a Rick Mears type of race, hanging around within range until near the end, fine tuning and saving the best for last. In 1993, the Penske-Mercedes was a strong, but far from dominant runner. Jerry

*The PC22 was another of Nigel Bennett's Indianapolis 500
winners with Emerson Fittipaldi in 1993 (Norm DeWitt)*

Breon recalls, "Paul Tracy was out of the race and I was over at the
Goodyear garage dismounting tires and packing our stuff up. I was
looking out the corner of my eye at the monitor and thinking, 'Well,
we'll have another load today as we aren't going to win this year.' Then
next thing I see, there's Emerson going for the win."

Taking advantage of an inexperienced Nigel Mansell on a late race
restart, Emerson flew past and was on his way. As with Jim Clark in 1963,
Mansell had almost won his first oval race ever, and 1993 is considered
one of the greatest Indianapolis 500 races in history. Perhaps one of the
post-race quotes that was most memorable was when John Andretti
said (I paraphrase), "If before the race you had told me I could finish
17 seconds behind the leader, I'd have taken it. Who'd have thought that
would be for tenth place?" Keep in mind that this was an era of multiple
chassis manufacturers, multiple engine manufacturers, and two tire
manufacturers. It was a world apart from the spec car shootouts of
2018. In this most level of playing fields, Fittipaldi and Penske Racing
had won again. Chuck Sprague reminisces:

We started ninth, there was a short green flag segment at the beginning and then there was a caution. Emerson came on the radio saying, 'Boys, I'm telling you that we can win this race,' which was a bold prediction after fifteen laps, or whatever it was. Emerson's was on his game at the Speedway, big time. He won in 1989. If you look back to 1990, Emerson was absolutely a rocketship and then started blistering front tires. In 1991 he was leading when the clutch failed at a pit stop and the gearbox got destroyed trying to get it into gear. 1992 was what it was. 1993 he won. In 1994 he had a lap on the entire field when the wall got in the way. What a lot of people aren't aware of in 1993, and even more impressively in 1994, with that big engine, is that we ran the entire day in both cases with making zero adjustments to the car. No tire pressure, no wing, no stagger. I don't even think he adjusted the roll bars in the cockpit. At Indianapolis that is an extraordinary accomplishment in both cases. Emerson's approach to setup, he was very fussy about tires and consistency, and he worked really hard at it.

Paul Tracy had five wins that season, including at Long Beach carrying #12, the number his fellow Canadian, Gilles Villeneuve, had carried when he'd won at that circuit in 1979. Fittipaldi and Tracy finished 2-3 in points, behind the rookie champion Nigel Mansell. Obviously Nigel was no rookie in the normal sense of the word, having been the 1992 World Champion, but it was still an incredible feat in an unfamiliar car, on unfamiliar tracks. Nobody would have guessed Nigel's lack of oval experience when watching the incredible wheel-to-wheel battle between Paul Tracy and Mansell at Loudon, New Hampshire.

Chuck Sprague recounts, "That PC22 was actually a really good car. We made a strategy mistake in the Phoenix race and crashed as a result and that is the only reason Mansell was the champion. You've got to remember that he only won by eight points. We lost twenty points when we didn't come in while being suspicious of the left front tire, and then stuffed it. It turned out he did have a puncture and crashed. That would have made all the difference, we were on Mansell all year. Emerson and Mansell at Cleveland was epic."

It should be mentioned that Teddy Mayer was in a management role at Penske Racing, working at the races. Crossing paths with Mayer in the paddock at Laguna Seca in '93, I blurted out, "It's the Wiener! I haven't seen you since the Bruce and Denny show," which brought a big smile to his face (although in hindsight we'd crossed paths in the McLaren Formula 1 pits during the James Hunt years as well). I asked if he'd been brought in to engineer a particular car and he explained that his was now in a more overall role managing the team. Teddy Mayer passed away in 2009, one of the great organizers in racing and a big part of the success for any team that was lucky enough to have him.

1994 was to see total dominance for Team Penske, the likes of which had been unseen since the PC17 of 1988. Jerry Breon says, "The PC23 was the car we used with the Mercedes [stock block] engine at Indianapolis and we had five 1-2-3 finishes with that car throughout the course of the year. That was probably the most successful year." Having won twelve of sixteen races pretty much covers the bases, eight of them courtesy Al Jr., except for the fact that it was also the year of the parallel development program for the surprise introduction of the Mercedes Stock Block V-8 for the Indianapolis 500. The car was an incredible challenge given its power, not unlike the supercharged Novi had been for a previous generation. Chuck Sprague describes it:

> We were starting to have balance problems with the tires on the pushrod engine cars when we started running harder. The guys would put white marks on the tire where the tire valve stems were and the wheels were spinning in the tires. We sent all the wheels out and had the beads sandblasted with very course sand, which immediately completely demoralized Newman-Haas when they saw it. Peter Gibbons told me a couple years later, 'When we saw you had to sand blast your rear wheels to keep the tires from turning, we knew we were screwed.' To tell you the truth, we never even thought to hide it, it was just an issue we were dealing with. For this to be happening at Indy of all places, a track where it's a hard low-grip tire and you are in top gear. It was amazing, and it's a good thing that we never tried to road race that engine, thank God. We would have broken the entire car.

The race was complete and utter dominance between Fittipaldi and Unser Jr., Al winning the race after Emerson threw the car away in turn four with fifteen laps to go after leading most of the race. Discussing this MB stock block program with Tony George in 2006, he was to the point, "There are still people today who talk about that, that it was an engineering challenge brought together to become the most insane thing allowed to happen at the Indianapolis Motor Speedway." That is a fairly accurate summary of the situation. Although Raul Boesel did well to put his car in the middle of the front row, it was the classic case of bringing a knife to a gunfight, and he never led the race (after having had the fastest car in 1993).

Chuck Sprague says, "In 1994 we gave a PC22 to Rahal and most importantly we gave him Emerson's race setup from 1993. I told either Bobby or Hogan that, 'The gift here is not the car you see in front of you, it's the piece of paper that I'm about to give you.' They just put that setup on there and went out and went, 'Holy smokes!' I think they did a little tweaking to accommodate the '94 tires, but he came from twenty-second to third. I fully expected those guys to be up there in the year, and that '93 setup was the best thing you were ever going to have if you had a 4-valve engine. It was a magic setup." Bobby Rahal adds his perspective:

> There were two of them for Mike Groff and myself. Frankly, when we got them there was a tremendous amount of work to be done to get them race ready. They had been sitting around as show cars, and it was a bit of the thrash to get right. The setup wasn't, to my memory, much different from what we had. I think we had the fastest lap on the second weekend of qualifying. It was a very stable and very nice car, even better than I remember the '93 Lola being. Coming down pit lane I said to one of the guys at Penske, 'If I'd had one of these, I'd have won four or five 500s by now!' It was a nice car to drive and we moved up through the field and finished third. Of course, that was the year of the Penske-Mercedes [stock block] and so nobody was running for the win, we were running for first in the conventional engine class. Roger and Carl Hogan were friends and the next year they needed us, so we gave them our cars and the setups.

Even more unbelievably, in 1995 neither car of Fittipaldi or Unser Jr. managed to qualify for the race in the new PC24. Jerry Breon still wonders about that, "Well, to this day I still don't know what happened. The year before the car was so dominant and the Mercedes engine was so strong the year before, we probably didn't concentrate upon improving the car for the next year as much as we should have. It still raises the question in my mind, 'Exactly what did happen there?' We were still on Goodyear tires at that point, which was causing everybody some problems. I'm convinced that it just comes down to complacency between '94 and '95."

The team had been thrashing between the PC24, the previous years' PC23, a Bettenhausen-sourced '94 Reynard (thanks for those 1991 PC20s, Mr. Penske), and finally a pair of Lolas from Rahal. No doubt the chaos of four different chassis being tried in a musical chairs situation contributed to the lack of results. It had to be nothing short of a data hurricane with bits and cars changing on a daily basis. Some of the blame could be set at the feet of Goodyear and their latest tire, however the car had reasonably good results elsewhere. It didn't help that the PC23 of '94 had raced at the Speedway in Mercedes/Ilmor stock-block form, making that data essentially worthless in 1995.

Chuck admits, "We borrowed a lot of cars from a lot of people. It was very important to us that we did not repeat the mistake of 1992, where we were dealing with both the '91 and '92 car, lost our focus, made a mistake and put Rick in the wall. We were determined that was not going to happen so were very cautious and diligent in not putting anything on the racetrack that we didn't have 1,000% confidence in. It was extremely hard, and certainly one of the darkest eras for the team. But that nobody crashed and nobody got hurt was far more important than any other outcome."

It is amazing that the Penske team didn't get horribly lost between all those different chassis. Bobby Rahal says, "Well, I think they did. It's hard to believe, right? It's Penske Racing, but they were pretty desperate. They changed some things and Emerson was on a lap that would have qualified and they waved it off, so neither of their cars qualified. It just shows you how competitive that series was back then."

Rahal could relate to what Penske was going through. "In 1993 it was a crushing blow when we didn't qualify after I'd won the championship in 1992. After I didn't qualify in 1993, Miller gave us a three-year contract extension within a week, and they were a phenomenal sponsor for us. When 1994 comes along, and we were obviously having issues and weren't competitive, I couldn't go to Miller and tell them that we weren't qualifying for two years in a row. That drove a lot of the rationale for reaching out to Roger to lease those cars, as we could not afford to not qualify. So, we had to withdraw the Honda cars before we could qualify the Penskes."

Chuck Sprague confirms that much of their unfortunate situation was the result of being on the losing end of a tire war:

You've got to remember that in 1995 we finished second in the championship. The big thing about 1995 through 1999 was Goodyear, and everything was academic when you'd made your tire choice. If you had Firestones, nothing else mattered. When our guys drove Firestones for the first time, they were astounded. In 1994, when everyone would be pulling out of the tracks after a race, the Firestone test team was coming in for testing the day after. They knew what they were doing, testing when the track was all rubbered in, and that was key. It did not go unnoticed. A long time ago I heard that their tires were manufactured overseas and they were not restricted to the same chemical limitations that the Goodyear guys faced. The compounds that had to do with the durability and consistency, which was everything.

When headed into the Speedway on race morning in 1995, I saw something that brought back memories of how Emerson Fittipaldi had refused to drink his milk in 1993's victory lane, substituting orange juice as a promotional stunt. The van in front of me had a sign in the back window that said "Emmo, you should have drank your milk."

The PC24 of 1995 was still effective for much of the year, taking four wins on road/street courses with Al Unser Jr. and second in the points. Al had also taken second at Michigan and Milwaukee, Paul Tracy taking wins at Surfers Paradise and Milwaukee. For some reason the car seemed to work reasonably well everywhere except for at Indianapolis.

1996-97 were the last Nigel Bennett cars. Breon explains, "By 1996 we were struggling, and we went two or three years there without a win [1996, 1998, 1999]. Paul Tracy ran off I think three wins in a row [1997 Nazareth, Rio, and Gateway], but that was about the only high spot we had then. By then Roger was looking for a way to stop the bleeding. At the end of 1996 is when I made the change from the race team to the business administration side of things, so I quit traveling. I was there with the race team all day every day in the shop."

For 1996, Emerson was racing the current PC25 for a new satellite team – Hogan Penske, and Fittipaldi had taken fourths at Nazareth and Milwaukee, which was not far off the results of Unser Jr. in the Penske Racing entry. However a major crash at Michigan caused career-ending injuries to Emmo and his seat was taken by Jan Magnussen. The Hogan Penske entity ended at years' end, with the retirement of Fittipaldi.

John Travis had left Lola to design the last Penske PC27 for 1998, and although it was not a success, it was a packaging marvel, which created its own large number of issues. The latest Mercedes/Ilmor was tiny in comparison to the previous versions, and Travis had placed Penske's latest gearbox ahead of the rear wheels, which proved to be a mechanic's nightmare. The high nose/low wing configuration mirrored the current design wave that had swept through Grand Prix, from Benetton's Rory Byrne. However, the car still had Goodyear tires in a Firestone world, and regardless of the effort and craftsmanship involved, that doomed the effort. Andre Ribeiro, a proven race winner with the Reynard or Lola/Honda/Firestone combination, joined the team for 1998 to achieve an utter lack of results, his best finish a seventh place from the entire season mostly strewn with DNFs. Ribeiro retired at the end of the season.

Chuck remembers, "When Andre Ribeiro came on board in 1998 and drove Goodyears for the first time, he was frightened to death. He got out of the car, pulled me around behind the truck and said, 'I don't even know how to say this...' I told him, 'I'm not surprised as you can see it in the data.' Even Mario talked about at the end of the '98 season where he said, 'We will never know how good the Penske PC27 was because it was on the wrong side of the tire war.' It was true, that's the

John Travis' PC27 was a compact design that was let down by tires.
This is the wind tunnel model (Norm DeWitt)

only part that touches the ground and it was a huge gap."

The PC27 had transmission problems early in the season. Chuck explains, "The gearbox was problematic, but it was eventually solved. The bottom line was that those were the days of no-lift shift. The team wasn't doing the programming. We were fighting issues with tearing up dog rings, failures, and so on. There were some early season development issues. We brought in Grant Newbury, our unassigned head of engineering who was without particular race responsibility, so he was available for projects. We assigned him to work hand-in-hand with the gentleman who was doing the gearbox programming in Toronto, and that day the gearbox problems disappeared."

Chuck continues, "The PC27 had an engine that was fifty pounds lighter. It was a very elegant design and John Travis did a terrific job. But again, if you were on the wrong tire it doesn't matter. All the engine programs were getting highly competitive, and we wound up with a peaky engine and a tire that didn't tolerate wheelspin. Now you've got a real bad combination as when the power is delivered abruptly, and if you spin the tires and they overheat and don't recover well. Road courses and street courses will be more than problematic at best."

Why on earth did Penske Racing stay with Goodyear? Chuck says, "K-Mart Auto Centers were Goodyear. It was an exclusive multi-year deal and he [Roger] was loyal. At the end of '99 when Goodyear pulled out, that solved the problem." The only team to do well with Goodyears was Derrick Walker's car driven by Gil de Ferran. Chuck explains, "Which is why my radio during practice was tuned to his car and not to ours. Seriously, because he would come in and Gil basically downloaded. If you listen to him on the radio, he's never out of breath, never passionate, just nuts and bolts. I would listen to his debrief and then when our guys came in I'd listen to ours. We've got different cars, we've got different drivers, different engines, and we've got the same problems. Do the math… so we learned that our approach had to be, 'Focus only on what the tires need.' It's not the car."

One would think that Penske Racing and Walker would collaborate on the tire situation and setup. Chuck didn't know, "If it happened it happened at a level above my pay rate. You might ask Roger and Derrick about that. I'm sure there were some prayer meetings."

Through 1999, the team switched back and forth between the PC27B and a Lola, finding no success with either. Al Unser continued to persevere with the team, his best finish a fifth at Cleveland. As the struggle continued, the effort hit the lowest of lows with the loss of Gonzalo Rodriguez at Laguna Seca when he cleared the gravel trap and impacted the wall in what little runoff there was atop the Corkscrew. The team had signed Gil de Ferran and Greg Moore for the upcoming season, and then another savage blow came at the season finale in Fontana when Greg Moore lost his life in an early race accident.

Sprague says, "My last year with the race team was 1998, so it was after my time, although I was still with Roger until 2010. But I don't miss having not been around for the 1999 season. At Laguna Seca it was a freak thing before the days of Hans devices. What little bit of saving grace that there is was that it wasn't a mechanical mistake on the part of the guys. It was a driver decision that led to a very unique set of events, tragically. But it doesn't change the outcome and it doesn't change the tragedy."

Jerry Breon recounts, "The big moves were made at the end of 1999 as we switched the engine and the tires. We brought in Tim Cidric to be the team President. We had hired Greg Moore along with de Ferran, but Greg lost his life at the last race of the year out in California. That opened the door for Helio." Given that CART had seen a Reynard/Honda/Firestone advantage for the previous four years, it was probably an overdue and understandable decision. Starting with 2000, the team switched to running Reynard chassis, Honda engines, and Firestone tires, and there was immediate success with Gil winning the CART title. Chuck Sprague recalls, "You know, Gil and Helio hit it off immediately. It was definitely dark days, but when we went to the Christmas party that year [1999], Gil and Helio transformed the entire atmosphere of the Christmas party and the entire team in that one night with their give-and-take and their spirit. It was an exceptional evening and it was exactly what the team needed and of course it paid off immediately."

For 2001, the team had used the knowledge and facilities of Penske Cars to heavily modify their Reynards for the better and Penske's de Ferran was to repeat as champion in what was otherwise a Lola year. Other than the two Penske Reynards, the best Reynard result was Franchitti who finished seventh in the points. Penske Cars DNA was still playing a strong role in the team results. Chuck Sprague says, "If you are in the off-season and you change everything, as you work through it you can figure out which one worked, or what made the difference. I firmly believe it would be 'all of them.' But I don't know that the team would have done any less well with a Penske PC28/Mercedes/

Firestone combination than they did with a Reynard/Honda/Firestone combination. There is no way to just ever know."

The team still used their facility in England to fabricate carbon pieces for everything including various bits for the pit carts and miscellaneous for their ALMS P2 DHL Porsches. There were also small carbon bits here and there, such as aerodynamic covers for the rear wing adjusters on the Penske IRL Dallaras in the mid 2000s. However, there just wasn't enough work to justify keeping the facility open given how everything has shifted into spec car programs. Breon reminisces, "Officially it would have been about March 2007 when we closed Penske Cars in Poole, England. I really hated to see that place close, that was a gorgeous facility. We had two places over there, the original Penske Cars where we did the Formula 1 program, and the early Indy cars. Those were in the small original facility, I think Roger bought the place from Graeme McRae. We ended up outgrowing that and moving a couple of doors down on the same street into an industrial park. I enjoyed it, as it was always the wrong time of the year when I was trying to get home to my family for Christmas, as the months of December and January were busy times at Penske Cars working on prototypes for the following year." Chuck Sprague says:

> That's the tragedy of their history is that the downfall of Penske cars began in 1995 and then over time the team made a switch to the IRL, where you weren't able to show creativity or manufacture your own stuff, then they were left with no choice. During my tenure at the team, particularly when I was the Team Manager, by far my most important connection was with Nick at Penske Cars. I was managing the parts operation personally because there was so much at stake. We didn't want to waste effort making anything we didn't need, but we also didn't wind up without key spares or development components. After each crash we triaged every part and if something could be salvaged, it was. They were farming themselves out, doing work for Formula 1 teams and various outside companies as they had the CNC shop and the carbon shop. Nick Goozee was very creative to keep the operation available. But in the end, the decision makes itself.

It was the end of an era, the end of Penske Cars. Although Penske was no longer a manufacturer, in the spec car era it is probably even more important to have the hive mentality on sharing data. Rick Mears explains:

> We do more of it now these days as there is a bigger con-glomeration of information, more data to go through with specialists in different areas, but the principle is pretty much the same. Now we can measure in areas that are almost beyond what we can feel. Before, sometimes you'd have to make enough change that I could tell what it did. Today, when the rules stay the same, you start in here [middle of the envelope] with your setup and start filling out the envelope. Pretty soon you get out to the corners and that's all that you have left if you are trying to find that last little bit. The corners are so small that you have to find two or three minute things together to make the change to make a difference where you can feel it. You know it is pointing in the direction you are looking for, but it still boils down to seat of the pants at the end of the day.

Time and time again, you see Indianapolis teams that let everybody go at the end of the season and then re-hire personnel in the spring for the next season. Jerry Breon admits, "Something I never had to worry about. Never once in those forty-two years did I worry about what I was going to do the next year or if I was going to get a paycheck the next week. That's a credit to Mr. Penske and the way he runs things. He's tough but he's very fair, and it's the integrity that kept me there the whole time."

Team Penske is much more a cohesive unit today than in the days of the late 70s and early 80s, defining the very concept of synergy, where the whole is greater than the sum of its parts. As for Roger Penske, he still can be found making the strategy calls from the pit box at race after race. Jerry confirms, "Roger is a pretty unique individual, still to this day. Eighty years old and still going strong. I couldn't keep up with him."

Team Penske's run of success continues, as they celebrate back-to-back Indycar Championships in 2016 and 2017. From that humble single-car entry of 1968, on through to 2017's four- and five-car armadas, across those fifty seasons there has been one constant theme, and that is Excellence.

Rick Mears is the most successful driver for Penske,
with four Indianapolis 500 wins and three Championships (Norm DeWitt)

The 'Baby Borgs' of Penske Racing.
Their success continues (Norm DeWitt)

CHAPTER 12

FRANK WILLIAMS (RACING CARS) LTD

*Piers Courage in the first World Championship Grand Prix
for Frank Williams Ltd, the 1969 Spanish GP (Bernard Cahier)*

Many are familiar with the story of success that is Williams Grand Prix Engineering, which recently celebrated their fortieth anniversary in Formula 1 racing, now known as Williams F1. What is a less familiar tale is that there were other Williams-led teams that had been racing at the highest levels for nearly a decade before. The first of those was moving up after some success in Formula 2. In their second Formula 1 Grand Prix appearance, using a heavily modified used Brabham, they finished second in the Monaco Grand Prix. The year was 1969 and the team was Frank Williams (Racing Cars) Ltd.

The story begins in 1961 when there were races at Mallory Park. A young racer of little means named Frank had crashed his Austin saloon car and met a driver named Jonathan, whose race had ended similarly. As it turned out, they shared the same last name – Williams. Soon after they met up with Jonathan's friend, Piers Courage. The three young men were to remain connected for the rest of their lives and, through sheer determination, all found their way into Grand Prix racing.

One would think the young Frank Williams had dreams of racing a Grand Prix Lotus 25 or BRM, but there wasn't a specific goal. Sir Frank Williams says, "Never a car specifically, I just focused on doing the best in the car I was driving. I never had anyone to operate the cars I drove and so had to work hard to make sure I could afford the best cars I could." For the 1963 season, he was to accompany Jonathan, helping in his Formula Junior efforts at various circuits in Europe. Their flat in London became a home for aspiring racers, including Courage, Bubbles Horsley, Charles Lucas, and others. Did Frank dare dream of the success that was so soon to arrive? Sir Frank recounts, "No, I wasn't that sort of person. I made sure we enjoyed the days that we were living. If we worked hard and took our opportunities then success wouldn't be too far away."

Continuing his efforts, Frank was racing in Formula 2 and Formula 3, funded through buying and selling parts and cars. The business side, Frank Williams Ltd., was becoming a success, and although there was never a specific moment to where this became apparent, Frank began to sense his greatest racing talent was in team ownership and management as that was where he was finding success. He explains, "It was my commitment to a new young driver. I took a year off racing [1967] to earn money so I could come back and buy another car, but that was before Piers Courage. I invested in him as I saw an exceptional talent. The moment I started to support him, all of a sudden my focus was not on returning as a driver, but making sure he had the best chance to win and fulfill his potential. What also helped is that the business side of things was going quite well and if you're succeeding at something you normally stick with it."

Jackie Oliver had moved up into Formula 1 by 1968 with Team Lotus, and was another from the London crowd of up-and-coming English racers. "I knew Piers very well as we both lived in London during the same period and I used to see him a little bit socially. Piers was a little different to all of us. He was a very quick driver, but we all came up from Essex and Piers was just a little bit more polished than we were. He never worked on his own cars, a bit of a silver spoon, but he blended in very well with us."

Paul Butler was the Dunlop tire representative for Formula 2 at the time, later to become the Moto GP Race Director. Paul reminisces:

Piers was the Courage Brewery family, he was a talent that came along a bit like Charles Lucas. That whole crowd in Formula 2/Formula 3 were a really good mix socially. Quite a few in the middle of the scale, and Piers at the upper end of it. He was a charming, go-for-it character. They all would stay in the same hotels and drink in the same bars. At the Reims event there was a famous bar called Bridget's and there was always a lot of uproarious fun going on. On one occasion, David Piper's mechanic was a bit of a ruffian and he knew everyone was in the bar. He came along and threw a thunderblast [a powerful firecracker] into the passageway, so everyone left and started fleeing down the street. The cops arrived and Trevor Taylor was running between me and somebody else, and crossing the road he tripped over the curb at the very moment the cops fired into the air. We immediately assumed he'd been hit and we froze. He scrambled up and said, 'What are you guys playing at, let's go!' That sort of stuff was relatively commonplace, and Piers was a pretty good party animal.

Frank was different though, as he was always serious and ran a very professional outfit and was way ahead of the game for his age. In every era, the competition and standards are different. It is hard to make comparisons except with the rest at that time and Frank was right up there and was doing it with limited resources. He was years ahead of Ron Dennis and McLaren in perfection of presentation.

It wasn't long before Frank Williams Ltd. was selling new racing cars, such as the Brabham. Sir Frank says, "What really helped was that I could speak a few languages. Being multilingual gave me the edge over Lotus, Cooper, or Brabham when selling to customers overseas. The company was really quite unique because of this. Having a secretary also helped as I was always contactable, and back in those days it was important." For the 1968 season, Frank was to enter a Brabham for Piers Courage in Formula 2.

Dunlop's Paul Butler (left center) with Piers Courage (right center) and a new Formula 2 tire in 1968 (Paul Butler)

Tire tech was a critical part of any successful team, and Williams had cast his lot with Dunlop. Paul Butler recalls:

It was always very experimental as tire development in that period was really quite exciting. People were trying stuff, they didn't have the technology and analysis methods available and you took a flyer on the compounds in particular, as they were all

all cross-plies. We'd run a compound 970 at Zandvoort in July. It was a wet-weather compound that had not been tried in the dry. It was wearing so well in wet conditions that the technical guys said we should really try it in the dry. Piers shot off into the distance and we were actually gaining two seconds a lap on the rest of the field, this compound was so good. There were only a couple of options and this wasn't a time when compounds would change from race to race, it was more 'season-long' in those days.

Piers retired in the final race that day at Zandvoort, but the speed was there and it was only a matter of time before Frank Williams Ltd. found their way onto the top step of the podium.

The Williams team was to score its first major victory at Monza in 1968. Piers Courage had been drafted onto the Reg Parnell BRM Formula 1 team, running the V-12 car, scoring a sixth in France, and a fourth at Monza, while gaining valuable experience. However, earlier in the season, the Formula 2 race at Monza had presented a conflict in the schedule with the Dutch Grand Prix, so long-time friend and associate Jonathan Williams was brought onto the team as the substitute driver. Frank remembers, "It was a very special race. I was of course elated at the time, but I understood that it was a first step. We had more races to win and so we immediately focused on getting the money ready for the next race."

At the end of the season, the Williams team had prepared their car for Piers to run in the Temporada Formula 2 series in Argentina. Yet another of this merry band of Englishmen running around the world chasing success in motorsport was the racing correspondent, Andrew Marriott:

Piers was a wonderful guy, and I'm proud to say that Piers was a friend of mine. We were all in that circle, growing up racing together, me as a journalist and Piers as a driver. Strangely, I never went to Harrow [the London racers' flat of legend], don't know why really. My favorite was the 1968 Temporada series, we were quite a close group, staying in the same hotels. I wrote quite extensively about it as I wrote a chat column about it each week in Motoring News. There was nothing else happening that time of the year and Formula 2 was very competitive. During the third round at El Zonda [San Juan] raceway, Rindt missed the first practice because he went on a sightseeing tour. He drove into the Andes to look and came back for the second practice, thinking it would be enough. There was a most massive gale sweeping down this valley to the racetrack, which was surrounded by these huge mountains. Only Rindt went out, because he had to do a lap to qualify with all this dust and grit swirling around.

Rindt had to start the race from last place, yet had worked his way up to third by the finish. It had been a costly sightseeing trip.

Formula 2 and the Tasman series were good testing grounds for new concepts that were similarly being adopted in Formula 1. Andrew says, "They were just trying everything, there were no wind tunnels to see what worked and what didn't. I think Piers was running a biplane in that series. There were a lot of good people that year. Techno had some good cars there, Winkelmann Racing was there with Rindt, and Jackie Oliver was there in the Lotus 48." Some of the cars sprouted wings while others didn't. In the first of the races, Piers' rear-wing strut failed on the Williams-entered Brabham, which was not an unusual fate for the winged cars in 1968. The wing strut on Rindt's similar Brabham also failed on the same lap as Piers', underlining that the failure was the design and not from car preparation. Paul Butler recounts:

As with the Tasman series, these races were the off-season playground of the up-and-coming next generation of Formula 1 superstars, as well as many regulars from Formula 1, such as Rindt, Siffert, Beltoise, Courage, Brambilla, de Adamich, Pescarolo, Rodriguez, Oliver, and Reggazoni, and Carlos Reutemann was driving a single-seater for the first time. We should have won the races, because Dunlop had won [most] all the races in the European championship [Formula 2 was for engines of 1.6 litre capacity], and we had everyone on Dunlop except Ferrari. They had a Fiat Dino-based V-6 engine that came in two guises. After this, they were off to the Tasman series, which had a 2.5 litre upper limit. I have it on good authority that they were using their bigger capacity engine. I'd been sent off with great fanfare that these races were going to be a great promotion for Dunlop in Argentina. I was miserable, most of the time I had Jochen Rindt in my ear saying how they've got to try Firestones. It was nothing to do with tires; it was that Ferrari was using a bigger engine. What do you do? You can't protest Ferrari in Argentina, so we only won one race, the final round with Piers Courage driving for Frank Williams in the Brabham.

The Ferrari Formula 2 cars had a mediocre European season, so skepticism about their pace in Argentina is understandable. Reports from the races made reference to the Ferrari resurgence being due to their performance at high altitude. With two of the four race weekends being held near Buenos Aires on the Atlantic coast, that conclusion makes little sense. At race one in Buenos Aires, the Ferraris finished first and second, the first of three wins in a row. However one must balance that against their improved late-season form, finishing the last two rounds of the European Formula 2 drivers' championship first and third at Hockenheim, followed by first and second at Vallelunga. Paul Butler continues:

The Temporada series finale was at a very bumpy circuit in Buenos Aires, it was quite old even then. That was the one that Piers and Dunlop won, and it was the only one we got a result out of. It was the famous race where Reggazoni had contact with de Adamich and shot up the pitlane. There was absolutely no spectator or personnel control, and the pitlane was full of punters and photographers. Several got thrown up in the air. Of course, all hell broke loose. The Army is out there with rifles, pointing them at people. They called for an ambulance; it was a big old Chevy ambulance that comes roaring up with medics hanging off the back. The medics put this guy onto the wheeled gurney, threw him into the ambulance, slammed the doors, jumped onto the fender at the back and banged on the roof. The driver who had been waiting for his moment revved the engine the engine and dropped the clutch, but they hadn't closed the doors properly. The driver took off, the doors flew open and the gurney came out the back and rolled down the pitlane. The crowd is going berserk, of course, as people are running after the gurney, trying to get him back to the ambulance. It was one of the most bizarre things I've ever seen in racing, it was Keystone Kops. Of course the police and the military were there and the guns came out because this was most unexpected and they panicked. But the race continued.

Such was the prelude to the second major Frank Williams Ltd. victory. Andrew Marriott says, "Piers won the last race in the series which was a complete fiasco due to a pit lane accident at the start. I actually saw it happen and I'm sorry, but you just had to laugh. I was probably standing next to Paul Butler when it happened." Piers had won by ten seconds over Rindt, who was also driving a Brabham BT23C powered by the Cosworth FVA, followed by Jo Siffert, Jackie Oliver, and Andrea de Adamich's Ferrari. That last race win moved Piers ahead of Tino Brambilla's Ferrari into third place in the final series standings, behind de Adamich and Rindt.

Frank Williams
(Racing Cars) Ltd
361 Bath Road
Cippenham
Slough
England

Telephone
06286 4646
Cables
Racecars Slough

The Frank Williams (Racing Cars) Ltd logo from their early days in Formula 1

Headed off to the Tasman series, there was no question that Courage was now on level footing with the Ferrari-Firestone combination, but there was an additional factor working against him, and that was a motivated Kiwi named Chris Amon, now driving the Ferrari Dino alongside teammate Derek Bell. Graham Hill and Jochen Rindt were teamed in 2.5 litre versions of the Formula 1 Lotus 49s. Frank Williams had acquired a former Brabham Formula 1 team BT24 from 1968, now powered by a 2.5 litre Cosworth V-8. In the New Zealand Grand Prix at Pukekohe, Courage finished third behind Amon and Rindt, but was to take second behind Amon in the following race at Levin. Race three went to Rindt, but at Teretonga Park in Invercargill, Courage had perhaps his greatest race. Passing Amon early in the race, Piers raced off to victory over Graham Hill's Lotus 49 by eighteen seconds. Given that the Tasman Brabham and Lotus cars were far closer to Formula 1 specification than the previous Formula 2 racers had been, this was a significant milestone for the Williams team. Piers also had set the race's fastest lap.

Unfortunately this victory was followed by a series of DNFs, Piers eventually finishing third in the Tasman series championship. However, bigger plans were afoot and those plans included moving up to Formula 1. Williams had managed to acquire a current specification Formula 1 Brabham BT26 chassis for 1969. Sir Frank explains, "I found a wealthy gentleman called Bridges who was based up north; he had a one-year-old Brabham that he had used for hillclimbs and other racing. We never met face to face, but I persuaded him to sell me the car. After the Tasman Series I had planned to get the car ready for Formula 2, but people kept telling me that I had a prepared Formula 1 car, and with the belief that Piers could become a champion, I entered him into Formula 1." Paul Butler recalls:

> Frank Williams was a good friend of Dunlop and Dunlop had been pretty good to him and it continued to be that way later. He was the first guy I know who did a full-color presentation of his team and their plans. In those days you had to get photographic separations to do color presentation and it was extremely expensive. He went to the trouble of doing that, and would arrive in a Rolls Royce and do the presentation in style. He knew what it took. He was the real deal, and he delivered.

For Frank Williams Racing Cars there was much to be done on the BT26, as it needed to be modified for the Cosworth DFV, that engine also being new to the Brabham factory team for 1969. The Brabham Formula 1 team ran on Goodyear tires, and one can only imagine the reaction from Sir Jack Brabham when he noted a similar current spec Brabham BT26-Cosworth on the grid shod with Dunlop tires. Fortunately it did not sour the relationship for Sir Frank as he continued selling new Brabham cars. Williams remembers, "Jack knew why we had bought that car, and he knew it was good. I had a good relationship with him because I had given him new customers by selling his cars." The news wasn't as well received by the Brabham designer, Ron Tauranac. Andrew Marriott says, "I thought Piers was really good, and he was going to be a front runner. Tauranac wasn't very impressed when they put a DFV in the back of that car."

Most people have an inner voice that moderate their hopes and dreams. Going to race Formula 1 had to be a huge leap of faith for Frank Williams. What was it that made him able to shed those doubts? "I had little doubt because I had a driver I had complete trust in. Piers' talent was obvious to me and that was what kept any doubts at bay. I also knew that if I worked hard then the chances of failure were lower. The factory Brabham was a very similar car, but Piers had a knack for looking after the car whilst still being fast. That was the key to the car, he made sure that the engine was in good shape, but still pushed so technical gremlins impacted us less than other teams on the grid. The only thing we had difficulty with were the engines."

After a promising fifth at the non-championship Silverstone International Trophy Race, the next race for Frank Williams Racing Cars in Formula 1 was the Spanish Grand Prix. This was a race forever equated with a number of high wing failures for many of the top teams, including Brabham. The race quickly ended in disappointment with engine problems leading to retirement on lap 18. The following race was at Monaco, the race where the high wings were banned after first practice on Thursday, requiring the teams to either forego wings altogether or scramble to fabricate suitable low wings for qualifying and the race. Courage scraped onto the grid in sixteenth and last place.

The race that was to provide a phenomenal result saw Courage tear through the field, up to seventh on the first lap. As the faster cars of Amon and Stewart pulled off due to breakages from the punishing nature of the street circuit, one by one Piers moved up in the standings until he reached second place, finishing seventeen seconds behind the King of Monaco, five-time winner Graham Hill. Sir Frank recalls, "It was a big day for both of us, although probably for Piers more than me. I remember the feeling well, but we still had other races that season and so we didn't let the success get to us. We both had bigger dreams in sight and focused fully on taking the next step forward."

The following races were a mixture of fifth-place finishes and scattered retirements. The qualifying was improving with each passing race, as Courage qualified seventh at the Nurburgring and then fourth for the Italian Grand Prix at Monza. At the United States Grand Prix, the race with the largest purse of the season, Courage finished second, bringing much needed finances into the team coffers. Sir Frank recalls, "Money was always an issue, and the race did offer a good prize fund I

*Piers Courage in only the team's second Grand Prix, finished a
shocking second place in Monaco (Williams F1)*

must admit, but it was towards the end of the season and therefore we
didn't have to think about too many more races. I also still had the sales
business so we had stable finances. It was a tough time to be racing if
you didn't have the amount of money some other teams had. I'm not
too keen on nostalgia, but the 1969 Brabham was the car that really
started it all for me. That car is very special, so it would be one that I
would want to find."

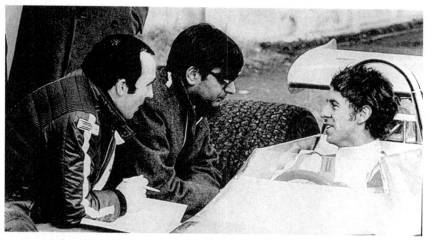

The 1970 collaboration of Frank Williams, Gian Paolo Dallara,
Piers Courage, and De Tomaso showed potential. (Dallara)

For 1970, Williams was to cast his lot with a new Gian Paolo Dallara-designed De Tomaso Formula 1 car. They had previously built a very compact Formula 2 car that had left a favorable impression on the paddock. Following miserable results from the first two Grand Prix of the season, the new De Tomaso Formula 1 car qualified ninth on the grid for the Monaco Grand Prix, so there appeared to be potential. Spa brought more disappointment, but at Zandvoort the car appeared to be closer to the pace, qualifying ninth on a track where Piers had shown great speed previously in Formula 2.

Jackie Oliver reflects, "There were a lot of us that had come up from racing in the UK. When we were at Zandvoort for a Formula 1 race in 1970, the morning before practice we were talking about his Formula 2 race there and the graduation into Formula 1, and that sticks out in my mind because it was the last conversation I had with him. It was dangerous, it was the 'killing years.' Sadly for all concerned, the Dutch Grand Prix was the race where Piers Courage lost his life in an accident.

Andrew Marriott reminisces, "I went to see Jonathan Williams a few months before he died [in 2014], and he spoke quite a bit about Piers and what a wonderful friend he was. For a long time after Piers was killed, Frank Williams ran a little cockerel on the cars. The Courage family brewing business, that cockerel was their symbol. It was a little metal badge on the cars, and it was nice. Frank and Piers got on very well."

Piers Courage, 1970 (Bernard Cahier)

Andrew continues, "At Cordova, in the middle of Argentina, before the race [December 1968] Jochen Rindt, Nina, Piers and Sally Courage and I were playing charades. Afterwards, Piers said to me that we were going to drive to round three of the championship and we got this car and had the most marvelous drive. In that journey, Piers wore this straw hat. When we got back to the airport he asked me if I wanted the hat. When I got home I chucked it somewhere, and that's where it sat for about twenty-five years. I've had it framed in a nice cabinet." In 2017, when he was again visiting Argentina, Andrew confirms that he had brought Piers' straw hat along on the trip.

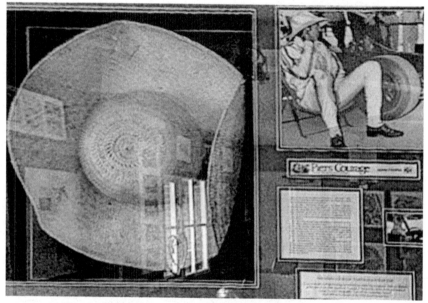

Piers Courage's hat and its place of honor at
Andrew Marriott's house (Andrew Marriott)

Williams continued in Formula 1, having entered a year-old March 701 and "tea tray" 711 for Henri Pescarolo in 1971, achieving a best finish of fourth at Silverstone. For the following season the team changed its name to Team Williams Motul, as they expanded to two cars for Pescarolo and newcomer Carlos Pace. Pescarolo drove the March 721 development of the "tea tray" car early in the season before switching to the Politoys FX3, a Len Bailey design that looked much like a McLaren M19 with the signature Tony Southgate shovel nose similar to a P160 BRM or Shadow. Pace had reasonable results with the older 711 March, taking sixth in Spain and a fifth at Spa, both results being far better than

Pescarolo's results in the 721 or Politoys.

Things looked promising for 1973, as the team had attracted Marlboro sponsorship, and the entrant was again Frank Williams Racing Cars. The Iso-Marlboro FX3B car was a modified Politoys car from the season before, Howden Ganley scoring a seventh in Brazil. The new Iso-Marlboro IR had its debut in the Spanish Grand Prix, but had an underwhelming season with Howden's highlight being a sixth in Canada. The team endured musical chairs all year with a variety of drivers. The following year saw Williams lose the Marlboro sponsorship while still scoring points in two races, the highlight being fourth in Italy. The cars became known as FWs, the beginning of that now-familiar model designation.

The early 1973 Iso-Marlboro livery on the
Politoys run by Williams (Norm DeWitt)

Frank Williams had a new car ready by mid-season 1975, the FW-04. As the team endured another season of musical chairs, Jacques Lafitte drove most of the season. There was one stunning highlight as Lafitte scored second place in the German Grand Prix at the Nurburgring, driving the FW-04 chassis to the best finish for a FW car, not exceeded until Regazzoni's historic win at the British Grand Prix in 1979. Having taken over the unloved Hesketh 308C from the team's closure at the end of 1975, the Hesketh was rebranded as a Wolf-Williams FW-05 for the the new Wolf-Williams Racing team as Walter Wolf had bought a controlling interest in Frank Williams' team. The car was hopeless, although Ickx had a strong run in the non-championship race at Brands

Hatch, qualifying fourth and finishing third. That was the sole high-light and Ickx had left for Ensign at mid-season.

It wasn't far into 1976 before Frank Williams knew he had enough of the Walter Wolf partnership and laid the groundwork for the formation of Williams Grand Prix Engineering with engineer and designer Patrick Head in early 1977. An old March, driven by Patrick Neve was the stopgap car used in '77, with a predictable lack of results. After the season the team introduced their debut design, the FW-06, which was the best of the non-ground-effect cars in 1978-79. Their following car was the stellar ground effect FW-07, unveiled at Long Beach in 1979. The FW-07B/C was raced to Constructor's Championships in 1980-81. The team endures to this day, having won a total of nine World Constructor's championships.

The FW-06 in its current place on honor at the
Williams factory museum (Norm DeWitt)

1996 World Champion for Williams-Renault, Damon Hill, sums up Frank Williams, "Frank is of a generation where people did it for the love of it. They loved the sport, they loved racing cars, and they loved the whole thing. He's also a responsible businessman, he knows he's got to make it work and can't just be spending money he hasn't got. He's been very successful as the proprietor of his own team. The sport has changed now where you've got owners that are distant from the sport and they have managers in there. The culture has changed completely

*The first of the Williams Grand Prix Engineering cars,
the FW-06, at Long Beach in 1978 (Norm DeWitt)*

from those times." Yet, Sir Frank Williams continues to be involved, sometimes to be found late at night in the garage studying his cars. His passion for racing remains.

*Sir Frank Williams late at night in the Williams F1 garage
studying his car, at the 2015 USGP (Norm DeWitt)*

*The Championship winning FW-07 and FW-08 cars brought
Williams to the top of the sport (Norm DeWitt)*

CHAPTER 13

BERNIE'S BRABHAMS

*The trapezoidal chassis of Gordon Murray's 1973-75
Brabham BT-42/44/44B was highly successful (Norm DeWitt)*

Bernie Ecclestone acquired Brabham (Motor Racing Developments) in 1971, and after a short period of time working with previous owner Ron Tauranac in his additional capacity as designer, Tauranac left the team. Making due with the previous BT-34 lobster claw car, Carlos Reutemann put the machine on pole at his home race in Argentina, but that was the highlight for the former Tauranac design. Ralph Bellamy's BT-37 was the replacement for the lobster-claw car, but it achieved even less, Reutemann's highlight being a fourth in Canada. Working with gifted South African designer Gordon Murray, Ecclestone brought the world some of the most innovative and successful Grand Prix cars of their era. The first of these cars was the BT-42, introduced at the Spanish Grand Prix in 1973. With many teams focusing upon the latest trend of wide and flat monocoques, Murray created a tidy machine that was a trapezoid in section (the following BT-44 and 44b were discussed at length in *Making it FASTER*).

The shape was essentially an angular version of previously seen concepts, going back to the late season 1968 monocoque Brawner-Hawk in America's USAC Championship Car series. The angular sides of the Hawk were rounded, much like Tony Southgate's BRM P-153 of 1970. There were others along the same lines, successfully explored by the World Championship-winning Matra MS-80 of 1969, with the bulging sides of the cockpit giving what could well be called the "pregnant guppy" look, versus the cigar-shaped Matra MS-10 of 1968. Sir Jackie Stewart explains, "That was the best place to put the fuel, right in the middle of the car. That's what the engineers at Matra did, they weren't racing people to much of an extent, but they were very clever and put together beautifully handling cars. It only became a real belly with the MS-80."

Derek Gardner used that same shape on his 1970 Tyrrell 001 to the point where one wonders if Tyrrell gave Gardner marching orders to design him an updated Matra. Jackie Stewart says, "He had his own head, but he was wise enough to know a good thing when he saw it. The MS-80 was the nicest car to drive. It was the right wheelbase, it was the right weight distribution. The Lotus 49 was probably a slightly faster car, but it was fragile."

The trend was established, and the two established design paths were "pregnant guppy," such as the aforementioned Matra/Tyrrell/BRM and Ralph Bellamy's McLaren M19A of 1971. The other design path was the wide and flat rear radiator rectangular monocoque, such as the

Lotus 72 or the M23 McLaren of 1973. Gordon Murray's car was unique in its crisp angularity, but otherwise followed the "pregnant guppy" path and was an immediate success. The 42 had conventional outboard coil/shock units, and the introduction to pullrod suspension to Grand Prix was with the BT-44 of 1974. The 44 was an improvement upon the 42, and the car was immediately on the pace.

Herbie Blash had his early roots as a mechanic for Rob Walker's Cooper-Maserati in 1967, driven by Jo Siffert. Moving on to Team Lotus in 1968, he worked with both Graham Hill and Jochen Rindt. Herbie reflects, "I joined after Bernie bought Brabham. I had first met Bernie when I was Jochen Rindt's mechanic. I couldn't get on with Ron Tauranac who was running Brabham, and who was the designer. I left for Frank Williams with the March, after the De Tomaso. Ron left Brabham, and I went back to Bernie. Charlie [Whiting] joined the team in 1978, and here we are in 2016 still together. I've done twenty-one years in Race Control and Charlie has done twenty years. I'm the oldest one, and I've now done over 750 Grand Prix. I haven't missed a Grand Prix since 1973."

Timing is everything. Herbie Blash had picked the right era for his return to Brabham, and results soon began to arrive. Reutemann in the BT-42 achieved a pair of third place finishes at Paul Ricard and Watkins Glen, the team achieving fourth in the Manufacturer's Championship. For 1974, Carlos won the third race of the season at South Africa in the new BT-44, also taking fastest lap. Later in the season he also was to win at Austria and the US, Watkins Glen seeing the team's first 1-2 finish. In the second car it had been musical chairs for most of the season, but late in the year Carlos Pace cemented his place in the team, taking fastest laps at both Monza and Watkins Glen. The team had slipped to fifth in the Manufacturer's Championship, but what mattered more was that the Brabham BT-44 was the car to beat by season's end.

Ron Pellatt was a mechanic at Hexagon running customer Brabhams. Ron says, "The 42 was a good car. It was pretty easy to work on from a mechanic's point of view, but it got a little dated on the aero package. The BT-44 came along and it was an improvement, a better car. The 44 had a different sort of suspension, it took us a little while to get our head around it as the pullrod had to be exactly right. If you got it wrong, it would upset the balance of the car, but you live with it and you learn what to do and sort of how to cope with it. It was the favorite, it just looked fabulous. It wasn't officially a works car, but we did pretty well with that."

A 1975 highlight was Carlos Pace winning his home Grand Prix at race two of the season, for what would be his only win. The circuit at Interlagos still carries his name. At race three in Kyalami, Pace took pole and fastest lap. Reutemann stepped up to win at the Nurburgring in mid-season along with five other podium finishes, but it was a Ferrari year, Niki Lauda sweeping all before him, leaving Reutemann third in points. The writing was on the wall, the team needed to partner with an engine manufacturer that could bring significant finances to the table. The option for running Alfa-Romeo engines was the route taken, as they had achieved much success with their endurance racing flat 12s in recent years. Given Ferrari's recent dominance, it would have been a reasonable choice as the flat-12 layout was the most attractive part of the deal.

Bernie Ecclestone recalls, "I thought it was good to be with a company like Alfa, rather than be in the hands of private people like Cosworth. It was better to be with a manufacturer." Gordon Murray created the BT-45 to house the flat 12, and much of the former tidy packaging went out the window as the team struggled with the sheer size and unreliability of the lump. Herbie Blash says, "The BT-45, to change the spark plugs you had to take the engine out."

Herbie felt like he had signed on for the BRM H-16 program. "Yeah, and the thing was so unreliable and the fuel consumption was unbelievable." Engines would show up with pumps in new and unique locations, causing panic within the team as both cars were endlessly modified to respond to the on-going engine revisions. At mid-season, Carlos Reutemann had seen enough of the Brabham-Alfa and left for Ferrari in the wake of Niki Lauda's horrible accident at the Nurburgring. Ferrari reasonably thought that if Lauda was going to return, it would not be in 1976, so they signed Reutemann which left Brabham-Alfa without their number-one driver. 1976 had been a disaster, mostly due to the Alfa-Romeo side of the equation. The best results had been a pair of fourth place finishes by Carlos Pace at mid-season.

For 1977 there was a new version of the car, the BT-45B. This car, driven by John Watson (newly arrived after Team Penske's withdrawal), and Carlos Pace, was immediately on the pace. Both drivers led the season-opening race held in the oppressive heat of Argentina, with Pace finishing second. At race two, Pace was battling with Hunt for the lead at Interlagos, Pace's home Grand Prix, when they tangled to the detriment of the Brabham. Watson crashed out while running fourth. There was no question that the second year BT-45 was a huge improvement. The cars that had been a weak proposition in 1976 were now challenging for race wins.

The 1977 Brabham BT-45B was a huge improvement on the 1976 car.
It battled for race wins all year (Norm DeWitt)

The next Grand Prix was in South Africa and it is mostly remembered for the tragic accident to Tom Pryce, who was decapitated when a marshal ran across the track carrying a fire extinguisher. The car continued flat out down the length of the long straight, flying into the side of Laffite's Ligier-Matra at the entry of turn one. Lauda was to achieve his first win since returning from his accident at the Nurburgring, bits of Pryce's Shadow wedged under his Ferrari. The Brabhams had again been contenders, Pace qualifying in second, Watson finishing sixth after setting the fastest lap of the race early-on. Sadly, the tragedies continued with the loss of Carlos Pace in an airplane accident two weeks later.

Brabham regrouped as best they could and brought Hans Stuck onto the team to drive their second car at Long Beach, race four. Hans Stuck remembers it well:

When Bernie called me after the crash I didn't expect him to call, and I was very pleased he did. For 1977 I didn't have any other thing to do as March did not renew my contract. I was very happy to be part of the team. Brabham with the Alfa engine was definitely a leading team. To race for Bernie was very exciting, it was an honor you know. When I was flying into London to meet with Bernie at his office, we sat together to talk about what kind of money. When we talked the phone rang and it was Arturo Mezario the Formula 1 driver and Bernie said '30,000 dollars… okay I'll ring you back in ten minutes.' Bernie turned to me and said, 'That was Arturo and he's willing to drive for me for 30,000.' I said, 'Oops, this is not so much what I had expected, I was thinking $70,000, but to drive for you Bernie, this is a big chance for me in a race-winning car.' I took the chance and signed the contract. Soon after we were at Monaco and after Thursday practice I was fastest, I took the provisional pole. That night we went out for dinner and Bernie said, 'You did a good job today and we want to be friends and honest. I have to tell you a story. Remember when you were sitting in my office and Mezario rang? I have to be honest, that wasn't really Arturo, it was my secretary in the other room!' It was my best year in Formula 1.

Bernie Ecclestone did not get a well-earned reputation as a master negotiator for nothing. Hans reflects, "Where would Formula 1 be without him today? It is just tremendous what he did, how he organized it. Even now it is amazing how professional he is acting, which is fantastic at his age." If you ever had any doubts as to how shrewd a negotiator Bernie can be, observe his recent negotiations playing Imola against Monza over the future of the Italian Grand Prix. Hans agrees, "Yes, yes, yes, he's also a real tough Backgammon player when I've played him, but you've got to be very careful that he isn't cheating you know." There are worse things than being on a team where the team principal always plays to win.

Hans continues, "The car, number one was the engine, it was fabulous. Especially at Monaco with the 12-cylinder in the hairpins I didn't have to use first gear because it had so much torque and it was wonderful to drive. Number two was to be able to work with Gordon Murray, who was the number one Formula 1 car designer. I learned a lot, he helped me a lot and I could help a lot as he one of the guys that was attracted to my style of driving, which was good. Number three was having Herbie Blash as Team Manager which was a big pleasure as he is a nice guy too. It was like Christmas in April."

Stuck immediately started up a relationship with Gordon Murray, to the benefit of both. "Before Monte Carlo, we had a lot of phone calls, and I told him, 'Listen Gordon, we need as much downforce as we can generate, with both new front and huge new rear wing,' and that's why we were amongst the front runners. My teammate, John Watson, he liked the car a little more on the push or understeering side. I was the guy who preferred oversteer more. Gordon was fantastic to work with as he understood that he was looking after the needs of the driver and he knew that the better a driver feels, the better he can perform. So at Monte Carlo we had a complete new aero." Obviously Monaco is a high-downforce, low-speed circuit, but Gordon was willing to listen to the drivers. That is very important and not always the case. Merv Wright managed Texaco Heron Suzuki in motorcycle Grand Prix racing (Barry Sheene, 1976 World Champion), and he famously once said, "If he wants a mashed potato sandwich strapped on his gas tank, and he thinks it's going to make him faster, just do it."

The flat-12 Alfa was a big thirsty lump of an engine, and having all that extra mass in the tail favored Hans Stuck's driving style. "I think so, I like the car. At Hockenheim where I finished third place, I ran out of gas coming out of the last corner. I just made it over the finish line, lucky that there wasn't pressure behind me." Similar was to happen to John Watson at the French Grand Prix where Mario Andretti passed the stumbling Brabham to win on the last lap.

There were a number of mid-season developments that occurred in 1977. Stuck recounts, "Aero… from race to race came new aero stuff and new dampers. They were doing those very expensive titanium springs with helper springs. There was a constant flow of development within the team. Also on the engine side, there was the Alfa Romeo factory near Milano. Carlo Chiti was the boss then and he was telling me how much power they had found and that they were working on the weight of the engine. After mid-season they had titanium exhaust pipes which were very expensive, especially when the car was run very low. It must have cost tons of money to replace them, but there was no issue. What we needed, we got."

As for engines, there was a constant revolving door of development, with qualifying specials and constant upgrades, which was creating a nightmare for Blash and the mechanics. Stuck explains, "Two different kinds of interest you know. I was interested in getting more horsepower, where Herbie was looking to fit them in. When the Alfa Romeo truck came to the racetrack, they always had fifteen to twenty engines in the truck, they were always over-engined on the weekends. Carlo would take me by the hands into the race truck and show me this newest engine with some titanium s**t that had this much horsepower and ask me, 'Can you take this engine?' Okay, I took the engine, went out in qualifying and, 'boom,' the thing would fall apart and I can't imagine that Herbie had to deal with all this you know."

The flat 12 Alfa-Romeo was turned from an endurance race winner to a Formula 1 winner by 1978 (Norm DeWitt)

It turned out to be a one-season deal for Stuck. "It all ended pretty sad. At Watkins Glen, James Hunt had pole position and I was on the first row. Bernie said, 'If you go win this race for us, I will extend your contract for another year.' I had a great start leading the race, but at the start the clutch cable broke. We had a cable clutch, not a hydraulic clutch. Of course you can change gear, but I knew that I'd have to make a pit stop and without a clutch, how can I take off from the pits?"

Stuck was running around the outside of the corners to get extra grip on the wet track. "Max Verstappen was doing that at Sao Paulo [2016] and he seemed to be finding so much grip around the outside, I was wondering why the other drivers weren't doing that. One or two laps before my pit stop the car jumped out of gear because of shifting all the time without a clutch damaged the dog rings. My dream of staying with Bernie ended in a Watkins Glen guardrail."

It wasn't long after Watkins Glen when Hans found out that he was being replaced by Niki Lauda for 1978. "Bernie was really helpful to try and find me another deal, which is how I got the Shadow. Niki was really important for the Parmalat sponsorship. If I had won a Grand Prix, it would be easier to sell me to Parmalat. I mean, who was better than Niki in those days?"

The BT-46 of 1978 brought back the trapezoidal monocoque design that Gordon had pioneered with the BT-42, and 44/44b from 1973-75. The car was designed with surface-mounted heat exchangers along the sloping sides of the cockpit, to take the mass out of the nose. Bernie Ecclestone says, "Gordon thought they would work, they were just radiators, that's all. We though it better to have the radiators there rather than in the front of the car, the weight distribution would have been better. But it didn't cool, it just didn't cool."

The BT-46 was also the first Grand Prix car to have carbon brakes, sourced from the Concorde program. Charlie Whiting recounts, "Starting in 1978, it was before anyone else. They were just beginning to use them, they were being tested but were not being raced." Bernie recalls:

> I remember when we talked about that with Gordon and I said, 'If it's good enough to stop a bloody Concorde, it must be good for us.' Then there was all the argument about the price, which is always there in the end. I remember Patrick Head saying he thought the steel brakes were more efficient anyway, and everyone would have to otherwise have the same brakes as we had. I said, 'Well, if they are no good, then why would they do that?' I was just saying to Charlie a few minutes ago how nearly everything we had done was to save money. Needless to say it cost more, it [saving money] doesn't work.
>
> About twenty years ago when everyone was using carbon brakes, there was some suggestion that we should go back to steel brakes to save money. It wouldn't have saved any money, it was just nonsense. But nevertheless, Williams did a test with the latest AP steel brakes and they were just as good as carbon brakes at the time but they just didn't last as long.

It likely stuck in Gordon's craw that he couldn't lose the front radiators as per the original design. The Brabham had struggled with side-mounted and front-mounted radiators, so there was certainly a well-documented history there regarding engine cooling on the BT-46, which might have bolstered the case for their need to cool the engine with a fan. Bernie confirms this:

> Absolutely, I think all these things happened a little bit by accident, because when we built the first fan car, it was a very basic car. Then it wasn't as sophisticated as we (later) made it. It was a general thought that this would work, but the problem was the regulations. I sort of – in my petition with the FIA – to say that we ought to make all these regulations not just for us, but that with all these bits and pieces had to have a primary purpose. What was the primary purpose? I said, 'As far as we were concerned, the primary purpose was for cooling.' The problem was sealing the car to the ground, we used rubber skirts. As soon as the car started, 'sweeeeeep'... they sealed automatically, it was run off the engine. The seal didn't take too much to keep it down.

The fan car was the next variant of the BT-46, the "B" model. It was unlike the previous Chaparral 2J of 1970 in how the Can-Am car had a separate two-stroke engine that powered the twin fans in the rear of the car. With the BT-46B, there was a single large fan mounted off the rear of the car that was powered by the engine. Tests showed some issues with the fan concept as in the beginning the fan car was shedding blades. Herbie Blash says, "Oh, yeah, but that was just initial testing, and we didn't do that much testing, but it's true that we did have problems with blades breaking. When we actually ran the car in Sweden, we didn't have any problems at all."

Niki Lauda and John Watson drove the fan cars in Sweden. The reaction from Colin Chapman and Mario Andretti was less than enthusiastic, as they both immediately saw the potential in the new car. Lotus released the following statement:

> Team Lotus International wish to state that they have today protested the eligibility of the Brabham cars numbers one and two, entered for the Swedish Grand Prix, on the grounds that

that the fan fitted onto the rear of the car constitutes a moving aerodynamic device in contravention of the regulations.

We believe that this fan has been incorporated into the general design of the car in such a way that its influence has a far greater effect in producing aerodynamic download on the car than it has of promoting airflow through the radiator, and so the production of download becomes its primary function.

The rules state that anything that has a primary function of influencing aerodynamic performance is an aerodynamic device. The rules also state that all such aerodynamic devices must remain fixed when the car is in motion. The blades of the fan are rotating aerofoils whose primary effect, we believe, to produce downforce on the car, and their effect on the cooling system is of secondary magnitude, which therefore renders the car ineligible to compete.

It will need a detailed and technical explanation of why we believe that the influence the fan has in producing download is more than ten times greater than the influence it has on cooling the radiator. It then becomes the fan's primary function. We would welcome the opportunity to explain these reasons to the technical committee of the C.S.I. as soon as possible.
– Colin Chapman

As the BT-46B had the fan tied to the engine, it would spin up with the revs of the engine. Niki describes it, "It was tied through the gearbox. It sucked the car to the ground and it was unbelievable. It was five seconds faster than the other cars." One wonders if the up and down speed of the fan as the revs climbed or fell would cause the car to porpoise along going through the gears? Lauda replies, "You had to keep the revs up, which we did. It did only one race and then it was off."

And what a race it was. The great adventure began in technical inspection where the official party line of "its primary function is cooling" was put to the test. Herbie Blash recalls, "When the FIA came, they were measuring the air going into the radiator, and the air underneath. But what they didn't know was that, while I was revving the engine up for the measurement, I was standing there with my feet under the skirt, keeping the skirt up. So as far as the FIA was concerned, it was legal [laughs]."

Bernie Ecclestone tells, "Colin Chapman came to me and said, 'Bernie, where are we going? We can all have this in two months and the cars can go on the ceiling if you'd like.' We would have lost all we had initially gained, so I withdrew the car, which upset a lot of people. It was never banned, I withdrew it. Gordon Murray was upset that I was not running the car. I wasn't interested, I could see the problem was going to be that everybody was going to have the same thing, and so what. And then it would be that everybody would have been quicker and quicker, pulling God knows how many Gs through the corners, just stop. Just stop it."

So, the team persevered with the flat-12 non-fan-equipped BT-46. The conventional BT-46 was to win the Italian Grand Prix, a Lauda-Watson 1-2 after Villeneuve and Andretti received one-minute penalties for jumping the start. It was yet another sad day for Grand Prix racing as both Ronnie Petersen and Vittorio Brambilla were seriously injured in a start-line melee. Ronnie passed away mostly due to medical malpractice.

Niki Lauda in the Brabham BT-46 on the pitlane at
Long Beach in 1978 (Norm DeWitt)

Charlie Whiting reminisces, "The BT-46 was a wonderful car. I arrived just as the fan car got withdrawn in 1978, and joined the Monday after Niki won in Sweden." The C model attempted to run in Austria with the radiators back on the side again, but that didn't work. When the Lotus 78 came out in 1977, the performance made it pretty clear that the ground effects car was the way to go. Was it just not in Gordon's DNA to just copy somebody else's car? Charlie answers, "I worked with him, but in those days I was a mechanic and didn't really help design the cars. But it was clear that he had gone done the surface cooling route, and if it had worked it would have been brilliant... but it didn't work so there were compromises that had to be made. With the flat-12 engine which it had, a proper wing car wouldn't work, so that's why the BT-48 went with a V-12 instead of a flat 12."

Herbie Blash concurs that the days of the flat 12 were numbered. "Ground effect came along, and because we were a flat 12 we didn't have any ground effects. So that's when, and I'd must say Alfa Romeo did a really good quick job, to make a V-12. You could see exactly what was happening with Lotus and ground effect." Was the withdrawal of the fan car when the need arose for the ground effect V-12 car? Herbie confirms, "Exactly, yes." The great irony is that Ferrari won the World Championship in 1979 with a flat-12 ground effects car. Charlie Whiting says, "Exactly... that's true, and I don't know what to say about that, really."

So a flat 12 ground effects car won't work?
Mauro Forghieri's Ferrari 312T4 that won the title (Norm DeWitt)

The Alfa V-12 did not appear until near the end of the year. Charlie Whiting remembers, "It didn't take that long to develop engines in those days." Piquet had to run the old flat 12 in the opening race in Australia. By the following race, the BT-48 V-12 cars were available for both team drivers. The BT-48 was to become one of the most unsuccessful cars

Rear view of the flat 12 Ferrari 312T4 that won six races along with the driver's and manufacturer's titles in 1979 (Norm DeWitt)

across Bernie's ownership of Brabham, despite having features that were years ahead of its time. Charlie continues:

The top half of the fuel tank was carbon, the top of the monocoque forward of the cockpit was carbon, and the rest was aluminum. I suspect that the ground effect wasn't particularly efficient. We did a test at Silverstone in April and the car was very quick, really good… amazing. We went back for the race and couldn't replicate it, we were two seconds slower and I don't think anyone understood why. In those days if you got it right, it was very right. There were lots of people doing things that they really didn't understand. Skirts were sticking and the efficiency of the skirts took some people quite a long time to master. The skirts of that BT-48 were a nightmare. Honeycomb was the main bit of it, there were springs and things sticking out… the wearing surface was like a hard polyethene or something like that, but then we obviously had to get into the likes of ceramics that we had to start putting in there, things like Lotus had been going through for two years. We had to go through a similarly painful learning period, in 1979. The BT-48 that won the non-championship race in Imola was against all the odds.

*Niki Lauda's 1979 V-12 Alfa-Romeo powered BT-48 in the
Queen's Hairpin at Long Beach (Gary Hartman)*

It certainly didn't help the relationship with Alfa-Romeo when Alfa started their own Formula 1 program, entering their new flat-12 car at the Belgian Grand Prix. By mid-season the Brabham team had begun planning their return to the compact Ford Cosworth. Charlie says:

> We changed engines for Canada and Watkins Glen, for the last two races we went to Ford engine cars. I was a lowly mechanic so I don't know what was going on there, but we were a bit fed up with Alfa Romeo as I understood it. The engines never came the same, you'd be waiting for one and it would sprout three new oil pumps on the side that no one knew about. So you'd have to cut great big holes in the underwings and things like that to accommodate it. Those were the sorts of things that were very difficult to overcome, but I'm not sure at what point that decision came. I suspect Gordon and Nelson were pushing Bernie to get rid of Alfa Romeo and, 'Let's just get a nice Cosworth engine so we don't have to worry about engines anymore.' Alfa wanted to do well, but they didn't for whatever reason.

For Gordon Murray it was Christmas in July, as once again he got to design a car around a small and tidy engine with moderate fuel consumption.

The Ford-powered BT-49 that was introduced at Canada in 1979 went from success to success. It achieved its first win at the 1980 Long Beach Grand Prix, and went on to challenge the Williams FW-07B for being the dominant car of 1980-81. Alan Jones won the 1980 title for

*The narrow but long V-12 Alfa in the BT-48. The learning curve
for ground effects cars was steep (Gary Hartman)*

Williams, but Nelson Piquet won the title scrap for 1981 against Carlos Reutemann, in what sadly passed for a championship-deciding venue – a parking lot in Las Vegas.

Despite all the success, Ecclestone could see the writing on the wall about the coming turbocharged era where the Cosworth V-8 was no longer going to be a competitive proposition. Bernie always wanted factory deals, it was the same with BMW as it had been with Alfa-Romeo. Charlie reflects, "I expect they were free with Alfa Romeo so that's a good deal. I suspect it was the same with BMW. We were having great success with the Ford engine and then we heard that Bernie had done a deal with BMW. We thought, 'Why on earth did you do that?' but it obviously paid off."

There were enormous struggles trying to get the BMW turbo engines to live for any time at all, but Piquet was stunned with the available power. Ron Pellatt was now a mechanic at Brabham working on the test team:

I got asked to do the Brabham with the turbine [turbo] in 1981 with the BMW. We were doing the car alongside the 49, it was all a secret project building the car and we did a lot of secret testing. I was in the background, really. In the early days the engine would only run for about ten minutes. We were running it too lean when testing at Brands Hatch, Ricard, and Silverstone. We got it right and realized the fuel was an issue. We were trying to copy the Cosworth car for fuel economy and there was no way we could match that with the BMW. We increased the fuel used by the engine to make it more reliable. We were only doing something like 3 miles per gallon, when the Cosworth was doing 4.5, maybe 5 miles to the gallon. They could do a full Grand Prix with what fuel they had, and we never could do that. So Bernie introduced fuel stops. We couldn't change the car, so Bernie said, 'Okay, let's do a fuel stop with the car,' and that changed everything. I don't think changing tires really came into it, we didn't have the speed to do that in those days.

Despite the difficulty, for most of 1982 the Brabham BT-50 ran with a massive fuel capacity needed to keep the engine alive for the full race distance. Nelson Piquet won the Canadian Grand Prix in 1982 with the BMW-engine car. Charlie Whiting marvels, "That was against all the odds, I can tell you. We failed to qualify in Detroit, then started fourth on the grid in Canada and won. We did three races that year – Monaco, Detroit, and Canada – with two BMW cars for Piquet and two Ford engine cars for Patrese. We won Monaco with Riccardo and we won Montreal on the day that Riccardo Paletti was killed. It was dark by the time we finished… just a horrible day, but we won."

The race at Monaco was one of the most bizarre in racing history. Derek Daly tells the tale:

In 1982, Frank said, 'I'm going to take a chance on a young guy,' and I was Keke Rosberg's teammate when he won his World Championship. That season is still regarded as the Wild West era of Formula 1. It was an unbelievably dangerous era and we lost Gilles Villenueve, Pironi [career ending injuries], and Paletti. They had 1200hp with the turbo cars in qualifying. Having the Cosworth was a big disadvantage, but Keke won the World Championship, so it was enough. The FW-08 was a shorter version than the FW-07B and was much more difficult to drive. At Monaco in the first practice session, I was P1 when my car broke. Instead of putting me into the spare car, they went and got another car and it was never as good. Midway through the race I was flying, and following Keke when he broke a right front suspension in the chicane. I had one of the fastest cars on the track, certainly in the top two when I lost it going into Tabac and broke the rear wing off on the guardrail, which straightened the car up perfectly. I kept going and never missed a beat. The only problem was that the oil cooler for the gearbox was attached to the rear wing, so when the wing went, it tore the lines off. So now as I drove around, the gearbox was pumping its oil out onto the racetrack. I was leading the race coming around onto the last lap at Rascasse corner when the gearbox started grinding and stopped. As I drove around, DeCesaris, Pironi, and Prost had all either crashed or ran out of fuel. Patrese in the Brabham took the lead, but spun, and he didn't know why for forty years until he read an interview with me and found out that he had spun on my oil. It was the most unlikely finish in Formula 1 history, all part of that Wild West era of Formula 1.

Pit-stop philosophy was all new to Formula 1. Charlie explains, "We tried to do it at Brands Hatch first, the first attempt. It was all going to plan, but the engine broke before we actually did a pit stop. Then we went to Germany and had a crash with Eliseo Salazar right before the pit stop. We had all the equipment, but I think people were beginning to think we were scamming them as we never actually did a pit stop. It did happen in Austria."

1983 brought the new car, the BT-52, and what a car it was. Whiting concurs, "A wonderful car, absolutely amazing. The concept was unique, just an amazing thing. I don't know what the weight distribution was, but it had a significant rear bias. The team was good, the refueling and all that sort of thing worked… it was a good year." The championship all came down the South Africa between Prost and Piquet. Prost was wearing his Nelson Piquet fan club T-shirt, which likely didn't go down well with the Renault board of directors. Piquet was to win his second World Drivers Championship for Brabham in three years.

Defending their title was proving to be anything but easy as their former reliability seemed to be gone. Charlie says, "In 1984 we had reliability problems, and didn't finish the first five races if I remember, all engine related. It was a derivative of the previous years' car. In 1985 we built the BT-54, which was a lovely car. We won Paul Ricard with it, which was the last Brabham win. It was the big car as we couldn't do

*Nelson Piquet in 1983 with the innovative BT-52 powered
by the turbocharged BMW (BMW AG)*

refueling anymore for 1985, which was a late decision."

Charlie Whiting recounts, "The BT-55 in 1986 was a disaster, the flying skate with the cranked over engine. We blamed BMW and BMW blamed us, there was blame on both sides really. It was basically discouraging. I believe BMW knew they had work to do, it was cranked over at forty-five, and at one corner the oil would go up all the way, and at the other end the engine would start to scavenge. It was a terrible chassis as well."

There was no depth to the tub so it was compromised in both torsion and bending, the result being a flexi-flyer. Charlie remembers, "If you were to build it to today's rules, it wouldn't have been a problem as using today's technology you'd have had a much more rigid and solid car. It wasn't like that in those days. The BT-55 was our first all-carbon chassis. The BT-54 was half and half." Herbie Blash confirms, "Half of the bits in it were just carbon panels which were just riveted. That's the last thing you would ever do, the technology just wasn't there at that particular time." Charlie comments, "McLaren... they were actually way ahead of the game." There is little doubt this is true when it came to composite fabrication.

Charlie says, "In 1988, Bernie stopped racing." Herbie Blash continues, "Bernie sold Brabham to Alfa Romeo. It was for a project called Pro-Drive, which was basically a Formula 1 car with a full body on it and two cars were built. Bernie was looking at setting up a new championship, but it never happened. Alfa Romeo decided not to continue it, but at the same time this Swiss gentleman came along as the team was still up and running and he wanted to buy it. It all happened very late and they asked me if I would run it. So I moved over to run that, but unfortunately he then ran out of money halfway through the season. It was left to me to be in survival mode."

Is there a favorite Brabham? Bernie waffles, "Probably not... but the 49, perhaps. Yep, probably." The BT-49 was the Ford Cosworth-powered car that debuted in Piquet and Lauda's hands for the Canadian Grand Prix weekend, when Lauda walked off the team into retirement. It was not great timing on his part as the BT-49 and its variants went on to win numerous Grand Prix in 1980-82, taking Nelson Piquet to his first Grand Prix win at Long Beach in 1980, and his first World Championship in 1981. The BT-49 was their most successful, and other than some clever hydraulics for 1981's BT-49C to circumvent the 6 cm rule, was the most conventional design across the Ecclestone/Murray era, and perhaps there is a lesson in that.

Chapter 14

Ensign Racing

*Morris Nunn and Chris Amon confer over the N176 Ensign,
from their most competitive year of 1976 (Morris Nunn)*

If there ever was an example of Formula 1 on a budget, it was Morris Nunn's Ensign Racing, a Formula 3 manufacturer whose cars raced in Formula 1 for eleven seasons between 1973 and 1983. It was the heyday of the "Formula Ford" era of Formula 1, where it was possible for a tiny racing manufacturer to buy four used Cosworth engines for £20,000 and actually contend for World Championship points at the highest level of motorsport. Morris recounts:

> After the Formula 3 cars, we did build a Formula Atlantic car and we did one race with that car at Snetterton with Mike Walker, and he won the race. At that time we were looking at building cars and selling them, but we didn't get any orders for Atlantic cars. We converted the Atlantic car to a Formula 2 car for a British driver who ran it himself; we only did the one car. We ran a works team for Rikky von Opel in Formula 3. There was a Bank Holiday race around August in 1972 and he won. Driving back he asked me what we should do for the following year. I said, 'What about Formula 2?' Rikky said, 'No, I don't like that.' I said, 'How's about F-5000?' And Rikky replied, 'No… do you think you could build me a Formula 1 car?' I said, 'WHAT??' By the time I finished that drive back home, we'd decided to go build a Formula 1 car. He would finance both the building of the car and the running expenses for the year, if I would give up the production side of Ensign.

Nunn had found his version of Lord Hesketh. Morris agreed, "Yep, I asked about what if he got hurt or something. Rikky said he would give me a contract that would finance me to get back into production. That was the deal we made, with no pressure upon him as a driver, and no pressure upon me in building the Formula 1 car. Our first race was Paul Ricard."

The results were mediocre at best. Fifteenth in France, and thirteenth in England were the season highlights for 1973. In 1974 the team went to Argentina and the experience was enough for Rikky. Morris explains:

> Ricky had practiced in Argentina and then he decided he didn't want to do Brazil and that we needed to go back to base. He couldn't tell me what was wrong, but he wasn't satisfied. I asked him if I could test the car with another driver. We called Brian

Redman and he came down to Silverstone and went through some changes on the car. Brian went round at the end of the first lap and said, 'Morris, I don't know about Rikky driving in South America, but I wouldn't like to put it into top gear. Give me more rear wing.' We gave him more rear wing, and he came in saying he couldn't feel the difference. 'Give me maximum rear wing.' He went out, came back in and said, 'The rear wing isn't working aerodynamically.' We had put an exhaust deflector onto the front radiator for the heat in Argentina. What it was doing was kicking the air right over the rear wing. We cut three inches off the deflector and Brian was able to continue testing the car and we dialed it in. We raced in the Daily Express Formula 1 race at Silverstone, with Brian finishing fifth after a race-long battle with Graham Hill.

The team did not do any tuft testing or wind-tunnel testing with scale models. Morris admits, "It was all seat of the pants." Soon after, Rikky von Opel decided to move over to Brabham. Morris says, "I think Bernie Ecclestone and Brabham were chasing Rikky for his sponsorship, so he drove the second car along with Reutemann. Rikky left us with everything to continue in Formula 1, but we'd have to find our own sponsorship. He had purchased six new engines and then took those engines to Bernie." Vern Schuppan, in his first race with the team, finished fifteenth in Belgium. That was probably the highlight of 1974.

Mike Wilds attempted to qualify the car at a number of races in late 1974. He says:

I had a tough time with the N174, struggling to actually qualify in Italy, Austria, and Canada. The chassis wasn't the best, but it wasn't so bad. The biggest problem, if my memory serves me well, Morris changed the fuel tanks in anticipation of an impending regulation change. These fuel tanks were quite complicated and we suffered from a drop in fuel pressure in left hand corners.

Finally, after a lot of frustration, things were modified after Canada and I qualified the car twenty-second, which is not a brilliant grid position, but there were thirty cars for twenty-five grid places. The car was much better and out-qualified such competitors as Tim Schenken in a JPS Lotus 72, Hans Stuck in a Works March, Vittorio Brambilla in a Works March, etc. I really enjoyed driving for Morris. Yes, we initially had a frustrating time but [Morris was] a great engineer and a pleasure to spend time with and drive for. I wish I could have driven some of his later Formula 1 cars.

Mike had finally been able to qualify the Ensign for a Grand Prix, lined up just ahead of the doomed Helmuth Koinigg (fatal accident on lap 10) in his Surtees. He recounts:

One story from that US Grand Prix... I was driving really hard to qualify, on my best lap I crossed the line to start my lap and Jody Scheckter was behind me in his Tyrrell. Throughout the lap, every time I looked in the mirrors Jody was close behind me. I finished the lap, which was good enough to qualify. Later Ken Tyrrell came up to me and said Jody was looking for me! I went and found Jody. Jody is quite an aggressive character so I thought attack is the best form of defense! I went and said very aggressively, 'ARE YOU LOOKING FOR ME?!' expecting him to complain that I held him up on that lap. Jody looked aghast and just said, 'I only wanted to tell how well you were making that Ensign go!'

At the start of the US Grand Prix the DFV fuel pressure relief valve failed and I struggled round the first lap and lost four or five laps while it was fixed. Eventually going out of the pits I spent a lot of the race dicing for fun with Chris Amon who was in the BRM P201. This led me to get the offer from BRM for 1975.

For 1975, Ensign had found a sponsorship, with Dutch driver Roelof Wunderink driving the single entry. Nunn explains, "We had a Dutch sponsor and they only wanted a Dutch driver. Wunderink had a bad accident in an F5000 car and was knocked unconscious for a period of time and was never 'right' after that." Plug in another Dutchman? "Another Dutchman, and that was when they brought Gijs van Lennep onto the team."

Germany brought sixth place and the team's first point. Morris says, "I wasn't very happy with the results of the sixth place, as we were last. At the end of the race the sponsors were in the pit and Van Lennep was going to arrive for the final lap. They said, 'Morris, slow him down because we are going to score a point.' I told them if he goes any slower he will stop. The sponsors fired him and we got Chris Amon."

Through Amon, Nunn had found his next engine supply, one that was almost unbelievably to last him through eight seasons of Formula 1. Morris remembers, "When we got Chris Amon, he had some shares in a garage in Australia. He sold the shares for 20,000 pounds and he gave me the money and we bought four used engines off Bernie, 5,000 each for them, and that was all the engines we had for the rest of Ensign's years in Formula 1. We were still running those engines in 1982." Were there updates continually across the years? "No, Cosworth was just maintaining them."

Chris Amon says, "I never realized that [the 20,000 pounds] actually paid for the engines." One can safely assume that things were a bit different when Chris was at Ferrari for three seasons in Grand Prix racing. This was providential for the struggling team, as Chris was always one of the fastest drivers in Formula 1. Morris concurs, "Absolutely, we kind of got him at the end of his career."

Chris recalls those early days of working with Morris in 1975, trying to bring his legendary talents at sorting racing cars to the team:

> I joined the team for Austria. One of the most significant things was when he talked to me on the phone. He made a big point saying, 'I engineer the car and you drive it.' He didn't want me doing any of the sorting or that kind of thing. When I got in it for Austria, it immediately didn't feel like that bad of a car, but it felt completely unsorted. It took me into the second day to convince him that we needed to change some springs and things. It was a good little car, but the engines weren't great. We didn't qualify very well, but I made up quite a few places in the race. I must say he did change his attitude after a while, and I think we ended up with a really good, sorted car.

When the race was stopped due to a massive cloudburst, Chris was in twelfth, which was also where he finished at Monza. Chris looks back on that early effort, "We ran a non-championship race in Dijon, France, and I remember having a big tussle with Hunt in the Hesketh. That car [1975] was not the original Ensign. It was similar to the '76 car, the one that the Dutch had paid for, and after Monza it became the Dutchmen's property. So for the first two races of 1976, I had to run the old 'von Opel car' at South Africa and Long Beach. I'll always remember that the pit crew told me after the race in South Africa, how Morris had said to the crew that, 'If he passes Andretti, the first thing I'll kiss his ass.' And I did pass Andretti, but Morris never did [the kissing]! It wasn't a bad car, although I think it was pretty heavy." Mario clearly didn't understand what was at stake that day within the Ensign team, but recovered to finish sixth for his last finish with the Parnelli Formula 1 car.

Amon had one last race with the original 1973 Ensign chassis getting eighth at Long Beach, but the new 1976 car brought a high level of competitiveness to the team. Morris describes, "The new car was a clean sheet design, new monocoque… the first car was overweight, but we could get right to the weight limit with the new car. We went to pullrod suspension front and rear, with inboard brakes all around. When Chris came, he wouldn't race the car with inboard brakes. When Jochen Rindt got killed, they reckoned it was the inboard brake shaft. We hadn't had a problem with it, but Chris wanted them off the car." Did they come off the car? Nunn says, "Yep, it was a case of that or he wasn't going to drive."

Amon confirms, "Yeah, I wasn't happy with those. I'd had them on my own car in 1974 as well, and had broken a couple of brake shafts. I had no time for them at all, really. Unfortunately, I didn't know that at the time, but that was one of the many fragile things on it [the Ensign]."

Nunn recalls, "We finished the car, did a few laps at Silverstone, then loaded the car and drove off to Spain. The fifth in Spain was the new car." It was the best finish for Ensign Racing to that point. Chris reminisces, "I was the meat in a Brabham sandwich, with Reutemann in front of me, and Pace behind me. I had a race-long go with those two."

At Belgium, the team was again on the pace. Chris continues, "I remember having a pretty good race with Hunt and Scheckter at the time, and in a split second I was upside down in the fencing. They found the wheel hundreds of yards away in a car park someplace." Morris says, "We had a rear wheel failure… actually the rear axle broke, the wheel came off, and he actually turned over."

The next race was Monaco, and past injuries caused a DNF. Chris explains, "The accident was in 1975 after I'd driven the Talon in the

Chris Amon in the N176 Ensign – 1976
(Morris Nunn)

F5000 race at Long Beach. This guy in a stolen car was being chased by the Highway Patrol, and he came through a red light and hit us. My foot must have come out the window when we were rolling down the road. I ended up in San Clemente hospital, and they shipped me back to England where I got gangrene and had to have skin grafts. In South Africa I don't think I was walking with a stick, but I wasn't much better than that. My elbow was sore from the crash in Belgium as well."

Morris recalls the unfortunate chain of events, "At Monaco we were in the pits right before the race and Elizabeth Taylor was walking by and the crowd was there with cameras. Chris was standing there and the crowd came back on him and a photographer stepped on his foot. He screamed and picked his leg up to grab his foot. So then the guy with the camera turned and hit Chris on the injured elbow with his camera. Oh Jesus, imagine starting the race at Monte Carlo after ten laps having to shift with your left hand. He retired, he couldn't drive anymore."

Sweden was the next Grand Prix, and Chris qualified in third, a shocker to the well-funded McLaren, Ferrari, Lotus, Brabham, and Tyrrell teams. He says, "I think I was running third when the steering broke, the six-wheel Tyrrells were in front of me, and they won the race. pretty good engines too. We were always playing catch-up in the corners. I ended up hitting the guardrail really hard. I was turning into a right-hand corner and the left-front wheel went left and I went straight ahead."

Morris recounts, "The left-front steering arm was bolted to the upright with helicoil threads. The mechanic that had built the front suspension didn't put the helicoil long enough, he'd put the short one in and it pulled out and his left-front wheel turned left. Chris was so quick; nobody could believe the speed he was doing when they were watching him because he was so smooth. As regards to all the drivers I've had, he was the best test driver." When you look at the long series of incredible drivers and World Champions (Andretti, Fittipaldi) that have been engineered by Nunn, that is high praise indeed.

Chris saw the failure as the likely result of crew fatigue. "It was Morris, his wife, maybe three mechanics, and myself. I think some of the problems were a result of the huge accident at the Belgian Grand Prix. With things like the helicoil, he was so short of money, and with a minimal crew… these guys were working three-quarters of the night, not just at races but between races. The accident at Zolder compounded the problem, as the car was quite badly knocked around and they were working all sorts of hours."

There were a number of other issues facing the drivers of the day. Morris says, "Remember, we used to get tire vibration in those days. Chris was getting really dejected because he said coming off the corners with all the tire vibration, putting the power down, that his vision would be blurred. It was no fun driving anymore."

The British Grand Prix brought another strong showing. Chris recalls, "In '76 we had some pretty good racing with people like Hunt, Scheckter, and those guys. Honestly, if we'd had equal engines it would have been no contest. I remember qualifying sixth at Brands Hatch, which is a pretty rough circuit, and being quite disappointed. I wasn't helped that I had missed first practice when I couldn't land my airplane in the fog."

It all ended with a DNF. Chris says, "A water leak was caused by debris from a first corner accident. I can remember seeing a photo of myself with Hunt's car about four feet in the air above me in the first corner. In hindsight I must have run over some debris and it split a water pipe or something." The team had called Chris in for the water leak, as they couldn't afford to chance losing an engine.

He continues, "By then I was fully aware of the situation and the finances. I was reasonably happy to accept those sorts of things, but I guess to a degree I was ready to wind down, as I'd pretty much made up my mind to give it away at the end of the year. I guess in a way that season was approached with the view of it being a last hurrah. If we'd had equal engines and reliability, I'm sure I could have won a race. The car had a lot of potential."

The next race was the Nurburgring, an undulating car-pounder of a circuit where durability is put to the ultimate test. Not necessarily a comforting thought for the driver of an Ensign N176. Amon remembers:

We hadn't qualified that well, it didn't feel very good over the bumps, we never really got it sorted. Teddy Mayer [McLaren] asked me to come and see him in the motorhome after qualifying on Saturday, as they were asking me to drive alongside James in 1977. When I got there, James Hunt was just coming out and said, 'Did you have problems in practice, or did you just not like driving that thing around here?' He was surprised at how far back I was. The key to the old Nurburgring was getting it to go over the jumps and bumps and things. If you can't get that worked out, you won't be going quick. Sweden was smooth, and Zolder was pretty smooth too. It worked well in Spain at Jarama, which once again was a smooth circuit.

It all went terribly wrong in the race when Niki Lauda had a horrific accident where he was trapped in the burning Ferrari. Chris explains his perspective, "I elected not to take the restart after Niki's crash, having seen how long it took for anybody to get anywhere near him. I thought if this Ensign has a suspension failure somewhere around the Nurburgring, having just seen what I'd just seen... I thought I could do without that at this stage of my career. I'd had two big accidents prior to that and I've got to say that once you lose confidence in the car, you do start to struggle. Any other track it wouldn't have been a factor, but under those circumstances on that track, and how it was so poorly marshaled... the whole thing was a real shambles. The fact is that the circuit had reached its use-by date." It is hard to argue with Chris' logic. Morris replies, "Yep, but now we were really scratching for drivers, looking at all kinds of new guys."

There is little doubt that when thinking back across the Ensign years, those stellar 1976 efforts come first to mind. Roger Penske says, "Well, Morris ran the red car with Chris Amon and obviously we were small teams working together to try and get a race win in. He and I became friends." Penske was to get that race win soon after, when Watson won the '76 Austrian Grand Prix, however Penske's team withdrew from Formula 1 at season's end.

Jacky Ickx had replaced Amon in the Ensign for the Dutch Grand Prix, qualifying eleventh and had a stellar race, moving up through the field. Unfortunately he was a late-race retirement with electrical issues after setting laps that were on pace with the fastest lap times of the day. Italy brought the new dark blue Tissot paint scheme, where Ickx finished tenth in another strong run, just behind Carlos Reutemann's Ferrari. Thirteenth in Canada was followed by a DNF at Watkins Glen, but the current Ensign certainly had the pace to worry the established teams such as Ferrari, Lotus, Tyrrell, Brabham, and McLaren.

Clay Regazzoni's 1977 N177 Ensign photographed in the modest paddock of the day (Morris Nunn)

1977 brought Clay Reggazoni over to the team from Ferrari, as Nunn was still running the same design he had run the previous year for Amon, although now rebranded the N177 (formerly the N176). Nunn reminisces about working with Clay:

> One of the most enjoyable seasons I ever had was with Regga. The first races he did for us was the Argentine. We finished

sixth, and then we went to Brazil. The first qualifying day he was either fourth or fifth quickest. The second day we dropped back to ninth for the grid. After practice he comes up with a couple of buddies and says, 'Okay Morris, see you back at the hotel.' I said, 'Hang on Clay, I need to talk to you about the car.' Clay said, 'Morris, the car is okay.' He puts his arm around me and says, 'Morris, you worry too much, I be fourth at the end of the first lap.' Sure enough he started ninth and was fifth at the end of the first lap. But then Jochen Mass went off the end of the straight and the catch-fence wire mesh swung across the racetrack and wrapped itself around our car.

Clay was always upbeat, he was great fun. He knew we were on a low budget. The first year with us he drove for nothing, he said, 'Put the money in the team.' When we went to South Africa he came with two rear wings that a couple of Ferrari mechanics had developed, this delta shape wing he had paid for himself and brought to South Africa. We built a second car, which we rented out to Teddy Yip that he ran for Patrick Tambay.

From this agreement, came another memorable highlight and disappointment in Ensign racing history. Morris continues, "Clay in Zandvoort... we were lying third and Tambay was running in fourth with our car run by Teddy Yip and Sid Taylor. The throttle cable broke on Clay's car and so his stopped, and Tambay ran out of gas. That was a real message that it's never over until it's over." Patrick Tambay still managed to be classified in fifth place.

The Derek Daly N178 Ensign was another points scoring design
(Norm DeWitt)

1978 brought a new N178 with rocker arm suspension and ground effects tunnels, as everyone was reacting to the success of the Lotus 78. A rotating door of drivers tried the Ensign that year, including the Formula 1 debut of Nelson Piquet. The best finish was scored by Derek Daly, who finished sixth in Montreal for a point. Derek recalls that time:

Morris was my first team owner in Formula 1, and I came after Jacky Ickx, midway in the 1978 season for the British Grand Prix at Brands Hatch. Mo Nunn was called No Munn because there was no money. So I traveled around the world with him, getting invaluable experience, but ran really well. How Dale Coyne ran in Indycar, we ran in Formula 1. All I know is that my contract payment for 1978 never got paid, and Morris promised instead of paying in cash, he would pay me by giving me the car I'd raced my first Grand Prix in. I was perfectly happy with that, and when I went to collect it, he'd already paid off another debt by giving my car to someone else. That was a laughing point of contention for the next forty years. I have no idea where the car is now, it was MN-07.

Then the team got it all wrong with Dave Baldwin's hideous N179 running a series of stacked radiators that extended vertically to the top of the cockpit opening. It didn't look altogether different from the 1978 Wolf WR5, which didn't exactly set the world on fire either. Morris recalls it, "We built a new car for 1979, and it was terrible. A front-radiated car that really didn't have any downforce." Derek Daly thought it was better than Nunn did, or at least felt the idea had potential. Daly claims, "I did drive that, and it was incredibly fast down the straight because there was so little drag. But that concept was much better than it looked, and if Mo Nunn had a budget and spent money developing that concept, that had real potential. But, he was too scared to develop it and didn't have the money for development. It looked like somebody bought a step ladder at Home Depot and fabricated it into a Formula 1 car."

For 1980 the team brought in Nigel Bennett, and the former Lotus and Firestone engineer was to get his first opportunity as a Formula 1 designer. The 1980 car was a full ground effect car, but proved to be a flexi-flyer. Morris says, "That car we built in seventy days, a ground effects car with a chassis that was so bad it would pop rivets under load.

*Nigel Bennett as Ronnie Peterson's Engineer at Lotus before
designing the Ensign N180, photo from 1978 (Norm DeWitt)*

It was weak, the understeer got worse and worse. We had to stiffen it up all over the place and every time we did that it was better, but not enough. Not a good car." Ninth in South Africa was the highlight of the year as at the following race it all went wrong in the Queen's hairpin at Long Beach. Reggazoni's brake pedal broke off at the end of the Shoreline straight and he had the awful choice of driving head-on into a wall, driving into Zunino's parked Brabham on the left, or Depailler's parked Alfa on the right. He chose to drive into the Brabham, and although he survived, he had lost the use of his legs as a result of the savage crash. The shocked team surveyed the damage and they discovered that the brake pedal had not been fabricated per the design drawings. It likely provided little consolation given that the team had missed how the assembly wasn't fabricated properly.

1981 saw an improved version of the N180, the "B" model for Marc Surer, which brought the team their best finish, fourth in Brazil. Morris remembers, "It was a reinforced 1980 car. We couldn't afford to keep him [Marc] because we didn't have any sponsorship, but it was the best result we ever had." 1982 brought the final and most impressive of the Ensigns, also Morris' favorite car. "That was the first car that was built

Surer in the improved N180B Ensign of 1981 scored a fourth place.
It was the highlight for Ensign Racing (Norm DeWitt)

with an aluminum honeycomb structure at the bottom and a carbon fiber/Kevlar top half. When we stress-tested the 1980 car, it was 4,000 foot pounds per degree, and when we tested one later on, it had dropped to 2,000-something. When we built the new chassis in 1982, it was in excess of 20,000 foot pounds per degree. We had Roberto Guerrero and Johnny Cecotto drive for us in 1983 and Cecotto finished sixth at Long Beach, his second Formula 1 race."

Roberto Guerrero explains, "Nigel Bennett designed the cars, and he was an incredible designer. Our total budget for the team was 400,000 dollars, and we were racing with Ferrari, Renault, BMW... so it was really cool." Amazingly, Guerrero qualified eleventh at the Detroit Grand Prix, and then finished eighth in the German Grand Prix. Shoestring budget doesn't begin to describe. Roberto continues, "We were just running the Cosworths. We'd go to the race with the one Cosworth that we had in the car and hope that it didn't break as we didn't have any spare engine. It was quite a feat, competing against those huge teams. It was amazing what we got out of the car with that team."

For 1983, Ensign Racing was officially gone, but Theodore Racing (Teddy Yip) ran a two-car team using the Ensign chassis. Guerrero says, "We became a two-car team with Johnny Cecotto, the motorcycle world champion." Cecotto managed to score a sixth-place finish at the Long Beach Grand Prix (US Grand Prix West), while Guerrero had qualified eleventh at Detroit, for the second year in a row. Those were the highlights. He recounts, "At the end of the season they closed their doors and I believe neither team raced in Formula 1 again. Nunn was able to get the most out of every dollar, and it was an honor being in

Jo Ramirez, Morris Nunn, and Nigel Bennett confer on improving the car (Morris Nunn)

Formula 1. It was a lot of struggle, but it was a lot of fun. We were really the final chapter of Ensign."

Morris says, "George Bignotti with Dan Cotter asked us to build an Indycar for them. We used the same tub as the Formula 1 car, only rebodied." That program brought Morris over to CART for the remainder of his career. Formula 1's loss was to be CART's gain, as it wasn't long before both Nunn and Guerrero were successful. Guerrero describes it:

Morris and I ended up in Indycar, from the chassis we used in Formula 1 for 1983 we built an Indycar and I tested it at Road America. It went really well, but Tom Sneva was driving for Bignotti at Indianapolis, and he hated the car. We came to the conclusion that it was because the sway bars that the Indycars had at the time were like 1-1/2 inches in diameter, and Formula 1 were a lot smaller. Putting bigger sway bars in the car would have probably solved all the problems, but Sneva wasn't ready to give it a shot. Of course he won the 500 with the March that year [1983] anyway, but that's how the association started. When Theodore Racing closed at the end of the year, Mo Nunn was offered the engineering job to go work with George Bignotti in

275

Indycars and Morris was the one who said they should have me as the driver for 1984.

Roberto Guerrero in the last of the Nigel Bennett
designed Ensigns, from 1982 (Roberto Guerrero)

Tom Sneva explains his perspective, "The Ensign was a car that we weren't able to spend enough time developing. It had too much Formula 1 input and not enough Indianapolis input, so it was at a disadvantage from the get-go. Morris was a road racer who became a good oval-track guy, but it took him a couple of years to develop the expertise you needed to run ovals as well as road courses. We just didn't have time and the March had a couple of years' development on it. We were the first one to have a March when it came out, and fixed the weak spots. George [Bignotti] probably had a lot of input on what was needed."

One finds advocates for Ensign Racing in the most unexpected places, such as a current Formula 1 pit lane. Robert Fernley, Chairman and co-owner of Force India, recalls:

The Can-Am car that Jim Crawford was driving in '82 and '83 was a baseline Ensign. It was the chief competitor against the former Al Unser Jr. Frissbee, which was Jacques Villeneuve under the Canadian Tire program. There were three really key cars that were competitive at that time. There was the Frissbee, there was the VDS [Tony Cicale design] that was operated by Al Holbert that became Michael Rove's car that became the championship car the following year. Jim's car was an Ensign 180B, the car that ran in 1981 which was reinforced – the former Marc Surer [Formula 1] car. We had a very close relationship with Mo and I ran what I would call the Ensign B team. I'm very fond of Morris, he and I had a long relationship. I think I'm the only person that has owned every Ensign that's ever been built. At some point the cars all have come through me, even the Ricky von Opel Batmobile.

There was a British Formula 1 series called the Aurora series and we won that championship in 1982 with the 180B and Jim Crawford in the car, against RAM racing – John McDonald and all that lot. It was my team, I financed the team, and there was no connection to Ensign financially. The connection was a technical relationship with Nigel Bennett helping.

The '83 Ensign that Sneva didn't choose at Indianapolis ended up back into the mix, only now at Long Beach in 1984. Fernley continues:

Sneva [rejecting the car]... I think it was more to do with the transmission, the Weisemann box and all the differences that they were running at that time. I took that Ensign back again, remodeled it, and Jim Crawford ran it at the first Long Beach Grand Prix for Indycars and finished fourth. We redesigned and remodeled it completely. Even today that is still the record for a rookie team with a rookie driver entering Indycars. By the time you got to the mid-80s, the evolution of Indycars with Lola and March competing with each other, obviously you had to have a new car every year. That car is now in Vijay Mallya's collection, along with the 180B, which is currently being rebuilt back to Formula 1 spec. It depends upon where Vijay wants to take it, but it's likely to go back to the Crawford car that won the Aurora championship, and will hopefully run in the Master's Formula 1 series next year. The Piquet 78 car, the MN-08, Vijay has got that as well, a nice looking car actually. The Piquet-Daly 78 car was the first Formula 1 car that I ran for Vijay, he has had that rebuilt and it's in Vijay's collection in England. He has two Formula 1 cars and the Long Beach Indycar, in the original livery that we ran it with in 1984 [Long Beach, Meadowlands, and failed to qualify for Indianapolis].

At that time, the Ensign was probably the best car to do that with, it was current, only a year or so old. The Williams and

Lotus from those days were very flimsy. Our record of finishing races was second to none. There is a big difference between running a Can-Am series with a car that has minimal development versus running a Formula 1 car with limited resources that needs to be developed all the time. Ours was a relatively easy program in comparison to what Mo was doing. A Formula 1 car very rarely runs the same way twice, and that doesn't really occur in any other formula.

This last chapter of the Ensign story certainly worked out well for Nigel Bennett, who had an advocate pulling for him to get one of the best design positions available. Mario Andretti explains, "I definitely was influential in trying to get Nigel hired at Lola. I knew he had that knowledge of ground effects from the beginning. I suggested Nigel because of my experience with him, as I had known him since he was a tire engineer with Firestone. He had in-depth practical knowledge of the business and to be in that Lotus fold with Colin and having worked with him there, I knew he understood about the dynamics of the car. I felt that he was going to be a pretty good bet. It was 100 percent a brilliant move."

Mario won the CART title with Bennett's T-800 Lola in 1984, and for 1986 Mario's engineer at Newman-Haas was Morris Nunn, again working with a Nigel Bennett chassis. Although they had wins at Portland and Pocono, it wasn't a great season by Andretti standards.

Following that stellar run with Lola, Nigel Bennett went on his own, and his designs soon arrived at Penske Cars. His first Penske design (the PC-17) went on to win the Indianapolis 500. Roger Penske recalls, "Of course Nigel was someone who moved over to our organization and 1988 was the first year we won the 500 with a Penske car designed by Bennett, so you can kind of see that history with Ensign and Mo Nunn obviously coming into Indycar. He [Morris] was a good friend, we raced against each other a lot, but never partnered. He would have been a great asset for any team."

Ensign Racing had brought a talent storm to CART. Morris Nunn was soon to engineer Emerson Fittipaldi at Patrick (and Ganassi) Racing in winning the 1989 Indianapolis 500 and CART title with Nigel Bennett's Penske PC-18, before later engineering Alex Zanardi and Juan Montoya.

CHAPTER 15

CHAPARRALS AND LONGHORNS

*Al Unser with the 1911 winning Marmon Wasp at the
100th running of the Indianapolis 500 in 2016 (Norm DeWitt)*

1978 had been a mixed bag for Al Unser and Chaparral Cars. As Jim Hall says, "The Lola that won the triple crown saw Al winning all the 500 mile races." The great irony of course being that, other than a single third-place finish early in the season at Ontario, there were no other podium finishes for the Triple Crown-winning Lola T500. Somehow, Al took the points battle to the wire and was narrowly beaten by Tom Sneva, whose nine podium finishes (without a single win) was just enough to bring the title back to Penske Racing.

Al was still running the first couple of races in 1979 with the previous year Lola T500 with some success, taking fourth at Phoenix, and sixth and third at Atlanta. The team then arrived at Indianapolis with the new and revolutionary Chaparral 2K, aka The Yellow Submarine.

Indianapolis in 1979 was the 500 where the breakaway CART teams had to obtain a court injunction to be allowed to compete in the race (due to restraint of trade issues). The Yellow Submarine had an obvious Formula 1 influence, but per Al Unser, that didn't extend to the details of the car. Unser describes it, "It didn't look like the Lotus, there was nothing that John [Barnard] took from the Lotus that I could tell. The man was a very smart man, an innovator."

The Chaparral 2K Yellow Submarine of 1979 is one of John Barnard's most famous designs (Norm DeWitt)

Rick Mears in the PC6 was the fastest car on track, but it wasn't long before the 2K hit its stride and was on the pace. After qualifying, Al Unser was on the outside of the front row, with Mears on the pole, and Tom Sneva alongside. Al led the first twenty laps of the race, and eighty-five of the first one hundred laps when it all went wrong on lap 103 and the car was retired with smoke and flame coming from the gearbox area, the victim of a broken gearbox line or fitting. Nevertheless, the Yellow Submarine had proven itself to be the class of the field under race conditions.

With a subsequent second at Trenton and a third at Michigan, the Chaparral seemed to be living up to its Indianapolis promise, but all was not well behind the scenes. Al explains,

> There was a lack of communication between John Barnard and Jim Hall; we just couldn't all get together. The car was a rocket and I guess I should have led every race, and I could say that I should have won everything. Where it ran really good was here at the Speedway, but at the other tracks we never could find the setup, until the last race at Phoenix. It just wasn't set up right, and I'd guess the underpans were flexing. Here I go... but they should have flexed more at the Speedway where the speeds were higher. If you understand ground effects, it doesn't make any difference once you reach a certain velocity of air passing, it's all the same. So if it was flexing on the short tracks at 150mph, it should have flexed here [at the Speedway] more at 225mph. So that's the trouble.

One has to understand that this was the infancy of ground effects in Indycar racing, and the learning curve was still vertical with everyone grasping to ascertain what worked and what didn't. It was also the era of the sliding skirt, which brought an entirely new set of variables into the mix. Although the Chaparral remained in the hunt for a while, the mid-season aero confusion had cost the team any shot at the title. For Johnny Rutherford it was much the same, and Team McLaren's early season pair

The Chaparral 2K ground effects tunnels were huge in comparison to 1979's Penske PC-7 (Norm DeWitt)

of wins at Atlanta were followed by only two more podium appearances, both third places. 1979 had become a Penske Racing steamroller with Rick Mears on top from his three wins and four second-place finishes. Bobby Unser was close behind on the strength of his six wins and two second-place finishes. Each of those Penske drivers had scored nearly twice the number of points as Al Unser in the Chaparral had, to win the inaugural CART championship for Penske Racing.

The Chaparral team had finally come to grips with the 2K in time to score a dominant win at the Phoenix finale, with Mears second and Bobby Unser third. All the hard work from that season's developments could well have paid off for Al Unser in 1980, yet at the end of the season, Unser left the team. Al recalls, "John Barnard quit, Hughie Absalom quit. Can Jim Hall beat the designer? I decided that there is a certain time in your career when you've learned what a team should be, or at least you think you have. One man can be the boss, yes, but one man cannot control all of it. Jim said, 'I can do it all.' Jim is a brilliant man, but he's got to have the personnel beneath him, and we didn't seem to have that."

Hughie Absalom gives his perspective, "Probably the main problem there was when Al and Jim had a bit of a falling out, so they lost a bit of continuity there. Al couldn't get any changes made, because they would have been Jim's changes. He would call Barnard up and try and get a

Johnny Rutherford confers with IMS Historian
Donald Davidson during pre-race festivities in 2016 (Norm DeWitt)

get a heads-up on what direction to go and John would be reluctant in doing that. They lost the continuity when John didn't continue with the program."

The great irony was that the Chaparral 2K was to have a stunning season in 1980 with Johnny Rutherford moving over from the disbanded Team McLaren in a classic case of right place, right time, right driver, and right team with the now-sorted Chaparral. Rutherford won the season opener at Ontario, then won again at the Indianapolis 500, from pole position. With seconds at both Milwaukee and Pocono, followed by two more wins at Mid-Ohio and Milwaukee, the season was a Johnny Rutherford and Chaparral runaway. At season's end, Bobby Unser was approximately 1,000 points behind, never having posed any serious threat to Rutherford's lead. I once asked Johnny Rutherford if he sent a thank-you card to Al Unser, who had turned the Chaparral into the weapon of choice, a weapon that Rutherford used to maximum effect.

Al could be forgiven for thinking that if the 2K was quick, just what would a currently dominant Cosworth-powered Formula 1 car be capable of? The Williams FW-07 was the fastest car in Formula 1 in 1979-80. One can certainly see why Al would jump at the opportunity to drive an FW-07 clone here in America's CART series. The resulting car was run by Bobby Hillin and it was called the Longhorn. Al Unser remembers:

Well, I got along with Bobby Hillin very well and he tried to get a team together. He thought and I thought that he had everything in place when I went there. It takes a while to learn who does what and if they are good at that job. We found that the designer that Bobby Hillin had, wasn't any good. You can't put the Williams team [into the equation] other than that Bobby Hillin paid for the rights to copy the Williams, period. Williams didn't control the team of Bobby Hillin, nor had they controlled any part of the running of the US team. It was a licensing agreement. People get carried away saying, 'It's a Williams.' Well, it was and it wasn't. Bobby bought the rights to the prints for the F1, but the designer who had been hired decided, 'That isn't the way to go, I'll do my own thing.' He built his own car, and the first Hillin car was no good. For two years we just dumped, and you can't imagine what that's like.

The cockpit of the Chaparral 2K
(Norm DeWitt)

Al's gamble with the Longhorn had rolled snake eyes. Bobby Unser concurs, "How wrong can you be? The initial plan was buying the original plans from Williams. The plans were all there for them to make it. It was Zink, as his interpretation turned out to be nothing like an FW-07. It was probably one of the very first honeycomb chassis that turned up at Indy." Ed Zink had been successful with his designs for Super-Vee and Formula Vee across the early-mid 1970s, the company turning into Citation Engineering, which then went on to success in Formula Ford. However, it was a big jump to go from those designs, to a full ground effects Indianapolis or Formula 1 car.

The Longhorn was known to be one of the most rigid cars ever raced at Indianapolis (pre-carbon era) and the moral of the story is that there is more to making a successful chassis than just creating a tub that is strong in torsion. The car certainly looked bulkier than the slender FW-07. It all went downhill from there as 1980 started with Al retiring the LR-01 in six of the first seven races, with their only podium coming as a third place finish in Mexico City. It was a complete disaster for all concerned. Al Unser tells of how the team sorted what had gone wrong at the end of 1980:

At the end of the season, Hillin and I took the prints of the car over to England and laid the prints out for Patrick Head [the Williams designer]. Patrick said, 'Oh no.' I said, 'What do you mean, Oh no?' Head said, 'I told him that his ideas wouldn't work.' I said, 'NO WAY!' I thought Bobby Hillin was going to pass out. I mean, he spent a lot of money. Regardless of what the figure was, it was a LOT of money for the rights to that. So there we sit. Then we took a Formula 1 car, a basic Formula 1 car, and brought it over here and started running it and it was so much better the next year. By then, every day the whole team was screwed up, we were replacing this guy, we were replacing that guy. Patrick Head came over to Michigan to see what was wrong with the car. It was pulsing going down the straightaway. We still don't know what it was.

Porpoising happened a lot in those early days of ground effect cars. Al agrees, "That's right, and they could probably cure it today, but in those days we couldn't and didn't know what was wrong with the car. The whole Hillin team was not happening right, we didn't have the right people in the right spots at the time."

Hughie Absalom says, "I came along when we actually bought one of the F1 cars. We did make two more chassis in America. But we still didn't get a handle on the aero package because the downforce on the Formula 1 car was much more towards the front. And that's what we were fighting for nearly half of the season was trying to move the center of pressure of the downforce backwards and reshaping all the under-bodies. Especially on the Speedways, it's the same old problem. You get so much downforce on the front that makes the thing want to be a little bit loose."

So you add wing to balance the car, which makes the car really 'draggy', the problem faced by Mario with the Lotus 78 previously in Formula 1. Hughie says, "Correct, so we kept working on the underbody. We were actually making some really good gains and then they went and hired Jon Ward, and he was one of the instigators of the Eagle with the BLAT system. So, he followed up, eliminating the Williams theory of underbody and followed the BLAT system, which was how the car was built for next year. So that kind of cancelled out the FW-07 when Jon turned up." However for 1981, Jon Ward stayed with the basic FW-07 aero package. Hughie laughs, "Under protest, because his philosophy was the BLAT and he didn't believe that the Williams interpretation was the best way to go. Jon Ward had joined after Indy sometime. We were pushing s**t uphill for a while there."

At Riverside in 1981 the Longhorn was running very competitively in the race, which one might expect given that this was a world-championship-winning Formula 1 car with Al Unser driving it on a road course. There doesn't seem to be a weak link in that combination, and it was very fast. Hughie says, "We were running second and I think

Al Unser with the 1981 Longhorn at Indianapolis.
Williams' Patrick Head stands alongside (Russ Thompson)

[Geoff] Brabham was leading in the Eagle. We broke an oil line to the transmission in that race."

The end result of the 1981 season was just as bad as the 1980 season had been, Al dropping from eighth in points (1980) to tenth in points. However there was an end of the season run that followed their strong outing at Riverside, Al taking third at Michigan and second in Mexico City, which at least gave the team hope that they had things turned around and headed in the right direction. However, for 1982, Hughie Absalom had moved on to greener pastures. Hughie recalls, "I was with Forsythe just that one year, it's what you feel comfortable with. Things change and if I'm not comfortable, I'm down the road, either before or after I get fired. I'd try to stay a step ahead." In this case Hughie had read the tea leaves correctly and separated from Longhorn Racing instead of enduring another frustrating season of modified FW-07s.

In 1982 the team again struggled for podium finishes, taking a best of second at Road America, with a third at Cleveland. Hughie says, "They should have won at Elkhart Lake in 1982, which was the closest that a Longhorn ever came to winning a race. As it turned out, by then I was with Forsythe and my car won the race, with Hector Rebaque. The Unser-Absalom connections continued as it was a Forsythe car in which Al Unser Jr. made his CART debut at Riverside in 1982.

To this day, the back of the business cards used by the Williams Formula 1 team have the outline of the FW-07B. Al Unser laughs, "I'll be darned." It is safe to assume that there is not a Longhorn on the back of the Al Unser business card. Al Unser confirms, "No there isn't. I'm still very good friends with Bobby Hillin today, and the guy put the effort towards it, but he didn't have the marbles in the right slots." This saga is proof that you can have plans to the greatest car in the world, but if you don't have the right people in the right jobs, you are screwed.

For the following year, Al Unser joined up with Penske Racing and showed that he had lost none of his speed, finishing in the top three at the first six races, including a win at Cleveland. By the end of the season at Phoenix, Al's fourth-place finish was just enough to hold off a flying Teo Fabi (winner at four of the last seven races) to become CART Champion for 1983. Al Unser was back.

*Al Unser came back in 1983 to win the title
with the problematic Penske PC11 (John Mensinger)*

CHAPTER 16

LITTLE AL AND THE BLAT EAGLE

Al Unser Jr. in the road course configuration Eagle at
Riverside in 1983 (Frank Sheffield)

Al Unser Jr. had come up through the ranks, becoming Super Vee champion in 1981 driving a Ralt RT-5, followed by a big step up to the Can-Am series. It was the twilight of the single seat Can-Am era, yet still with a handful of great cars. Trevor Harris had waved the magic design wand over a Frissbee for 1982 and created the Galles GR3, updated in many ways over the standard Frissbee. If you wanted to win the Can-Am series, your car had better have Trevor Harris suspension on it.

Trevor Harris discusses the evolution of his designs:

In 1981 I was designing the rear suspension for the VDS, it was very much like the Galles car. There were parts that could interchange, and that car won with Geoff Brabham in 1981 and again in 1984. It won the years when the GR3 didn't. I, Joey Cavalieri, and a guy named Jessie Suzuka were responsible for the first Frissbee, and that was a development of a 332 Lola, a Can-Am variant. We had one of the original Frissbees on which I completely re-did the front suspension and rear suspension with steel uprights and different suspension geometry. The monocoque was basically the same and the way the engine mounted was the same. The GR3 bodywork was different with a number of detail improvements that gave more downforce. Instead of being a slab side on the original car, we did a radius on the side of the car so the airflow was hopefully better, but we never knew for sure as we never wind-tunnel tested it. A lot of it was guesswork in those days, with tufts and oil drops, and we would check where the airflow was going. We were doing that, along with seat-of-the pants testing. We also did pressure testing under the car with very sensitive magnehelic air pressure gauges so we could kind of do some comparative work. We built one brand new car, the car that Al won the championship with in 1982 and Jacques Villeneuve [Gilles' brother] then won with it in 1983.

Trevor initially thought a re-bodied Formula 1 car was the route to Can-Am success. Trevor explains, "Gary Gove from Seattle was in the converted F1 car based upon an FW-07 Williams back in 1980 or '81. He was a pretty quick driver, and he was a key guy for Pete Lovely's outfit up in Seattle. It ran a limited number of races, but Gary had a bad crash and was injured in the car and that was the end of it. The thought at one time was, 'That car was going to destroy everything,' but it really didn't. Converted F1 cars really didn't outperform a reasonably well done full-body car."

There was some family history on the Unser side with the Frissbee Can-Am cars as well. Al Jr. says, "Trevor was the engineer on my Can-Am car in 1982. He redesigned the front suspension of the Frissbee with a pullrod system and that was when it got real fast. In 1980 Dad won at Laguna with a Frissbee for Brad Frisselle at Laguna."

In 1982, Little Al also had his debut at Riverside in a March 82C, and passed his Dad in the early laps on his way to scoring a fifth place finish for Forsythe. Al Jr. "That wasn't at Indy though, it took me a while at Indy." This is certainly true, as Al Unser (Sr.) was still at the top of his game, and was to win the CART title in two of the following three years. However it was a promising debut and for 1983, Galles was to step up into the CART series bringing Al Unser Jr. with them.

For 1983 the Galles team was also the factory AAR race team, running the latest BLAT Eagle (see *Making it FASTER* for the story of the '81 BLAT Eagle). The Eagle was a unique car that was quite unlike the normal sidepod tunnel ground effects car, and this one was also different from both previous and subsequent BLAT Eagles in having a Cosworth engine. For Al Jr., it wasn't just trying to adjust to a unique car, it was trying to sort out everything else as well. He recounts, "It took me a while to figure out Indycar racing. I think that was the biggest thing. Trevor did a fantastic job with the car, it was competitive, and it was strong. It didn't have a lot of downforce to it, but it didn't have a lot

The rear view of the 1981 Eagle with the lower suspension arms interrupting the airflow (Norm DeWitt)

*The rear underbody extension of the Trevor Harris modified
1983 Eagle (Alex Lond – Can-Am Cars, Ltd)*

of drag to it. Once we figured that out, that was when the car got quick. We stopped making mistakes like running it out of fuel and s**t like that."

Trevor Harris tells about the specific technical improvements made to the BLAT Eagle for the Galles effort:

> I was the designer at All American Racers in 1983, and that car was a serious update, or at least I regarded it as that, over the previous Eagle. I had worked with All American Racers running the previous BLAT eagle that Jon Ward had designed. I thought the front suspension was okay on that car and actually the original rear suspension was okay too, but the only catch was that the lower 'A' frame was under the underwing in the back and upset the airflow there. So I did an entirely different rear suspension on that car and put the lower 'A' frame above the underbody so you had a much cleaner airflow at the back of the car. There were several other improvements and there was work on the monocoque, which stiffened it up. We ran a Cosworth in it, and never had a Chevrolet.

Before taking the new Eagle to Atlanta for its debut race, the team had been testing at Phoenix. Al Unser Jr. recalls, "We had problems with the brakes pulling to the left. He couldn't understand what I was trying to tell him, and so I said, 'Well, I'll just give you a ride and show you.' Trevor said, 'Okay.' [laughs] So I got going and put on the brakes and it pulled to the left, he almost fell right off, I mean he came so close. We were running 20-25 mph and he was riding it like a bull. Foolish, that was what it was all about."

Team Manager Hughie Absalom is still equal parts amused and stunned. Hughie asks, "It's a bit crazy isn't it?" Trevor Harris confirms:

That is correct, it was at a very early test at Phoenix. We had a problem and the car was pulling in an odd way no matter what I did. So I thought, 'Maybe I can see something.' I couldn't find anything odd going on, and it turned out that we weren't the only ones with the problem. Goodyear had made some sort of error and other teams were having the same. We never put it together, it was some sort of construction problem. Years later running Firestones in the IRL series at Mid-Ohio, there was a race where a very similar problem came up. It was the same sort of situation where we tore our hair out trying to figure out what in the world was wrong with the car, and it turned out that again, it was a tire construction problem. There were other people doing the same thing, we weren't the only team fighting

Trevor Harris wins the 'fearless engineer' award for his riding atop Little Al's Eagle at Phoenix (Galles/Joe Cavaglieri)

the problem. I've seen it twice in my life. We checked the mechanicals on the car, tore the car apart, went crazy, and wasted a whole day.

The new Eagle had advantages over the competition. Trevor continues, "The entry to the tunnels underneath were so much bigger than your regular square entrance to a tunnel venturi. With the tunnels angled down and the way the air came in and rolled underneath, that is where the downforce came from. It was the vortices and how the wind rolled underneath. It was a great car."

1983 marked Little Al's debut at Indianapolis. The Eagle qualified in the middle of the second row, between Tom Sneva and Bobby Rahal, both in March 83Cs. In the race, Al Jr. had dropped to five laps behind, and with twenty-four laps to go, un-lapped himself at the restart. Letting his father through (it was Dad's birthday, after all). Al Jr. then drove a fast but wide Eagle for lap after lap, holding off the surging Sneva. After a few laps, the blue flag was waved at Unser Jr., who ignored it. After being stuck behind Junior for fourteen laps, Sneva managed to outfumble him in traffic and was finally past and on his way to eventual victory.

Al Jr. recounts, "I tried to make it very difficult for Tom to pass me. But Dad wouldn't go, he was fast all day, but then all of a sudden he was slow. That is what caused that whole thing because he should have checked out. They got a bad set of tires and his handling went away." On the TV coverage the commentators were saying, 'Al Jr. is just a rookie, he probably doesn't know what he is doing.' But Junior says, "I knew exactly what I was doing. If I was blocking him, he would have never gotten by me."

Post race, after being found innocent of intentional blocking, Al Jr. was eventually deemed to have jumped the re-start on lap 176, and was handed a two lap penalty. His finishing position remained unchanged, which was tenth place. It was a memorable debut. Hughie Absalom remembers:

I spent two years at Galles in 1983-84. We ran Junior in his very first Indycar race at Riverside [1982], when it was a Forsythe car. Galles ran the modified BLAT Eagle and it was good, Trevor Harris did a good job. I think what happened was that Galles had his 'spy' if you'd like. He had some big goals to meet, he had made his commitment. They had done very well in the Can-Am with the Frissbee, and coming into Indycar then they did okay, we did pretty good at Indy. After that it was a bit flat, but it was up and down, so then we bought the March. I couldn't tell you what the exact reason was, whether it was Junior's inexperience, but that probably was the most likely thing. Galles wanted results and he had the cubic bucks at that time to try and make whatever had to be done, done!

As a result, the team bought a March. Trevor Harris was of the opinion that Absalom had been behind that decision. Trevor says, "Hughie Absalom engineered my demise. He managed to convince Galles that the Eagle didn't work and that we really needed to have a March as some sort of backup. So Galles bought this March. It wasn't the matter of the March being a bad car, it was that we were a very small team trying to have spare parts. As for me, I was NOT eager to engineer a March. I'd worked on Marches before, but I was the Eagle designer having to set up a March when the Eagle was running well."

The Galles team had run the March in the race after Indianapolis, at Milwaukee. Starting fourteenth and finishing thirteenth, it certainly wasn't much of a successful experiment. For Cleveland, it was back into the Eagle. Trevor continues, "We were on the pole at Cleveland until the last minute when Mario got us. So, that wasn't too bad for a brand new car that had never run anywhere, considering that Cleveland was our first slow course." Little Al had finished ninth, but the speed on the flat airport circuit was undeniable. Next on the schedule was the

Al Junior in the Blat Eagle from the inside of Turn 6 at Riverside
(Frank Sheffield)

Superspeedway at Michigan, and again they ran the Eagle.

Al Jr. recalls, "In 1983, it was a fast car all year. We were super competitive at Michigan during the race, we were the quickest car out there, but I ran out of fuel twice. My team was a rookie team so we made rookie mistakes." The team eventually finished seventh at Michigan. That was the end of the road for Trevor Harris. Trevor explains, "It was a pretty effective car. Unfortunately I did not stay with the car all year, I got fired by Galles, the owner. It was a very interesting political situation that came up with somebody who thought my job was his. Unfortunately Little Al and Galles agreed with it, and after the Michigan race, I was out."

The next race was at Road America, which was a track that would likely suit the Eagle, and it did. In fact, Al Unser Jr. finished second to Mario Andretti after having led six laps of the race. The Eagle had landed, and one would think this performance would have guaranteed its place on the team versus the lackluster performance of the March in Milwaukee. However, that was not to be the case, and for the next race, Al Jr. was again back in the March for the superspeedway at Pocono. This time running the March proved to be highly successful, Al Jr. qualifying sixth and finishing second, led twenty-eight laps, and equaled his best performance of the season in the Eagle.

Al Jr. claims, "We were having trouble with the Eagle, and March was the car to have in '83, so Rick Galles bought a March. I raced the March and after the race when we got back to Albuquerque I said, 'I'm done with the March, the Eagle is a faster car than the March,' and so we finished the season in the Eagle. We would have been the quickest car at Indy if we would have known what we had, and we'd known what we were doing. It was just a rookie driver with a rookie team. If veterans had been in that car..."

Trevor Harris remembers:

It wasn't that the March was a bad car, it's just that if you are March, you don't go out and buy an Eagle because your March might not work. I was totally against it. Hughie was the Team Manager at that point. Joey Cavalieri, who was our Team Manager on the Can-Am car, got fired. It's a long, kind of unfortunate story. Even though the Eagle had run well on short courses and fast courses, Hughie's plan had us running the March at Milwaukee, and remember, this is the Eagle Factory

Team running a March, which is just unbelievable. The interesting thing was that Junior ran the Eagle very briefly. My recollection was that the time he ran with the Eagle was faster than he ever did with the March, but he raced with the March. He didn't qualify well, it was not a good weekend.

I respected Hughie as being a reasonable mechanic and a reasonable team manager, but the only thing was that he had a whole different scheme in mind for what he was going to do and how he was going to operate. His scheme and my scheme were not the same at all. It was unfortunate, but Al Jr. apologized for being part of the firing squad for me, and in later years Galles was trying to hire me in '95-'96, but I was busy and didn't really want to go through that again. It believe it took Al about five years to qualify at Indy as high as he had with the Eagle.

Side view of the standard road course configuration
1983 Eagle as run at Riverside (John Mensinger)

Actually, 1994 was the first year Al Jr. qualified any higher on the grid than he had in 1983 with the Eagle, taking pole position with the stock block Mercedes for Penske. Trevor adds, "It's a funny thing, Al Junior got a reputation for being a slow qualifier and as far as I could see, that was never the case. I don't think that was a fair commentary. If we got the car right, Junior was really quick everywhere we went. Al did a really good job with that car, but they ran the wrong configuration of the Eagle much of that season."

Countless times, Roger Penske would purchase a March if the latest Penske car didn't work out, but it helps if you've got a Penske-sized budget. Hughie reflects, "Basically that's what Galles did. You do whatever it takes to run at the front. That was what Galles did to pacify the sponsor and Junior, he went and bought a March. We were the only Eagle with the Cosworth in it. So that's how it all goes sometime, when you have to chase the front. Once you get caught in the middle of a situation like that it is hard. They were two completely different cars and one doesn't respond the same way as the other one does." For a small team, going in multiple directions at once is normally not the key to success.

For the next race at Riverside, again it was musical cars and Al Jr. was back in the Eagle, qualifying on the front row alongside pole sitter Teo Fabi. Little Al led more laps of the race than anyone else (twenty-seven) but in the end came up a little short, finishing fourth. There was little doubt at this point that the Eagle was as sharp of a weapon as the best March on the road courses. At Mid-Ohio, Al Jr. qualified the Eagle sixth, leading fourteen laps, but was out at half-distance with engine failure. Michigan brought a tenth-place finish, and the Eagle never was in contention.

One would have expected better things at Caesar's Palace in the twisty parking lot next to the casino, but again the Eagle was off the pace. Fabi then dominated the race at Laguna Seca, but Al Jr. did finish fourth, equaling his best finish since shelving the March after Pocono. After an eighth-place finish in the finale at Phoenix, and having scored seventh in the championship, it was the end of the Eagle program at Galles.

The team tried running the superspeedway rear
bodywork extension at Laguna Seca (Frank Sheffield)

Hughie says, "For 1984 we gave back the Eagles and decided to go with the March and we had two Marches from the beginning. At the end of the 1984 season I left." However, it was a better season for Al Jr. and Galles, as they got their first win at Portland, along with two other podium finishes (second and third) versus their achieving two second-place finishes in 1983. For 1985, Al Unser Jr. left for Shierson Racing, and eventually racked up a tally of thirty-four wins, two Indianapolis 500 victories ('92 and '94), and returned to Galles to win his first CART title in 1990. His second CART championship was with Penske Racing in 1994. His six wins in the streets of Southern California (would have been seven if he hadn't been punted out of the lead by his teammate Sullivan while leading near the end of the race in 1992), means that Al Unser Jr. will forever wear the crown "The King of Long Beach."

CHAPTER 17

THE ACTIVE GRAND PRIX CARS

The active front suspension system of the 1992 FW14B
(Norm DeWitt)

If there was ever a time when active suspension could have been useful, it would have been the 1983 Long Beach Grand Prix. There was a point when the cars hit a huge bump near the end of the back straight, which the hydro-pneumatic Ligier JS-21s floated across in comparison to most. That suspension system had been pioneered at Ligier by Gerard Ducarouge in 1981, a glimpse of active things to come from him when he joined Lotus. The Lotus team were pioneers in active suspension with the Camel Lotus 99T of Ayrton Senna designed by Ducarouge for the 1987 season. It was bulky and heavy, reportedly over fifty pounds for the installed system. The car won in Monaco and Detroit, showing some superiority as a street fighter. But much of the reason for their perseverance with the concept was Senna's desperate quest to find any technological advantage over superior cars such as the Williams-Honda or the McLaren-TAG. Ayrton in his pursuit of active suspension was ahead of his time, however by 1988 and the arrival of Nelson Piquet, the Lotus team had moved back to conventional suspension with the model 100T. Benetton and Williams were to pick up the torch.

Dickie Stanford was a mechanic at Williams when they made their first active–or 'reactive'– car. Dickie says, "With our first system we were controlling it wrong. Halfway through the year we ran the 'reactive' ride for a while, and used an American company to control the electronics. This was just after the Lotus, and we called it reactive ride because somebody owned the patent to 'active ride' suspension." Frank Dernie's FW11B of 1987 was built in both conventional and 'reactive' versions, and the Monza victory was the first and only win for the Williams-Honda reactive car, as well as being Nelson's last victory for the team. The tub of that race-winning FW11B currently sits on display at the Williams factory museum.

Benetton had begun their active program, working hard upon what they saw as an opportunity. Pat Fry was an engineer at Benetton during 1988-92, often in the R&D department for the team. Fry recalls, "We started developing with the '88 car, and although it didn't actually race, we did loads and loads of testing. When I got the numbers wrong on the Benetton active car in 1989-90, it was going quite quick at the time, but had the skirts sealing the front of the end plates to the ground and we were wearing all the end plates away up to the main plane. It was a relatively small team for the time with a limited budget so it was the middle of the year ['89] by the time we were trying to put active car parts onto the race car."

Pat Symonds was another engineer for the team, having come back to Benetton from the Reynard Formula 1 program that never raced (along with designer Rory Byrne) of 1991. He had been part of the team as far back as the Toleman days of the early 1980s, and was director of R&D at Benetton by 1993. Symonds says, "We were running an active Benetton back in the late 80s, in 1989, doing a lot of testing but not racing it. Then John Barnard came on board and he didn't like it at all, he said to stop it. The 1990 car was John Barnard's, the later high-nose cars were Rory [Byrne]. During our time at Reynard we actually learned an awful lot about the active system. We did use springs, they were pretty soft. The thing is that if you use a passive spring it actually helps, you reduce the bottoming of the car." The drivers saw the passive spring as having different advantages. As Benetton test driver Allan McNish explains, "I think there was a little bit of a spring in there (on the Benetton) for safety, an 'Oh s**t' spring in there."

Symonds continues, "Of course if you go the way Lotus did with their high-band width system [1987], you use an awful lot of power. Our system used about 7 horsepower whereas Lotus probably used four times that. So why use all that power to do the things that needed doing? We had all our clever electronics for weight distribution control, but took the heavy lifting out of it with the passive springs. McLaren never went quite as far. We were very much between the McLaren system and the Lotus system from what I understand."

By 1991, Williams-Renault had begun to challenge the all-conquering McLaren-Honda, clearly having the fastest pace by mid-season. While the team was racing this passive car to great success, Paddy Lowe was heading up the active program at Williams, and this was the turning point where the active suspension program began to match or exceed the capabilities of their passive car. Paddy Lowe recounts, "You've got to bear in mind that in those days we had very little data compared to now. So a lot of things we learned by accident or by some sort of observation. When we developed the Williams active car, and it was developed over a number of years."

Lowe confirms, "Williams had been racing the FW14A, which was quite a good car actually, the first car with the semi-auto box and Adrian's aerodynamics. We should have won the Championship, but didn't for a number of annoying reasons. In parallel, I was running the active car and we had put together a prototype toward the end of 1991. We did a number of tests and the whole point was to bring them along, do back-to-back tests, and be sure that the active car was quicker. Bear in mind that we didn't have the simulations to prove that, 'Yeah, that's what we want to do.' The active car required an entirely different way of engineering it. There were no springs, dampers, roll bars in the conventional way. It was difficult to do that and to be fully confident."

Undoubtedly it required a certain leap of faith for all concerned, both on the engineering side and in the cockpit. Paddy recounts:

We went to test at Paul Ricard and one thing we noticed when running late one evening. Renault had a garage around back straight, just before the corner. You could see all these sparks on the passive car and not on the active car. And then we were looking at data and the active car was slower at the end of the straight, but quicker on the lap. When we put all of these things together we realized the passive car would run all the way down onto the bump stops and skids, and the floor was completely stalled and that gave it more straight-line speed. The active car was constantly compensating, keeping the car at the perfect ride height for downforce, but not for drag. So, next the driver had a button for a low-drag setting, where we would put the front up and the back down. We probably achieved more drag reduction with the active car than the passive car in the end.

Of course, the driver needed to know what to push and when to maximize the effect. Paddy continues:

There was a problem you had, and this was very similar with DRS. The problem with Nigel, he is such an enthusiast, if you gave Nigel any button that made the car quicker, he would think, 'The more I use it, the quicker I go.' So you'd have to educate him that you wouldn't go quicker if you held the button on longer. He only discovered this the hard way at Estoril where there is a long straight and there is a sharp left hander at the end where he went into the gravel there one day. He had held the low drag all the way into the braking zone. The point is that you let go the button BEFORE you hit the brakes, not after. We saw it with Lewis in the Japanese Grand Prix just gone [2014], in the race when he almost lost it in turn one. He only just collected it due to massive skill when he left the DRS into the braking zone.

Allan McNish explains, "One thing about the active car was the increased forces on the [driver's] body. It worked the downforce in a different way and the high-speed G was a little bit higher. Physically on your body you were just hanging on. That's where Michael Schumacher came into his own, as he took his training to a different level, to a higher capability. Nigel was built like a brick s**thouse, and he just muscled the car around. His style worked for the active, because it was just fire it into the corner, drive it through, and I'll sort it out no matter what happens, and that worked. He was able to do that longer over the course of a race than Riccardo Patrese."

The FW14B at the Williams factory museum
(Norm DeWitt)

There was more to that story, as Patrese found to his peril. With the Williams system, if the system failed, there were no 'Oh s**t' springs to keep the tub from instantaneously becoming a carbon toboggan. Dickie Stanford tells it, "When they were testing this in early 1992 there was a glitch in the software, and Riccardo was driving it at Imola. On the straight it picked all four wheels up off the ground and it was just on its belly and did about six rotations. Obviously we cured that problem pretty quick. Patrese didn't really trust the car after that and it really dented his faith in the car. Whereas Nigel just put his faith in the system, but people forget that Nigel did all the development work at Lotus with the original active car, so when he drove this he knew what the car should feel like. From what I remember there is less feel with this system."

Allan McNish confirms that the lack-of-feel issue was also very much the case for the active Benetton:

You have to have a leap of faith with the downforce, because it had an unknown sensitivity. It wasn't a mechanical system like you'd expect from a spring and damper and the feedback systems from it. It was anesthetized in terms of the information the car was giving you. It was either grip or no grip, it was going around the corner or not. I didn't have any major failures with that, like I did with the four-wheel steer where we had a technical issue at one point, where the rear wheels were doing whatever they wanted to do, and the front wheels were not exactly in connection with the rear, but thankfully I survived that. It's motorsport, and that was an era of pushing the boundaries without limitations. Then the limitations came in when they banned it all. Now that I'm thinking about it, I've been very fortunate to come through that era. I did forty-five days of testing with the Benetton. They were good and it was also Williams' high point. If you did an active system now, it would be absolutely blinding.

Dickie Stanford remembers, "When Paddy was here, we had this very similar system to this on the FW13, which was the test car they used to take off about twice a month. Mark Blundell did the majority

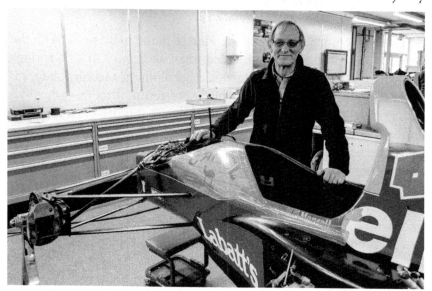

Dickie Stanford assembling a FW14B for the Williams Grand Prix Engineering fortieth anniversary (Norm DeWitt)

of the test-driving at little cold circuit in Wales called Pembrey. He must have done thousands of laps there. The major improvements were the electronics, the control systems. Paddy would work with a chap named Steve Wise, who is still here and is the head of our electronics."

As with most major advances on the technical side, the active car exposed a loophole in the regulations. Paddy Lowe explains:

I don't think there was any legality concern around it. We did actually get into a little bit of trickiness at one point with the way that they measure cars. Conventionally with the passive car, with the car stopped they measure the wing height, overhang and everything. The rules don't spell out that these rules apply when the car is stationary at static in the garage on tires at normal pressure. In those days the rules were measured to the ground. Nowadays they've done it all different, where they've done it in relation to the plank, datum points on the sprung chassis. When it came to an active car, it came to a whole different question such as, 'Hang on, this car can move around in the garage and what is the legitimate wing height?' If you have low drag settings, is the rear wing in a legal position? There was nothing to stop you, and it just exposed a weakness in the regulations.

The FW15 got put off a year (to 1993) and the '92 car was the 14B. Dickie says, "The difference between the 14A and the 14B is that we had to have these little panels just to cover the actuators. If you look at the 14A that is more straight down there along the nose as it has got shock absorbers. The weight of the cars is about the same. We went onto a weight-saving exercise because before the 14B was heavier, but we got it back to about the same. We still have the FW13 mule test car in storage."

In the US, the electronic advances through 1992 were primarily concerning traction control. Penske Team Manager Chuck Sprague explains what they were using:

That was a 'fox in the hen house' kind of thing for a lot of the teams, but it was perfectly legal then in '92. We had it for the last half of the season and developed it with Ilmor and Chevrolet. In our case it was extremely simple to implement, as we were the only team there with a transverse gearbox. At the back of the car you had this spur gear and all we did was drill and tap a hole into the back of the gearbox for a hall sensor that could count the teeth on the spur gear and then we knew the average wheel speed. In the front you had normal wheel-speed sensors. Once it got to be a certain percentage faster, it starting cutting cylinders out. It was very rudimentary, it wasn't based upon speed at all, it was just based upon wheel-spin percentage. I think we used 2.5% where it started to engage and by the time it reached 4 or 5%, you were only firing every third cylinder.

When we first raced it at Mid-Ohio, some of the teams commented on it being a 'funny sounding rev limiter.' We wanted to control wheel spin, as we could then run an open diff and Mid-Ohio rewards running that, but if you keep spinning that inside rear wheel, the tire is going to go away. Emerson being Emerson with all those years of experience, he could drive around there all day right on the edge of the system, at like 99.999% of engagement. When he did tip into the system, it was only a little bit. Guys with that much experience get it, they feel it. Tracy would just put his foot to the floor and let the car do all the work. Because it didn't cut fuel, it only cut spark, he'd be getting about 1.4mpg with raw fuel dumping out the rear of the car. Everybody agreed to ban traction control for 1993, but Nigel Mansell insisted we still had it as he said there were chatter marks in our tires when we pulled out of the pits, so that turned into a big pissing fight.

For 1993 Penske Racing responded to the ban on electronic systems in CART with a mechanical active suspension system that substituted mechanical bits for the electronics used in Grand Prix. Sprague explains:

It was just a hydraulic valve with an arm on it that read suspension position. It was just old school and instead of using electronics there was a hydraulic platform under each spring. The position of that was determined as it tried to compress the car down and the length of the shock changed, that platform would raise up to control the ride height. As the car was raised up, it would lower the hydraulic platform to keep the car at a certain overall position by adjusting those four hydraulic collars.

The car was assembled in the shop and the first time I saw it we were loading the car for its first test at Sebring. The engineer who was working on it was new to the team, and a large proportion of the fittings in the system were aluminum. I asked him if he checked the pressure ratings on those fittings. He told me that the fittings were good to 3,500 pounds and I told him, 'I don't think so, they were probably more like 300 pounds and weren't designed for hydraulic pressure.' He had looked at the catalog, but read it incorrectly. Had we run it like that, all the fittings would have broken, and that would have been a major crisis. We had to go, so we placed an emergency order for a boatload of stainless steel fittings and had to change every single one of them out. There were a lot of elbows and adaptors, but there was one fitting we didn't have and in the test that one broke.

The idea was to maintain the ideal ride height for the aerodynamics, and aero trumped all. Sprague continues:

If you make a car that is particularly good at a given ride height, but has high sensitivity to pitch, then you will be compromising. That is why everybody was running such stiff springs, because there was so much downforce available that the loss of mechanical grip from running such a stiff spring was more than overcome by the aero. With the PC10 as we went through the '82 season, we found that it was all about the ride height. Bump rubbers were part of the equation, but overall suspension stiffness was lesson one in ground effects. With the PC9 in '80 we went from having a truckfull of 300- to 600-pound springs to having 1,200- to 1,500-pound springs. By the time the whole thing was over we were at 3,500- to 4,000-pound springs. That was when it dawned on us that, although the electronics used in Formula 1 were illegal in CART, it could be done mechanically. Which brings us to the next evolution, which was to control the thing based upon speed in addition to ride height. Then the braking balance could be adjusted based upon the loads at each pushrod. You don't even need a track map [like most Grand Prix car systems], as the car can get lost if it spins.

Later on when we were testing at Firebird, we had the thing

come in on fire once when a fitting did fail or we blew a hose apart. The car comes in trailing black smoke, and we didn't even let the driver bring it into the tent, as we didn't want to burn the tent down. We got the fire out pretty quickly, but it was just a big mess. It was going to be a giant project if it was going to be ready to race. It's another thing to go wrong, and if it fails, what is your fail-safe mode?

The Penske mechanical active system did have secondary springs, much as they used at Benetton and McLaren with their active cars. Mechanical active suspension quickly became CART's version of the Brabham Fan Car controversy of 1978. Chuck Sprague recalls:

The program came to an end politically. Other teams got word of it and it came up at the CART board meetings. Barry Green said, 'Well, we can all go down this road, we will spend all this money and develop it, and all end up right back to where we are on an even footing. So what are we going to do?' What the other teams didn't realize was that we saw it as coming to a dead end because you could not get the resolution that you wanted from these mechanical sensors, as they were not sensitive enough. We also had to drag around this huge supporting hydraulic unit that looked like a five-kilowatt portable generator. It was a gasoline engine with a hydraulic pump and a reservoir so you could set up the car without having to start the engine, which had a hydraulic pump. We would also chase this thing around on the setup pad, playing with the car and applying load to it. Each one of these parts had enough play and enough random motion and stickiness to them that the weight distribution would change dramatically as the car tried to respond. You'd see changes in cross weight on the order of 150 pounds, which was huge. You could move anything a fair amount, move the ride height 20, 30, 50 thou and it wouldn't respond. Then the sensor would snap and overreact. The system was certainly promising, but in terms of being consistent and being something you were comfortable putting on the racetrack and turning a guy loose in it all day long? We were a long ways from that.

At Team Lotus, there was no current active development program such as that which was proceeding furiously at Williams. Mika Hakkinen arrived at the team in 1991. He recounts:

> When I joined Team Lotus it was really interesting. In 1991 they showed me some data from Ayrton when they were running active suspension at Monaco, how it was working and what it was doing. It was fascinating, unbelievable already at that time. Obviously at that time the aerodynamics were not that developed, so the active suspension never could give the great and best benefits. Center of gravity is always important in a racing car, so active suspension was able to control that a lot. Also, of course, we could control the stiffness of the front, stiffness of the rear, move the roll bars differently at low speed and high speed. Lotus did that already with Ayrton, but when I started running with Lotus in 1991-92, the active suspension was not being used. When I came to McLaren I was aware of active suspension, and I had seen some data, but I couldn't bring anything to McLaren to help them develop theirs.

Pat Fry had left Benetton and was in his first year as an engineer at McLaren in 1993. Pat says, "It was a ride-height controlled car, rather than a full active-ride system. At the start of the year it was purely just front and rear ride-height control. The aerodynamics on that car were not that great for a passive car because the diffuser stole quite a lot. Early on, the Benetton had a lot more sensible rear ride height at that stage. However, when you go nose up, rear down to stall the floor, the McLaren would gain a huge amount of speed, 7 or 8 maybe 10 kph doing that sort of thing because our diffuser lost more. Benetton had a diffuser designed with a much greater range as their passive car was much better, so when they tried to stall it they only gained 1 or 2 kph."

Fry continues, "The McLaren system was fairly simple in that it was driven by lap distance. We had a pretty good model of calculating where you were around the lap and effectively change the front and rear ride heights, eventually also the roll function in the end. We didn't have full roll control until about one-third of the way into the season. The driver would say that it was understeering in this corner. I'd ask at what point, so I'd lower the front ride height a little. The aero platform side of it was so easy to control. As soon as you knew where you were not limited for grip, you would instantly go into low drag mode, lift the front and drag the tail, and drop it right before the point of braking. The main thing came from sorting the rear." Obviously, dragging the diffuser on the track wasn't a great idea. Pat agress, "Well yeah, if you got the numbers wrong it would do that."

Mike Andretti had come across from CART to Grand Prix to drive the McLaren alongside Ayrton Senna. Mike explains that he had never driven an active car before:

> That was the first active car that McLaren had as well, their first and only year before they banned it. Feel-wise it wasn't anything different, but in setup you could make the car do whatever you wanted in terms of changes through the corner, which was really interesting. You'd say, 'Well, in this kind of a corner I need the front to work a little bit better after this point in the corner,' and you could have it move cross-weight or lower it. You could have it do what you want, and the setup for every corner was different. In a fast corner you needed it to do this, or in a slow corner you needed it to do that. So you would be able to adjust it, and the interesting thing was that debriefs were for hours as you literally would dissect the corner and go over exactly what you wanted the car to be doing. It was fun from a driver's standpoint, as you'd say, 'I want this here,' and you get it there. In a [passive] setup it's all compromise.

Setting up the car was a unique experience. Andretti continues, "We would start from braking, such as, 'Well, I'm locking the rears, or bottoming while braking.' So you'd get the ride height higher when you first started braking. Then when you got past a certain point you could drop the front or drop the rear, and then as you turned in it would be a little too pointy turning in, so you'd keep the rear lower to make it more stable. Or you could do left front weight and as you started to turn in and understeer you could start to get right front weight. There was so many of those things you could do."

Pat Fry recalls the challenges of the day. "Mike Andretti... he sort of struggled as the whole concept of it was quite a bit different from what he was used to, and we were all learning at the same time to be honest. All the drivers there are massively talented and on top of their game." The McLaren could not automatically keep the proper heights to maintain a certain optimum aerodynamic profile, so it was an exhaustive process.

Andretti concurs, "You had to manually tell it where and what you wanted it to do. The debriefs were mega-long as you had to dissect every part of the track and tell it what you wanted it to do. Me and Ayrton would leave after maybe three or four hours. Maybe the Williams or Benetton was 'smarter' [chuckles], but ours wasn't smart. I don't think our car was a good as the Williams and maybe that was part of the area where theirs was better as it was 'smarter.'"

He continues, "Our car was different as it had springs, and I think the Williams car had no springs [entirely hydro-pneumatic suspension]. We first had the soft springs at Monaco, and there we changed the springs a lot as with the really soft springs, yet you could control the ride height better. Softer was better as there was more grip, but normally you'd give away ride height. But with the active system you didn't have to, you could keep it low. So at all the circuits we ran it very soft. You had to be careful when you dropped the tail and where you were bottoming, such as turn one at Imola where you'd be flat and you'd have to watch that as it could be tricky. You determined where to lower it, and you'd only do that at certain corners. Ayrton and I would debrief together and work on this side by side all the time."

Everything was being pushed to the limit. Mika Hakkinen was a test driver at McLaren for most of the 1993 season, and he recalls the frustrations:

It was amazing technology, everything was being discussed meter by meter. Debriefs were quite interesting as it was very detailed on corner entry, mid-corner, and exit, but what often happened in the debriefing was the computers would crash. Oh my God, you should have seen the engineers, they would get so pissed off that they were ready to throw the computers out the window. The time was limited, but you would have to wait and reboot everything. There was so much information even at that time, and the computers were steaming hot as they were on the limit. Pat Fry was scratching his hair [makes a face like he's pulling his hair out]. It is so incredible what they did at that time with active suspension. Everything is handmade, so every development took a longer time to do. We were developing constantly, all the time. Suspension, the hydraulics, and all the time we found something new in the system to improve and it just got better all the time.

I will always remember in high speeds what we did from the steering input. The steering input was moving the chassis balance like this [leans over like a toboggan on banking], so when you would turn to go into high-speed corners it would lean in like this. It was amazing, in some corners we would go flat in sixth gear and without that active suspension there was no way that could have been possible. With the active suspension they moved the center of gravity and they moved the angle of the car so it was able to perform in unbelievable ways. It was really good fun. The funny thing is that straight away, you went flat. It was unbelievable as the car was able to go so quick, both aerodynamically and mechanically, but the engine wasn't able to increase the acceleration, and we just needed more power. I said, 'Give me more horsepower, give me more power.'

Mid-season that possibility at McLaren became a reality as Mauro Forghieri had brought his latest Lamborghini V-12s to the team for testing with the new "B" model of the McLaren MP 4/8, to replace the underpowered Ford V-8. Mika recalls:

The engine was great, it sounded great, it looked great, and it made a LOT of power. It had some negative of course… the fuel consumption, the oil consumption, the cooling… a lot of these issues were there, but the power was there which we were missing. We were very quick on low fuel, and with 20 kilos of fuel we were really very fast. Then it was the time to run a race distance with the car to see how it performs, so naturally they put 200 kilos of fuel in the car. We were at Silverstone and again when I went out, I started pushing flat out and it was incredible. The weight didn't really influence the acceleration because the engine had so much power, it was brilliant. But Silverstone has corners called Becketts where you go a little bit left right, left right and you put a lot of load into the car and the engine was part of the chassis.

Hakkinen was sensing the engine straining through that high-load section and it turned out that the engine was bending under the loads. "Definitely the loads were massive and the engine was really long. On the Hangar straight it just exploded. Pieces came through the floor and

The MP4/8 active McLaren was used by Ayrton Senna in some of his greatest drives (Norm DeWitt)

into the tarmac and, 'Hmmm, maybe this engine is not ready yet,' you know. But it was good experience."

Pat Fry says, "The Lamborghini was half way through 1993 when McLaren were testing it. The Lamborghini was running in the back of another F1 car, and we had good day to be sure, but we had a lot of blowups as well. When Honda left in 1993, we were using a customer Ford engine and I'm sure the other Ford teams were saying, 'Well don't give them the latest stuff.' With the active car that year with Ayrton, it was a bloody good little package. How many races did he win that year? Five, with that fantastic effort at Donington. With the Lamborghini there were days where it had a huge amount of power. It used a huge amount of fuel and it wasn't very reliable, but those are the sort of things you work on."

However, there was a fundamental aerodynamic problem with the MP4/8 that wasn't solved for years. Hakkinen describes it:

Even when we changed to the Lamborghini we were still having the same balance problem with the car, and for me that was a bigger issue than anything else. When you reached a certain speed with the car, it always would do something that made me have to lift the throttle and I wanted to keep my foot down all the time. There was some issue that we were not able to fix. Certainly we were studying it all the time, so they really started looking for that. We didn't get it right in 1994, but then came more understanding that it was coming from the aerodynamics, that the front wing was not able to perform at certain angles in the corner. They really started looking for that, because that it is the first thing that touches the air and it directs the air around the car.

The McLaren had a button that would raise the nose and drop the tail to take the angle of attack off the wings on the straight. Hakkinen says:

Later on we did it automatically and when it came out of the corner automatically the bottom would go down and raise the nose. You didn't even have to push a button. Actually the angle on the car wouldn't come down until you got onto the brakes, the car was staying up going into the braking. It was a massive effect, unbelievable. Automatically it was also moving the brake balance, it was not constant brake balance. When the speed goes down the aerodynamics is less and often what happens is that you start to lock the front tires. So it would automatically move the brake balance towards the rear under braking. The angle of the car was changing, the roll bar was making it softer on the front so that way you would get the front end going into the corner.

Hakkinen confirms it was the most interesting time of his career. "We tried so many things, we tried a fully automatic gearbox so that you don't have to shift anymore. We went to fly by wire, we moved the throttle from the pedal to the steering wheel to make it quicker. I said that every time when we come out of the corner I feel like I'm late onto the throttle as your foot is heavy. Paddy Lowe said yes, so we moved it to the steering. They were ready to do anything to be better than others."

There were also experiments with blown diffusers. Mike Andretti recalls, "Those were introduced part way through the year with the exhaust into the diffuser and you had to be really careful in fast corners, because if you had to lift, you were losing a couple hundred pounds of downforce. When you committed you had to stay committed. If you got it trouble and had to get out of it, it got hairy. It was really challenging for me as I didn't know any of the racetracks and qualifying laps were limited. If you got in trouble during practice, which I did probably at half of my races, I was learning the tracks during the race. Another problem that I had was that they could do whatever they wanted to the car and you would never know. Ayrton's car was never having the problems mine was having." Mike got the podium at his final race for McLaren in Monza. "Yeah, it was good because it pissed Ron Dennis off... so it was great. He was mad that I was on the podium." Mike Andretti was hardly the first, nor last, to find himself facing the scorn of Ron Dennis. However, Mike absolutely cherishes his year working with Ayrton Senna. "We worked really well together, we were friends. He said that I was one of the best teammates he'd had. He thought I really got screwed and even had a press conference at Portugal at the first race after I was let go. He was the first guy to call me when I won in Australia [the Surfers Paradise CART season opener in 1994], he'd stayed up all night watching it."

Somehow, when in his first Grand Prix with McLaren, Hakkinen was flying the car over the curbs at Estoril, when one would expect an active car to just soak up the bump while keeping the chassis static. One wonders if it was it because of the static element with the passive suspension (springs). Mika explains, "Yeah, it's true, but it has to be not 'too big' [laughing] as it was a big curb like this [holding his hands apart] that I went over." Hakkinen was right on the pace at his first race, out-qualifying teammate Senna, which was a very uncommon situation.

The previous year, Johnny Herbert had been struggling with the non-active Lotus. He says, "If the aerodynamics are not very good, then it didn't matter what you did with the car. Which was sadly what we had. We had a good car in '92 when I was with Mika, but we were two and a half seconds slower than Nigel. Two and a half seconds! It was a different world. Back then with all the testing that went on – Williams, McLaren, Ferrari – they did all the tire testing so all the tire construction was done for them. When it was passed to us, the construction never worked because we didn't have the downforce to use the tires properly. The active era for me was the best time as it was just brilliant working with the engineers to try and understand what we needed."

So, for 1993 Johnny Herbert had stayed with Lotus and with a comparatively small budget (compared to Williams or McLaren). He remembers it was quite an experience as they struggled with trying to develop an active car:

We did indeed have an active era in '93. I have to say it was the most interesting time I had in F1, because there were thousands and thousands of different setups that you could do. You never went through the whole damn thing, but it had the ability to ride the curbs, and to have the complete right poise when you were under-braking or under-accelerating. The three best laps I probably had in my life were at Barcelona in '93. As I went out onto the circuit, the car had a bit of a twitch. So we went around and when I stopped on the grid, they plugged in the computer. The car was twitching on that [right] side, going 'Ch-ch-ch-ch-ch.' I did the parade lap, came around and stopped on the grid, they jumped over again and plugged in the computer. 'Ch-ch-ch-ch-ch.' They couldn't fix it. As the lights started to come on, the car stopped twitching. So I took off and passed a couple of people on the first lap. Then the team got on the radio and said,

'Come in now, the system is failing.' The left rear had stopped working completely, it was literally like a dog with a broken leg, as it was just there. The other three corners compensated for that one. So, although I didn't have a left rear, I was able to pass three people in the first few laps. You could still drive it, and it was brilliant how the system worked.

The rear active suspension of the FW14B
(Norm DeWitt)

Dickie Stanford had a more reliable system over at Williams:

Well, if there were errors in the programs, or the sensors got a bit noisy, then it would develop problems. If it was okay in the garage it should be okay on the track. If there was a failure, it was a failure [unlike a twitching Lotus scenario]. It doesn't run springs, this was full-on hydraulic, but it has special concave washers, which you put this way and that way, so that the washers flex. These cones have different thicknesses so you can run different spring rates. There were two on the rear, one on the front. Alain Prost would go out and set a time, and then he would sit in the garage and wait for somebody else do go quicker. Then he'd go out and pop another lap and then come back into the garage again. We never knew if he was on the limit or not. I always said that if he'd had somebody else like Senna in the 15, we'd have found out how good it really was, because it was way much faster than that. It was easily the quickest car.

Johnny Herbert remembers it fondly, "Back in the early 90s it was basic data back then. You had a line of what it was doing, but you couldn't analyze it in a close way because the data wasn't good enough. The Lotus active though, it was okay. I think (teammate) Alex Zanardi may have had a time when the system was blinking, but it was quite rare. Compared to Ayrton Senna's Camel Lotus, this was a much smaller mechanism, the whole system was way smaller. With ours it was lovely."

There were a few memorable races for Herbert with the active Lotus:

Just before the race at Imola, we had a surfboard that we nailed to the pitlane outside the garage. When we started the race I came in for the first pit stop, I drove over the plank, pushed the button and the wheels came up so we could change the tires (as the tub dropped down onto the surfboard). Unfortunately the only problem was that when the guns were going off, the car was rocking about, so we only did one pit stop with that. My best circuit for the active car was at Monza in '93, at the end of the long straight, still using the old chicane back then. With the active you'd just throw it to the curb, the suspension absorbed the curb, the monocoque didn't move. The car didn't jump or fly, it would just absorb it. It was just an awesome piece of kit, how it could launch over curbs without disrupting the chassis.

Herbert looks at where we've come from that active era. "It's a shame we don't have it now. Now we have heave dampers trying to replicate such things and yes, it's very clever. [Simulated] traction control was what we had a couple of years ago in F1 effectively, which was one clutch drops that was 50%, and you'd go for 50 meters and then you'd release the second one. It was almost automatic, where now it's just one paddle where the driver has to get it correct, which is where the human becomes the problem because they can't get it right all the time."

Alex Zanardi was to have an enormous accident in the Eau Rouge/ Radillon sector of Spa in his active Lotus, from which he was lucky to escape alive. That did not affect Johnny's assessment of the active Lotus, but there were other issues. He recounts, "Unfortunately in '93 the 107B was not a good car, aerodynamically it had a lot of drag. However, the physical system, the full active system, I thought was very good. We had some small springs [200 pounds or less than that] in case the system failed. Of course we had the button that dropped the tail and raised the front. It lifted the front more, the tail went down a bit so the diffuser didn't touch. Zanardi went through Eau Rouge flat out on the first or second lap on cold tires, from what I remember, and it [the active] didn't fail."

Meanwhile, Benetton had been making huge advances on their active program for 1993. Steve Matchett (Benetton team mechanic) feels that the Benetton was the class of the field:

> I would say that by the end of the '93 season we had the best active car out there, but it took a lot of work to get it there. When we first tried to introduce it, we couldn't get it to go down the pitlane. For an hour, the active car would sit on the weight scales and effectively have a nervous breakdown. It would just start shuddering and shaking as soon as we'd fire up the active suspension system to level it. Getting new software sent over, changing the moog valves, changing the physical components on the car, trying to get the corner weights right, we were just trying to measure it. By accident one evening I had a 10mm combination spanner in my hand and just waved it across the moog valves as I was moving from left to right in front the car and it immediately cured it. It was like a magic wand, just waving the spanner across the moog valves cured this nervous-breakdown shaking that the car was undergoing.

Allan McNish reflects, "Some of the things they were doing were exploring the boundaries and what the technology could do then. The consistency of the Moog valves were a main factors of the active ride, how it was controlled and how it was working. I was 23 and always wanted to be a racing driver and I was climbing into anything. If you want to be logical and sensible and you saw Steve doing that, would you climb into the thing and go fire it off around the circuit? If you had any sense you wouldn't. I learned a lot that year and I learned a lot from Pat. To be honest, I still think Pat is one of the best and most instinctive engineers I've worked with, ever, and I've worked with some good ones. I worked with him ten years later at Renault and it was very interesting. He had an engineering brain, but he also had a real-world-scenario brain so he could adapt numbers to reality."

Steve Matchett says, "That issue with the Moog valves was just one example. Slowly we eventually did get it to work. The joy of it was that everything that the mechanics and engineers would do when the car came back to the pits after a practice session, you could do in a heart-beat with a laptop. Once we got it, it saved a huge amount of setup time. It was a wonderful toy."

Symonds is another who thinks that Benetton had found a magic bullet by the end of the year, and the end of the era:

> I think it was, we had spent that time at Reynard developing it. As I said, the hydraulics had been the really difficult part. While at Reynard, we actually went to a Royal Air Force base and had taken a look at the hydraulics on a Tornado and determined where we had been going wrong and got that sorted. At Reynard we did all the development of the system. Also the '93 Benetton was the only car that has ever had four-wheel steering, and between that and the active suspension it was really quite sophisticated, very much an integrated system. We could measure effectively what the car was doing and there was an electronic arbitrator. There was no arbitrator on the McLaren system. The arbitrator would understand what the car was doing. It was incredibly sophisticated, it would understand exactly what the car was doing on turn-in, handle a steady state of understeer, and it was doing all these things.

The goal was an Artificial Intelligence (AI) presence to the point where the car actually knows what it wants. Symonds says, "That's what the arbitrator did. Of course, Michael [Schumacher] loved it and really exploited it. Patrese struggled with it, it wasn't a natural thing. I wish that we had done a fully integrated system. It was rather difficult." Allan McNish agrees:

> There is no question that they were going in that direction. It started with the automated gearbox, and basically it would shift up and down within the parameters that you preset, to what was right. It wasn't GPS based, it was distance based, so it reset when it went across the line. Now everything runs off GPS, but back then it wasn't quite enough. On the gearbox side of things you could decide that you were outside of the correct powerband and needed to downshift, and therefore it would shift without knowing if you'd run wide. But you had overrides on these

things so you could hold it and override what it wanted to do. It was a leap of faith from a driver's point of view. Downforce, for example, is something that is a feel more than anything, when you fire into a corner. Back then it wasn't as consistent or sophisticated as it is now, in terms of how the downforce acts. You would go into the corner, like Bridge at Silverstone, which was a 180mph compression right hander, and you went in just thinking, 'Hopefully it sticks, and hopefully I'm not going to get snap oversteer in the middle of it.' It was without the suggested little indicators you normally get from a racing car when it starts to slide or move. Those cars were either grip or no grip, and there was nothing in between. You could quite easily fire yourself off backwards at a rate of knots. With the active car, it was at a higher performance level, but was desensitized to the driver.

The four-wheel steering box was on the left-hand side and the active suspension controls were on the right. In the maps you could raise or lower the whole map, and adjust in terms of its activity. You were in a situation where you'd be going down the straight and be adjusting with one hand and then immediately go to the other hand to adjust something else on the other system. Then you'd have to predetermine what you wanted in the corner from the gearbox. Most of the time the straights were actually busier than the corners. Sometimes at Silverstone going down Hangar Street you'd be doing two or three things at once, and racing drivers can normally only do one thing at once, we're not exactly that dexterous when it comes to multi-tasking.

As a result, the active cars could approach sensory overload with the driver trying to fine-tune all the parameters while flying around a circuit on the limit. Allan continues:

You had so many different options to tune it all the time. You would switch it on, and it would rise itself up and then settle down into its preset ride height, moving up and down until it recalibrated itself to its static state. I don't think it used lasers, I think it was pushrod sensors and things like that at the time. The little black box, which had front ride height and rear ride

Michael Schumacher and the dashboard of the 1993 active
Benetton at the Belgian Grand Prix (Paul-Henri Cahier)

height adjustments, which you could adjust corner to corner if you wanted. The understeer oversteer balance, was basically just a balance shift as to how you wanted that. There was the four-wheel steering on the other side and it became a bit much because we also had automated gearbox settings and various overrides for that. The push was for everything being automated, and the gearbox was the first part of that. But you've got the parameters and once outside the parameters you get problems. We had a year at Audi where the car shut off when Loic [Duval] was leading at Le Mans in the dark, because the computer sensed something was wrong when it wasn't. Sometimes you need a Control-Alt-Delete, as it needs a reboot. When I jumped from McLaren to Benetton, the system wasn't in the, 'Let's hope it doesn't fail today,' sort of thing. It was a wee bit more correct than that.

To raise the front and drop the rear, you pushed a button, which was to stall the floor at the rear to take the drag off the rear. You couldn't actively move the wings, so what it did was drop the rear and lifted the front, but it was also a way of making the wings less efficient. You'd come around Copse, which is a long accelerating right-hander and you'd be coming out of the corner at 160mph and thought, 'Okay, I think I can hit the button now,' and then it would go up and start understeering. It was very exploratory. I remember we went to a test and just tried a basic change and it was like an instant gain of lap time straight away. 'Why have you gone quicker?' 'Well, the car is better.' Back to back, back to back, trying to sort why it wasn't the intended result from the change they made, but they were treading into the unknown.

We were the only ones with the four-wheel steering and it only raced once. It was a bit odd, because to some extent it was contrary to your natural instincts as a driver. The rear would steer in the opposite direction to the corner. Say for example, in a left-hander, the rear would steer right, so you would have less stress onto the front tire and have less understeer. But, you'd originally felt that as oversteer, so you had to stop your brain from counter-steering when what the car was actually doing was trying to help you. It was also to help with tire wear as well. It was more than a degree [of steering movement], but it wasn't much. It had some improvements, and was a mechanically based system. However, when it worked it was a lot of fine-tuning as for the high speed or different angles of corners and to get the benefit, it was slightly different in terms of the way that it worked. You had to have the right amount of movement depending upon the type of corner. Silverstone, the old circuit in the 90s, was a good circuit for it because you had lots of transitions, you had bumps, you had flat out corners, you had corners with increasing speeds and decreasing speeds. I did the signoffs for it to be used in the last two races, Suzuka and Adelaide, but it was only used at Suzuka. We pushed hard to get something out of it before the end of the season, because it was known by mid-season that it was going to be banned for 1994. There was a whole understanding going on, and even though they banned it, you can't un-learn what you've learned.

From 1994-on, teams were finding ways to engineer passive systems to utilize the advantages they had achieved with the active cars. The heave damper system was a way to interconnect the front and rear suspension when it is not allowed for them to be mechanically connected. Pat Symonds explains, "For years they produced anti-squat, anti-dive suspension systems and the drivers complained they couldn't feel what was going on with the car." Ferrari in 1979 introduced a system of cables to keep their car from pinning the nose. Williams adopted the heave damper to achieve a similar result. Symonds explains the issue, "The history it is that we did a linked system on the 1981 Toleman. It never worked, but we tried. Then of course Renault was running the interconnected suspension in 2007. The interconnection got banned, so now everyone is using a combination of heave dampers and various things to try and limit the movement. They are tied together electronically, they are not allowed to be physically interconnected. What we try to do is to try and concentrate on tuning our pitch closely."

Allan remembers, "Many other things were banned around that time as well. Tire-pressure release valves were banned as well, where the valve itself had a release point on it so that if the tire pressure got above 20psi it would automatically lower the pressure. If you had a failure, then you had a puncture."

In many ways the cars hadn't changed, despite all the progress in certain areas. Allan says, "We'd just gone to radials in some categories, we were still running cross-ply tires in 1990. You didn't have the diffs like you do now. If you had hydraulic or electronic, or even a viscous diff like you would later on to have some compliance in it, but we just had a standard diff back then. To be honest, we didn't do much with the diffs back then, it wasn't a big development point yet. We developed the systems around it."

Soon the great experiment was over, and the active era was legislated out of existence for 1994. Paddy Lowe recalls, "As far as I'm aware, it was simply banned as there was a lot of lobbying by the teams that didn't have it, and it was Formula 1 politics. Williams had won in '92 with active suspension and traction control and these things so in '93 you had a lot of teams trying to copy it. Probably everybody might have been having a go [with active suspension], but there were teams without it. Ferrari were the principal protagonists. We felt they were behind the curve and Ferrari felt they had been outsmarted on all of these new technologies. We ended up in the position at the Canadian Grand Prix in 1993, where all of the active cars and those with traction control were all protested as being illegal. Max Mosley was behind it, but presumably he had been lobbied by a number of teams, most likely Ferrari."

Paddy Lowe continues:

I was at McLaren by then, so we went to the World Motorsport Council. I was there, and the main defenders of active suspension were McLaren, Williams, and Lotus. On the traction-control side it was also McLaren, Williams and that was it. As a result of this court case, both traction control and automatic suspension were both found illegal. It went to the court of appeal and they agreed, but none of these cars could be changed easily. So a deal was done by Bernie, which was famously called the Hockenheim agreement, as he went around the paddock in Hockenheim. All the teams agreed to ban a whole list of electronic systems starting with the '94 season and there was a single bit of paper that listed them. The deal was that we can carry on as we are, but we are going to have simple cars for 1994, there will be none of this electronics crap. There was a belief that the cars would remain broadly mechanical, a piece of paper that was going to keep the cars in the dark ages! It didn't work. The cars just carried on getting more complex.

That said, there were realistic fears that the active cars could be capable of speeds that the drivers were no longer capable of withstanding, nor the circuits being capable of containing.

These limits to what the human body could endure came to the fore again a few years later in 2001 at a CART race held in Texas. Mike Andretti recalls, "With a combination of vertical and lateral G's, we found a point that was past what a driver could deal with. They had this special drivers' meeting and I was thinking, 'What is going on and why are they doing this?' They asked, 'Who had these symptoms?' and I was one of the three guys that didn't raise their hands. One of the drivers didn't remember going through three and four, he had lost consciousness. I didn't have the problem, but I was thinking, 'Ooh man, this is bad.' That was the limit and we found it. Luckily I was probably in the best shape that I was ever in during my professional career that year, so I didn't notice it." Everybody has different personal limits for such things and most of the racers that day had found theirs.

Gary Olvey was on the CART safety team at the time and was there at Texas as well:

They were having problems getting an adequate blood supply to the brain, no different than a fighter pilot that doesn't have a G suit on, or hasn't been acclimated to handling those kind of G's. The situation there was high vertical and high lateral G's combined. The day after the race I talked to a whole bunch of people and one of them was the Secretary of the Navy. They were trying to develop gimble fighter jets where the aircraft would never get level and that would benefit radar [signature] from what I understand. They had their pilots in gimble centrifuges and they couldn't get anybody to tolerate it. They bagged that whole thing and aren't doing that anymore. In the centrifuge, as soon as they went beyond a couple of revolutions they went blind in one eye and lost their depth perception. We showed that we [CART] were way beyond what they were trying to do, and had to call off the race. It's a good thing that we didn't run that race as it would have been catastrophic.

Another issue with some of the cars, especially in the ground effect era, happened when going around corners. There were often situations where everything was dependent upon you being able to turn the car under load. Martin Brundle struggled with that exact condition, saying the '92 passive Benneton was nearly impossible to steer under load, while adding that complaining about the car is in their job description. He says, "The Benetton was a very simple passive car. I think every driver has said, 'The car is a bitch to drive.'" This wasn't just a case of complaints as usual, as this problem limited the driver's ability to steer the car. Pat Symonds concurs, "Yes, that was truth as we didn't have power steering. Benetton first put power steering on the car in 1997. In those days the driver would commit to the corner and would put the steering into lock, as the steering was so heavy you'd just locked your elbows into it."

Grand Prix racing is currently considering bringing back active suspension systems. Dickie Stanford says:

I know, the engineers have already come over, I've been looking after them. They are lucky that we've got this here [Williams FW14B, chassis #6] and we've been getting it running. It's all big and bulky, but in its day it was the height of technology. These connectors, it's all 3,000psi stuff. You wouldn't have anything that size on a car today. We are pretty sure that Mercedes last year or the year before were running some sort of limited system, it's not allowed to be electronically operated so it's some sort of manual operation. It creates more pressure as one wheel is being loaded and pushes back up. There is talk of Red Bull and Mercedes running some sort of steering system as well, there is always one wheel turning much more than the other. I don't think it's anything to do with Ackermann, it's the way it's been designed. Pictures of the cars on a tight corner, it's too much for Ackermann.

Then there have been some examples that challenge the known boundaries of technology. In the 2010 America's Cup, the helmsman was able to pull up data such as rudder angle, which is then printed down the inside of his glasses, akin to being a Terminator. Rick Mears has a driver-centric view of such advances, which he sees as reducing the difficulty of the driver's task at hand:

Active suspension… see I don't like all that stuff, to me it just takes away my tools. Any driver aid, any time something makes it easier for the driver, I don't want to do it. Even say 'paddle shifters' today, as an example. Yeah, they are better and are easier on gearboxes, and probably save money in the long haul. But to me as a driver, if I go out there with an H pattern and I can do a better job at shifting quicker, I can gain some time. If I shift five thousand times during a race and miss a shift three times versus eight times, I get an advantage. With traction control, all of a sudden now my computer guy is racing their computer guy instead of my foot being the traction control. ABS braking, I just stomp it down and let it do the work. That I might be better than the other guy at modulating my brake pedal, it doesn't matter as I lose another tool. Bumps on the track, 'Oh, we've got to get rid of that bump, it's a safety thing.' Any of that stuff that comes in, I lose tools. Wake up, anything you do that makes it easier for you, makes it easier for everybody else out there.

Mike Andretti has the opposite opinion, based upon his personal experience with the active cars. "Oh, it was an amazing era. I'm glad I experienced that. I just wish I'd had more time with it and in a different situation. From a driver's standpoint it was amazing. You could ask for the world and you were getting it, which was pretty cool. They say, 'Well, all these tools it took away from the driver,' and that was bulls**t. The driver was even more involved because he was helping them by saying, 'I need this here or that there,' and you would tailor it."

Mears counters, "Difficulty is one of the main tools that you have to work with. I think we are headed in the right direction and the guys are starting to realize that there are limits to everything. But the difficult part is, 'How do we get there?"

And thus the philosophical discussion over what is best for the future of open-wheel racing continues...

CHAPTER 18

DALE COYNE

Dale Coyne in the 1986 DC-1 alongside Emerson Fittipaldi's Patrick-March at Sanair. Dale finished 12th (David Ayers)

A bunch of college guys rented a motor home and went to the June Sprints at Road America. One of them was in his second year of college for golf course design and maintenance. He was sick as a dog all weekend and didn't emerge until at the end of the event when he heard the F5000 cars hit the track. The young man was Dale Coyne, "I thought those were way cool, and that was it. I came back here to Joliet and hooked up with my dentist, Dick Dejarld." Dejarld had run a McKee Mk18/Chevy for three years in the L&M Formula 5000 series, and was later to campaign the Schkee Can-Am racer. Dale says, "I talked to him about how I wanted to get started. He told me to start with Atlantic because it's a faster car and has the right amount of tire for the car."

One might expect that Coyne would have been driving a staggeringly fast street machine to prepare himself for stepping straight into Atlantic. He explains, "I had a Cosworth-Vega. The Atlantic wasn't a dependable car, so I raced the Cosworth-Vega while I was getting my license stuff out of the way, as you needed two Regionals or whatever the rules were, and then went back to the open-wheel car."

There were no years of pounding around autocross circuits or road racing FV or FF. Road America in 1977 was Dale's first Pro Atlantic race.

I actually started with a March 722 running amateur Atlantic with a Lotus twin cam that was nothing but problems. We finally got a BDA and that worked pretty well, but I never really learned how to drive in that car, so I went back to air-cooled Super Vee. We ran a Lola 328 and ran every race we could in the Midwest. It was a nice car, it was dependable and it just ran, and ran, and ran. We ran SCCA Regionals, SCCA Nationals, and we ran a little bit of everything with that, whatever was at Road America.

We ran the Autoresearch car but only ran a couple of races. At Milwaukee we ran against Michael Andretti, so it must have been 1982. Michigan sticks out in my mind, as we got an electrical problem and went all the way to the back, lost a lap early-on and worked our way back to the front. It was probably the most fun that I ever had in a Super Vee. The whole field was bunched together and you really had to trust each other. It was pack racing at its highest.

The Autoresearch car was uncompetitive on most circuits as it was a ground effect Ralt RT-5 dominated era and Dale's Autoresearch was flat bottomed. Dale says, "The Ralts were killer and I got a RT-5, the ex-Galles car. It was my first ground effects car and we were highly competitive with that." Derek Daly, Rosberg's teammate at Williams Grand Prix Engineering, recalls, "In 1982 we did the USGP in Las Vegas, which was the end of our season. A week or two later I got a call and was asked if I wanted to do the Indycar race at Phoenix. I knew nothing about Indycar or Phoenix, but agreed to do it. I have a picture me on the starting grid beside Johnny Rutherford, and the mechanic on the left rear wheel was Dale Coyne. He was a mechanic on my very first Indycar race."

Dale confirms, "We had bought an Eagle from Wysard, and they ran Derek for his first race in a March. The Eagle was the car they had run the year before, and I was getting that car ready. I think that was the only time I was on somebody's pit crew. 1983 was a little bit of the RT-5 and we won the SCCA National at IRP. To go on a tour with the Pro Series was expensive, so I just ran multiple races where I could. I didn't run the Runoffs because you need to follow the Nationals to get enough points to qualify. We had our eye on Indycar at the time and we picked up the Eagle."

That's a pretty bold move, going from winning SCCA Nationals straight into CART. Dale agrees laughing, "Yes, it was! I had a landscape business in commercial maintenance and that's why I didn't put a lot of effort into the Indy 500 because May is the busiest month of the year for a landscaper. We'd make enough money to run the rest of the year."

He continues, "We were running Super Vee at the time and Indycar had just started their CPI program, it was important financially to get there when we did. That was when their franchise program hit in 1984 and we looked at that and thought, 'We can do that.' So had we bought an old '81 Eagle and built our own stock block for it, as Gurney was still playing with those."

Mike Hull is now Managing Director at Chip Ganassi Racing, but in 1983, he was with Arciero Racing running a PC10 for Pete Halsmer. Mike reminisces:

In those days, CART had initiated the franchise system and they were trying to get people to fill the grid. Frasco (CART Chairman) was trying to get promoters interested in the series, and he had to go to promoters and say, 'I have a guaranteed field of twenty-four people.' Dale was one of the guys who said 'Okay, I'll do it.' He had been racing Super Vees and he bought an Eagle Chevrolet stock block thing and ran around at the back before he got his s**t together and helped CART establish itself as being one of the entrants. You have to realize that if you committed yourself to running those days in the franchise system, you got paid to show up, and that did help him a lot. I thought he did a good job of representing himself and understanding the importance of what he was trying to do and not get in the way of other people in the process. I thought he did a really good job of that at the time.

Coyne at first put Jim McElreath in the car for race two at Phoenix, a driver both at the end of his career as well as having a long successful resume. McElreath was unable to qualify. The next race was with Tom Bigelow at Milwaukee, again unable to qualify. At this point Dale climbed into his car and clicked off a series of five straight DNQs. After skipping the Pocono 500, Dale next arrived at Mid-Ohio, qualified for the race, and finished fourteenth. Phoenix and Laguna Seca were the next races where Dale was unable to qualify the car. The grand experiment wasn't looking particularly grand at that point.

"We had the Eagle for a year and then in 1985 we had an '84 Lola. It was Mario's championship car from the year before. They had resurfaced the track, but when we went to Mid-Ohio we ran fast enough to where we were only one-tenth [of a second] off Mario's time from the year before." That was the highlight as Dale had started the season a non-qualifier at Long Beach, followed by four more races where he either didn't qualify or didn't start. However, at Michigan Dale finished twenty-fourth, and had four other similar finishing positions before not qualifying for the last two races. The twenty-third place finish at Road America was as good as it got for Coyne in 1985.

At some point, one would think that discouragement would enter the equation, but there was no quit in Dale Coyne. He remembers, "In 1986 they changed the rules tremendously so we built our own car. We took the gearbox and the tunnels and corners off that Lola and we fabricated our own bodywork to build the DC-1 in 1986. We had the same car, but we had to rebuild the whole bodywork, they were all carbon tops with aluminum bottoms." Coyne was unable to qualify the

DC-1 at Phoenix, but did manage to scrape onto the grid at Long Beach, which better suited the stock block engine.

The crowd would roar as Dale came by, as he was running far behind the field. No doubt everyone could relate to the guy who probably was in way over his head running that old heavily modified Lola-Chevy and was probably on the knife-edge financially, but was out there living the dream. Dale says, "I remember driving Long Beach and everybody was standing up to where it was like driving through a wave and I was beginning to wonder what was going on? It was like at a baseball game, you'd drive down there and everybody would stand up. It was the same impact when we ran at the Meadowlands, everybody liked that thing. It was a small block Chevy, but it was all aftermarket stuff. They had tons of torque, but once you got hit third gear it was all over. They would pass like you were standing still. The tighter the street circuit, the smaller they were, the better. The Meadowlands was my favorite circuit because it was a tremendous amount of work, you almost never went straight. There was one little straight, but everything else was turns."

Arie Luyendyk recounts, "He had no chance but he showed up all the time. He looked like a duck out of water because he was totally different in his whole setup than everybody else. His was the only stock block engine running, when we all ran turbo engine cars. He obviously ran on a shoestring, and you can only do that for so many years, but that's why he became a fan favorite. The whole field would come by and then Dale would come by with that different sounding engine and for some reason people would just embrace him. Not to mock him at all, but it was kind of like Eddie the Eagle, who was a fan favorite too." Derek Daly concurs, "The stock block... the whole field would go by and this rumble would come down through the streets and Dale would get a bigger cheer than everyone else. He had the passion, and he had more commitment to the sport than almost any owner that I know here, so good for him."

Results were beginning to improve, and Dale took twelfth at both Cleveland and Sanair. For the first time, the number of finishes exceeded the number of DNQs. Coyne explains, "For 1987 we bought an '86 March. We were still always running the small block Chevy, we always had to convert the cars from the Cosworth, that's how we could afford to run. We were one of the last guys who were still trying to run that engine. When Genesee beer's team went away, we were the last ones to run it. We ran Barnes cylinder heads with different valve angles than the normal stock block, and Cosworth pistons had to be made custom, but were better to be sure." His best finish in the March-Chevy was fifteenth at the Meadowlands, with seventeenth at Portland and Road America, which appeared to somewhat suit their new car.

Dale continues, "We went to Road America and Sheirson was building his own engines for Al Unser Jr. I remember asking Little Al about the new cams that they had ground for Road America and how it was coming off the corners. Al said, 'I can beat every car off the corner except one, and it's that f**king Coyne. That thing loves Road America.' Yeah, until I hit third gear, but in second gear, that thing was a rocketship. We used to break parts, like at Road America going up the hill from turn five we snapped both half-shafts like twigs. We had tons of torque. We bored and stroked the thing to try and make horsepower and get rid of the torque, but the architecture of the engine created a lot of torque."

1988 was nearing the end of the trail for Dale Coyne the driver and he decided to put somebody else in the car. Dale admits, "We tried everything while we were wondering, 'Is it me or is it the car?' We put Dominic Dobson in the car at Long Beach with all the best stuff, putting in the best motor. We gave him the best of everything, as he was a pretty good driver at the time. His first pit stop his eyes were huge and he said, 'How in the hell do you drive this thing? You touch the throttle, it goes BOOM!' Until you hit third gear and the Cosworth cars go singing past you something fierce. He qualified last and finished last, so I said, 'Okay, I guess it's not the driver.'" Coyne was back in the cockpit and there was a bit of optimism following a thirteenth place finish at Milwaukee. Later in the season they got another tight street circuit and finished twelfth at the season finale in Miami. That was the last race for the Chevy stock block cars, and Dale regrets having not made the switch sooner:

In 1989 we started with the Cosworths, and that was when I realized that I should have always been running Cosworths. It takes more man-hours to build, but you don't need to be as finicky and I built the engines. You don't have to do all the massaging that you had to do running small block Chevy parts. The thing I didn't realize was that guys like Shierson and Patrick said, 'We would have given you our old parts. We run our parts hard, but if you just drop your revs 500rpm you could run on parts for another 1,000 miles.' It was frustrating, plus when I raced the car in races during 1991, I realized how easy that car was to drive compared to the stock block. I wish I had used old Cosworth parts that I could have gotten for free. If I'd bought two-year-old Cosworths with old parts, I would have been ahead of the game.

Coyne only drove in one race that season, scored twenty-seventh at Pocono. It was musical chairs in the team Lola/Cosworth between Guido Dacco (most races), Fulvio Ballabio, Ken Johnson, and John Paul Jr. Dacco scored a pair of twelfths at Detroit and Cleveland, and once again twelfth was the high-water mark for yet another season.

Dale had other interests at the time. "I got out of the landscape business in 1989-90 and my racing helped the Chicagoland program. There was a bunch of local guys with stars in their eyes and not realistic numbers. I brought some reality to that and jumped in with both feet." Originally a drag racing venture, it eventually turned into a paved oval.

We got the entitlements done for the dragstrip insanely quick as we had a friendly City of Joliet, which we had annexed into the City instead of dealing with the County. Cary Agajanian was asking, 'How did you get that done so fast?' We knew Bruton Smith was looking around to build a track near Chicago and was looking around over in Gary. Menard and Tony George were looking somewhere else and were all getting push-back from their communities. We had built a short dirt oval along with the dragstrip and Tony George and I were sitting at the top of the bleachers on a quiet day. I said that we can get all that land, we can get the entitlements done and to go forward so we decided, 'Let's do this.' I spearheaded all that, Tony was a partner. ISC was a third partner and they bought us all out several years ago and they still own it. It has NHRA drag racing and Cup racing.

For 1990 Dean Hall, a former Colorado downhill-skiing champion, drove almost the entire season, scoring elevenths at Cleveland and Mid-Ohio. 1991 saw Coyne behind the wheel again for three races, finishing at Cleveland and Portland, and missing the cut at Detroit. That was the end of Dale's driving career. The most notable driver to find his way into Coyne's car was the debut of Paul Tracy at Long Beach, where the car

Dale Coyne with his team during qualifying at Fontana in 2013
(Norm DeWitt)

suffered engine failure after thirty laps. Randy Lewis' twelfth at Toronto was the best finish.

For 1992, Dale teamed up with Chicago Bears legend Walter Payton to form Payton/Coyne Racing, running most of the season with Eric Bachelart and Ross Bentley. Bachelart broke through for an eighth at Long Beach, then seventh at Detroit to take eighteenth in points. Ross Bentley's best result was eleventh in the opening race at Surfers Paradise. 1993 was mostly run with Robbie Buhl and Ross Bentley. Buhl took sixth place at Long Beach, for the team's best finish to date. The following year it was complete and utter musical chairs, with eight different drivers in the two team cars, the only constants being Bentley and Zampedri driving much of the schedule, Zampedri scoring a seventh place finish at Portland and ninth in the Laguna finale.

For the inaugural US 500 at Michigan, run directly against the Indianapolis 500 in 1996, Payton-Coyne achieved their first podium finish, third place with the former F3000 champion, Roberto Moreno. Mostly, the team continued to struggle with mid-pack drivers, at best. Dale recalls, "In the late 90s we started taking drivers that had some skill. We had Cristiano da Matta and Bruno Junquera and tried to make it work with guys like that. I built my business up so we could afford to fund the drivers." Oriol Servia was to take third at Laguna Seca for Coyne, but Bruno was the driver that proved Coyne Racing could be a threat. Bruno had a run of three consecutive podiums in 2007, with the highlight a second place at Zolder, Belgium, followed by thirds at Assen and Surfers Paradise.

*Justin Wilson was the most successful driver in the history of
Dale Coyne racing, and a great guy (Norm DeWitt)*

Bruno Junquera says, "I had two great years with Dale Coyne, and in 2007 I almost won Zolder. I had a lot of fun, Dale was a nice guy. Many people don't talk about Gail Coyne [Dale's wife]. She is behind the scenes and works very hard as well." There was only one thing that Coyne's teams had yet to achieve, and that was the top place on the podium. When Champ Car folded early in 2008, the team moved over to the IRL/Indycar series. 2008 wasn't to see much success, but that was to change for 2009.

Dale says, "Justin Wilson was the breakthrough guy, we won two races with him." At Watkins Glen in 2009, Wilson had brought the team their maiden win and at Mid-Ohio the entire transporter was painted up with the victory lane celebration graphic from the Glen. It was no fluke, with Justin having led forty-nine of sixty laps. After twenty-five years of continuous running at the top level of American open-wheel racing, Dale Coyne was at long last in victory lane.

That seemed to break the jinx and from that point on, Coyne's team were contenders. Wilson scored another victory in 2012 at Texas. Dale recalls, "We were lucky when we won in 2014 with Carlos Huertas [Houston], and that weekend with Mike Conway in Detroit was a magical weekend." In 2013, Conway was a last-minute substitute for Ana Beatrix at Detroit and utterly flattened the field in Saturday's race one. Race two, starting from the pole, appeared to be more of the same until a pit strategy call resulted in his finishing third. Simon Pagenaud had his first win, and the first win for Schmidt-Hamilton motorsports.

Schmidt's and Coyne's teams had won the first two races of the 2017 season, winning against the other teams that could be described as juggernauts… Penske, Ganassi, and Andretti. Sam Schmidt remembers, "We didn't really cross paths until I became a team owner in Indycar. He's a super nice guy and given a tremendous amount of opportunity in this series to people that otherwise would not have had that opportunity. I've always said that the teams with the lesser budget want to win just as bad as the other guys do, and they put in as much time and effort as anybody else, if not more. It's an ongoing battle and we worked really hard with Simon and now James to get to where we are. We are still considered a pretty small team compared to the big guys, and I know our budgets are smaller. It was great to have both these teams win, especially us at Long Beach, I think it's one that everybody wants to win. It was nice to check that bucket and now it's time to check the Indianapolis bucket." No doubt, Dale Coyne would concur with that conclusion.

For the past seven years, Coyne racing has been sponsored by the Boy Scouts of America. Dale says, "The little town I was in had a Cub Scout pack but did not have a Boy Scout troop, so I was only in Cub Scouts. It's been a real good relationship, we picked them up in their one-hundredth-anniversary year."

Conor Daly had his first ride with the team at Long Beach in 2016, eventually taking the ride at Foyt for 2017. "I needed that to really get my Indycar career going, and that vaulted me into my next ride. It was one of the things that will always have been an important part of my career. I really learned a lot from working with those guys, especially my engineer Mike Cannon, I learned a lot from him. They do a lot with very little." Derek Daly reflects, "Years later, to see my son driving Coyne's car, shows you how the wheel keeps turning and that it's a small family."

Michael Cannon trying out his latest innovations in
bicycle suspension (Norm DeWitt)

The highly experienced Michael Cannon has been an engineer with Coyne for four seasons. He says, "The desire to win is there. We may not have quite as many luxuries as we did ten or twelve years ago when I was with Forsythe, but one thing about Dale and his wife Gail is that they don't have any children. We are their family and they definitely treat us like family. I thoroughly enjoyed working at Andretti, I thoroughly enjoyed working with Forsythe, but I love working for Dale and Gail."

For 2017, Newman-Haas 2.0 is back together at Dale Coyne Racing, as Sebastian Bourdais was back with Coyne, only this time with his old Newman-Haas engineer Craig Hampson. Their previous partnership had led to absolute domination of the Champ Car series for years, and the reunion started off on the right foot with victory in race one at St. Petersburg. Dale says, "We've been working on it for years. We've had Tom Phillips, the chief mechanic from there, and a couple of other mechanics from the team. It was a natural for those guys when that closed down, they wanted to live at home and we were the logical choice down in Plainfield. We've been talking to Sebastian for over a year about coming back, and when KV folded, then it made sense. And we've been talking to Craig year after year about coming back."

Craig Hampson gives his perspective:

> I don't necessarily want to say it's a Newman-Haas vibe because I don't want to take away from Dale and Gail, and all the guys who have been long-time Coyne people, because they are very good people and they work very hard. We've got four guys who once worked at Newman-Haas, but there are a lot more guys who were part of the culture here [at Coyne]. The biggest thing I can offer is that everybody is a racer and they are all working extremely hard. With such a small crew, you end up having to work substantially more hours and wear a lot more hats, compared to working at another place. In addition, not having a lot of spare parts means rather than having next weekend's pieces already built up, you have to take apart this week's pieces. We have two chassis per driver, but we only have enough uprights and suspension components for four chassis. So, if you want to roll them around on something, we can't leave anything back in the shop being built in advance for the next event. We just can't do that, we have to keep rebuilding it all. For the mechanics, this is all really hard. Not so much for the

engineers, even though there are things for me, like sims that we might have previously done that we can't do, or wind-tunnel tests that we can't do. Everybody is wearing three hats.

Sebastian Bourdais confirms, "I've been personally trying to get Craig back for years now, almost ever since I came back to the series in 2011. When I saw what was happening at KV, the inconsistency of the effort, the uncertainty, and the tough winters of having built something and watching it crumble during the winter, it was not positive for me and I had to find a way to do something else. They [KV] were budgeting their thing from February through September. If you are trying to build a serious program, you can't do that. That's what Dale brings to the table. He has a small and reasonable team and he builds his budgets from January 1st to December 31st, so you have a stable platform that you can build off and make things better, and that's the plan."

Late in the opening race of the season in St. Petersburg, Pagenaud tried to run down Bourdais and reduce the six-second gap. Bourdais recalls:

You burn a lot of fuel when you run full-rich hard, max attack, something around 2.75 on fuel, and going all the way to 3.3 or 3.4 if you really had to save fuel. The pace is not dramatically different, but to gain that little extra it was a lot of fuel you had to burn. When Craig told me we were good on fuel and then six or so laps later he comes on saying 'Uh… dude, here's the new fuel number.' All of a sudden we started burning fuel like there was no end. The manufacturers in their war of trying to get a few more horsepower, have some maps where fuel pours in and it drains the tank pretty quick. You are picking your strategy based on hypotheticals, and whether it is going to fall your way or it's going to bite you up the ass, you can never know which one it's going to be. Obviously at St. Pete there was a bit of luck involved, but you need some when you start from where we started. The good thing was to pass the Penske car to get the lead and then run away with it. If you qualify in the top six, you put yourself in the position to be a real contender, and that's all we can ask for.

Don't think you can ever tell what is really going on from some of the television coverage. Sebastian continues, "They are trying to explain things they don't understand to the public. When they are trying to explain what we are doing on fuel strategy, we go by codes and Craig was telling me 46 15 22 whatever. They are trying to understand and compare that to the 3.7 that Simon was getting, which was obviously an offset from the real fuel numbers that he was getting. They thought they heard 1.8 from what was '18' for me and then were trying to explain that I was getting 1.8 mileage while Simon was getting 3.7. That would have to be a different category [of car]."

After the race Craig spoke to Sebastian over the radio, telling him to cool it with the celebration. Sebastian admits, "I've fried diffs before doing donuts and doing things to the engine that people didn't like. We know each other very well." Craig Hampson confirms, "In Champ Car with Newman-Haas, when he got a win, he would do donuts. He didn't start that, Zanardi's the guy who started that, but I knew he was going to try to do a donut. Honda was not going to like that because we have to use an engine for 2,500 miles, so I told him, 'Don't do that', but he did it anyway. It's different now, that you only get four engines a year and they are expected to go for a long time. If you do anything to overheat it or over-rev it, that's a problem, but it was okay and we were able to use it for Long Beach. It's just good business not to waste money for no reason." Craig commented to Sebastian in victory lane, "It's like the good old days."

Back in 2007, Dale Coyne and I were taking in the beauty of the lake behind the paddock at Mt. Tremblant Champ Car, and we concurred that it was the most beautiful paddock anywhere in racing. So, now that Bourdais had won the 2017 season opener, is St. Pete now the most beautiful paddock in racing? Dale replies, "It is, and hopefully Long Beach tops it." This remark after the win in St. Petersburg was followed with a second-place finish at Long Beach and an eighth at Barber.

Craig Hampson says, "I started in the series after Dale had stopped driving. For a long time Dale was the one with the driver who brought funding, and they were constantly in your way while you attempted to lap them. I can honestly say that I've lost a race because we tripped over the Dale Coyne driver trying to do our in-lap into pit lane. But clearly, once Dale had a sample of having a good driver in the car, he said, 'You know, this is a lot more fun when you have a good driver.' So year in and year out, he stepped up the program. He signed Justin [Wilson] and then he won a race. A few years later they finished sixth in points.

Sebring 2017 with Dale Coyne, Mike Hull, Bob Mayes,
Sebastian Bourdais, and Patrick Bourdais (Norm DeWitt)

That's the kind of thing this team is capable of, and we will try to do that sort of thing."

Headed into the first oval race of the season at Phoenix, Bourdais was leading the points over Dixon, Newgarden, and Pagenaud. Sebastian could almost do no wrong at this point, having won the 24 Hours of Daytona for Ford, St. Petersburg for Coyne, then taken second at both the 12 Hours of Sebring and Long Beach. Craig recounts,

We are off to a flying start, but they are going to get us, right? The resources of the other teams, particularly on the ovals, where they have done the wind-tunnel testing, the upright testing, the gearbox testing, and they've got the spare parts, they are going to outrun us at those places. We just need to make sure we don't make mistakes, we don't crash, and we deliver the best job we can at those [ovals] and we'll play to Sebastian and Ed Jones' strengths, which is clearly on the road and street courses, and we'll see how many points we can score. I'm not saying neither one is a good oval racer, but I really don't think we have the simulation ability or the hotrod parts to be able to contend on the ovals. The fact is that Sebastian finished in the top ten of every single oval race last year, on a team that really was underfunded, so he can do it.

Sadly, at Phoenix, Bourdais was collected in a first-lap melee when a car in front of him spun and took him out. Indianapolis brought Bourdais nothing but grief. First in the road course Indy Grand Prix, Bourdais had an engine failure on the fourth lap. In practice for the 500 on Fast Friday, Sebastian turned the fastest time of 233.116 mph, nearly one mph faster than anyone in the field. The following day, the car came around on Seb and turned him straight into the barrier, leaving him seriously injured with a fractured pelvis and hip, and Bourdais was fortunate to have survived. Coyne's championship-contending driver was out for most of the season as a result, although his spirits remained high when he returned to the Speedway before the race. Being from Le Mans, Sebastian also sent greetings for his many fans in France, which was shown on the many screens around the circuit just before the start of the 24 Hours of Le Mans, where he had won the GTE Pro class for Ford in 2016.

Sebastian Bourdais on his way to second place in the
2017 Long Beach Grand Prix (Norm DeWitt)

Bruno Junquera says, "I've seen his team improve so much in the last ten years, and with Sebastian he has a really good team. When Sebastian crashed at Indy, that car was the ride that everybody wanted, they wanted the Dale Coyne car." James Davison drove the Bourdais car in the 500 and ran as high as second place. However, after Servia and Davison split Tony Kannan going down into turn one, a touch between the two drivers mid-corner triggered a multi-car accident. Ed Jones also put in a stellar drive, and was also a contender for victory in the 500, eventually finishing third, only a half second behind Sato and second place Castroneves. Many felt that Jones was every bit as deserving of the Rookie of the Year honors as the award winner Fernando Alonso. Dale Coyne cars were now contenders along with the best in the business, at the biggest race in the world.

Other 2017 highlights were Tristan Vautier's stellar drive in the crash-fest known as Texas, where Kannan, moving up the track, caused a multi-car pileup which wrecked both Jones and Vautier, earning Kannan a few choice comments from the normally composed Coyne during the ensuing red flag. At season's end, Ed Jones was awarded Rookie of the Year by Indycar.

Arie Luyendyk compliments Coyne, "Dale realized that he'd better do something different and he got somebody to drive his car and what he did later was great. He always pulls off a win on a yearly basis. Now, this is the big leagues, and [at Phoenix, 2017] he's leading the points. It's pretty awesome when you think about it."

Bruno Junquera agrees, "I think he is one of the most special people in racing. When you work for a team owner who is a racer, that's really good. Carl Haas was a racer, and Dale is a racer. I remember in the 2008 crisis he took a risk and was buying Sonny Barbeque [restaurants] and now they have about three times that many of them than they had then. He's a pretty smart guy, and with Gail, they are a very nice couple."

And Conor Daly adds, "They do a very good job. It's good and I'm not surprised at all by what is going on, because I know how good our car was a year ago. With Bourdais, and Hampson together there, and Cannon all working together… that's a really good combo." Craig Hampson says, "I think Dale puts more of his own personal money into this than any owner out here except for maybe Roger Penske. Dale puts millions of dollars of his own money into this every single year, and that level of commitment is really impressive. The fans of the series ought to really support and appreciate him for that. Year in and year out, he puts two cars out there, three for Indy, and he does this on his own dime. I think that's a really important part of Indycar racing that maybe people don't acknowledge. It's lean here, but we are doing the best we can with what we have."

In closing, Mario Andretti weighs in, "He's one of those that survived, and he was just hanging on by a thread. But perseverance worked for him, and that's why everybody respects him. He's so well liked and the best part is that he has set an example to a lot of people. He was the underdog, and that was what he was considered for so many years. As an owner, he was still pretty much the same way for a long time, I think he should be commended. It was David and Goliath, and he was a David. Now he's getting close to being Goliath and everybody loves that. I do, for one, and that's a wonderful success story."

Dale Coyne calling the shots (Norm DeWitt)

CHAPTER 19

AERODYNAMICS

BC Comic appears courtesy John Hart Studios

Other than the electronics revolution of the past few decades, nothing has had a bigger effect upon the design of the modern racecar than aerodynamics. The quest for low drag combined with stability was very much the previous order of the day, such as when Gian Paolo Dallara was hired as the aerodynamicist for Ferrari's Grand Prix program in 1959. As adjustable rear wings showed up on innovative cars such as the Chaparral 2C, the awareness that one could both have low drag and high downforce from a single design started a revolution. By 1966 the Chaparral 2E with its rudimentary, but moveable, high rear wing awakened the world to the possibilities. In 1968 this concept arrived in Formula 1, and by 1969 the entire field was composed of high-wing machines, faster to the point that even the typically conservative McLaren introduced a biplane at Monaco (with large nose-mounted dive plane wings as well, making it a triplane). Given the then-recent catastrophic failures of many cars with high wings (see *Making it FASTER*), the high-wing era of Formula 1 was outlawed by emergency decree, starting with the Saturday of Monaco weekend.

The Chaparral 2E of 1966 wasn't the first high wing car but it made the concept operational and effective (Norm DeWitt)

The ground effects era of 1977-82 was pioneered by Team Lotus (see *Making it FASTER*), and saw ever-increasing levels of downforce achieved from endless track and tunnel testing. One assumes that ever since that time, every team would have the wind tunnels and simulations humming along 24/7. The reality is somewhat different.

In 1988, Stefan Johansson had arrived at the Ligier team, teamed with Rene Arnoux. Stefan says, "It was a Michele Tetu design. The poor old guy, he came up with that brilliant idea of putting half of the fuel tank between the engine and the gearbox and Guy [Ligier] was, 'Yeah, OUI!' and that was the end of the discussion. Arnoux and I were joking that the car had aerodynamics by Sir Isaac Newton, as the only thing that kept it on the road was gravity." The Ligier-Judd JS31 had a sadly predictable lack of results, neither driver qualifying in San Marino, nor scoring a single point in 1988. Normally aspirated cars had a difficult enough time in 1988, but to put that engine into a chassis without downforce was engineering suicide.

Five years down the road, things hadn't changed much amongst the smaller teams. Johnny Herbert provides a good example of the gap between the haves and have-nots, from his time at Team Lotus in 1992-93, "At Lotus we used to do aerodynamic testing on an oval with a Corvette towing the car around. That was aerodynamic testing. I'm not sure how the hell that worked with a Stingray in the front. They had a 40% wind tunnel but they never really used it. There was no computer-based software for it, they'd have a bit of an idea, stick the new front wing on the car and ask if it was better."

Amidst all this seat-of-the-pants engineering, the larger Formula 1 teams were exploring some of the most engineering-intensive concepts, trying to find a way to achieve what the Chaparral had explored so effectively in the mid 1960s, that of changing from high downforce to low drag while on the fly. Out of this came McLaren's elegant solution, with a driver-actuated method of stalling the rear wing known as "the F-Duct." The name had nothing to do with the inventive method of ducting past the cockpit, it was because the snorkel air intake emerged from the "F" part of the word "vodafone" atop the tub just ahead of the cockpit.

McLaren engineering director Paddy Lowe explains:

Well, it wasn't me who came up with it, for a start. It was some colleagues at McLaren, there were a number of key individuals that worked on it. There was more than one key idea in there. The first idea was that we could bypass Article 3.1.5, which doesn't constrain the driver. Actually I don't think that came first, to be honest. There were three key ideas in there, one idea being to stall a wing by introducing a flowing slot, which is something that is probably used in various aircraft. The second idea was to be able to switch flows using the venturi switches, which is solid-state aerodynamic switch. The whole way it worked was there was a hole that the driver blocked with his elbow. Interestingly as other teams copied it, which depressingly was copied more quickly by the other teams than we'd have liked, they had ones where the driver took his hand off the steering wheel and blocked the hole with the back of his hand on the side of the cockpit. We thought that was a bit wrong really, because in that sense the driver was less under control of the car, essentially driving one handed. You only needed to do it in a straight line and with ours the elbow came out at exactly the right place on the straight and the driver kept both his hands firmly on the wheel. Blocking this hole changed the pressure in the control duct that came from the bonnet scoop, and blocking the hole, that fed into the high pressure in the signal port of the first stage of this venturi device. And that switched the air from the main inlet, the power circuit, which came from the roll hoop. That either allowed full flow to a regular exhaust in the back of the car, or when switched to this controlled flow, you could blow the slot in the wing. It was a separate port, I think it came in below the main air entry, it had nothing to do with the engine. The key thing is that the venturi switch had no moving parts, which is necessary to comply with Article 3.1.5. The only moving part was the driver's elbow and the Article 3.1.5 doesn't talk about the driver. The driver can move his helmet and affect the aerodynamic performance and that's clearly not part of the outlawed circuit. These were the three innovations that occurred.

McLaren engineer Pat Fry adds:

The F-Duct... that was a nice concept I have to say. I was involved in the F-Duct development at McLaren in 2010. I was the one that solved it, along with the aerodynamicist who got it working at the Jerez test. It was one of those concepts that was at least a year in development, playing around in the lab getting the two-stage fluidic switch working on the McLaren. It was a great concept, but when it comes to the practicality of getting it working at the track, it took quite a lot of inventiveness at the track to get the basic thing to work. How it worked in the wind tunnel, it was a bit more of a challenge as we had to get the intake duct up and out of the boundary or surface layer, get it up into clean air flow. Then we had to design all the ducting in the chassis to get it working.

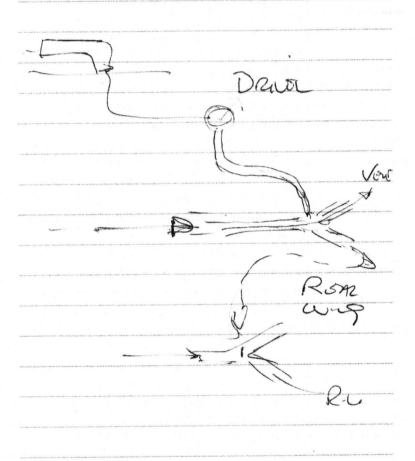

The "F Duct" was a system that gave the driver control over airflow used to stall the rear wing (sketch by Pat Fry)

In some ways, McLaren was creating a monster, requiring a driver to somehow block a hole in the cockpit, while twiddling knobs on the wheel, perhaps moving their head lower or to the side to enhance the airflow into the airbox feeding the engine, and of course drive the car. Pat Fry weighs in:

> It's one of those things isn't it? It's worth lap time so you try to work out how to do it. The Ferrari concept was a good way, and a huge amount of work went in. But, the Ferrari one had the single-stage fluidic switch and with a single-stage switch it is never perfect. So, when you blank the duct in the chassis, you blow air in from the side of the switch. We had a double-stage switch to make sure we got the full effect to the rear wing. You've a duct on top of the chassis that fed down to a hole that you put your elbow over. When you blocked it the air came down to another fluidic switch, a clever Y piece basically. The main flow comes out through the vent tube. The whole thing was what we called an F-Duct but the main [air] flow that was going to the rear wing was coming through a separate entry up by the roll hoop. That air would be directed to a slot on the back of the lower wing, and the switch changed the direction. There was a 2nd switch [Y] in the McLaren [ductwork] just to make sure you got 100% switching of the flow to the rear wing. It was an interesting, fun concept and the hard thing was developing it. McLaren had been looking at this since the winter of 2008 with the initial bits, as a concept to get into the 2009 car. But we were in so much trouble with the 2009 car and the aero program for the 2009 car had started so late and we were so far off on the total downforce, we just had to put all our effort onto that, rather than into the things which were icing on the cake.

Jenson Button says, "We would actually use our arm. We had a pad on our arm that we would use to cover a hole in the side of the car where you would push against it, keeping both your hands on the wheel." Pat Fry concurs, "It wasn't too bad for the driver to seal, to be honest. It didn't have to have a perfect seal to get the thing to switch. I think everyone got excited about it, but all of McLaren development was working down that single rear wing. The delta CD that we had on it wasn't massive, it was 8-10kph top speed gain. It wasn't a monumental step because we put all our bets on that wing working, we ended up with quite a low-downforce wing. At the first race when we were really quick in a straight line, everybody was thinking it was worth 20kph, but half of that was the DRS effect and half of it was not having the right wing. We were running lower downforce than we'd otherwise have run. The Ferrari one actually had a bigger DRS effect."

Paddy Lowe reflects, "It was overall almost a two-year project, perhaps a year and a half. A thing like that is pretty difficult to put together, particularly the venturi switch took quite an awful lot of R&D to make that work. The nature of these things is that you don't put ten people on it when you don't know if it will work. First R&D normally runs quite long and thin until you see that it works, and then you put more resources on it. Of course the people that are copying you don't have that problem because they know it works, they've seen it working. So they can afford to put ten or twenty people on it straight away and then perhaps deliver it in two or three months." Imitation the most sincere form of flattery? "Yeah, absolutely. It's just slightly frustrating to see that it was copied it so quickly. We had it on for race one in 2010 and I think the first other team had it around race eight or something like that."

The system created a significant advantage on the straights, akin to a DRS effect in 2017 terms. Inevitably the drivers began to wonder if this could find a use in high-speed corners where perhaps the downforce wasn't so critically important. Jenson Button recounts:

There was only one corner really where I did it, and it was for one lap in qualifying and that was at Abu Dhabi in turn two. We had so much downforce back then, and it was just about flat without the F-Duct. On my last lap in qualifying I used the F-Duct and went through there. It was very loose, but I gained tenths of a second as you carried that speed all the way down the straight. It was scary definitely, but it was the same as when we could use DRS around the whole lap. You ended up using it in places that were so dangerous to do it in, but you had to try. At Suzuka, you saw a Red Bull do it through 130R, but the Red Bull had so much more downforce than anyone else did. So you'd try it and lose the rear and have a massive chance of putting it into the wall. It was a massive step in downforce with the DRS open, and it was the same with the F-Duct and you took more risks than you should have done, so what they've done now was definitely the right thing with the regulations.

The biggest problem that Formula 1 may face today is the hyper-critical aero packages that are seriously compromised the closer one gets to the car ahead. To try and help 'the show,' Grand Prix racing adopted the Drag Reduction System, or DRS. This allowed cars to open a wing element when they were within a second of the car ahead in designated DRS zones, typically on the longest straight or two. The idea was that the reduced drag could assist the car behind to overtake before arriving at the next corner where the airflow to the front wing had to be restored. In the area of the nose or front suspension, everything is aero driven. As Paddy Lowe says, "Anything in that area is determined by the aerodynamic benefits. It's a quite dominant aspect."

The pullrod front suspension 2015 Ferrari. Advantages were aerodynamic, which is everything (Norm DeWitt)

When James Allison was the Technical Director at Ferrari in 2015, he brought me over to an enormous wall mural showcasing the front of Vettel's Ferrari and explained the goals of the aero program on the modern Formula 1 car. The f250 point is the edge of the front wing on the inside point. The main idea in the area of the front wing is to direct the turbulent air around the outside of the front tire and not the inside, so it does not reattach and screw up the rear aerodynamics further down the side of the car. The advantage of the pullrod is that it attaches high at the inside of the wheel versus low. The net result is there is less to interfere with the aero at that critical point to underside aero. The pushrod is the opposite, it attaches high at the tub, which is a less critical area, and then is low at the wheel hub. That is an aero negative in a very important area, and explains the pullrod Ferrari in a nutshell.

Pat Fry was the Engineering Director at Ferrari in 2014, and he explains exactly how the nearly horizontal pullrod operates. The key is in the front-suspension droop angle when the car is at static load. He explains:

> Pullrod versus pushrod… pullrods have been around for ages, like back when I started racing about twenty years ago. They came back into the rear of the car to make the packaging tidy and clear it up around the lower part of the brake duct area. Red Bull were the first to do it in 2009, which just makes the back neater. So when you see an almost horizontal pullrod in bump it actually is pulling because the car is getting wider. We are trying to get that all clean and we are trying to minimize the effect of the suspension area. The pullrod was not a dramatic game changer when we went down that path in 2012. When you look at it, you go, 'It's impossible, that would never work, that is such a shallow angle.' But when you really look at the full geometry of

Details of the 2015 Ferrari showing the pullrod routing and aerodynamic enhancing assemblies (Norm DeWitt)

it, you realize it's not as bad as it looks. There is also some small CG benefit of moving 2.5 to 3 kilos from the damper units.

The difficulty at the front is that we run the car much stiffer, so therefore we've got to be careful with the stiffness of the pullrod. In reality the concept is not driven mechanically, it's driven aerodynamically. The reason for it is the aero benefits and what it does for you. The pushrod gets quite challenging as the higher you push it up, the more you have the similar sort of problems that we have with the pullrod. The last few years, people have been pushing the lower wishbone up and up, to try and make space for aero devices. Various people actually have the pushrod going through the wishbone to try and get a lower point to get a sense of the mechanical advantage. We have exactly the opposite problem with the pullrod going through the top wishbone... we want the top wishbone as high as we can. It's all driven by that blockage of the junction of pullrod and wishbone, you don't want it anywhere near the bottom of the car. Even when you've raised the lower wishbone up, there is still an aero penalty of having all that stuff meeting in the middle.

In 2016 things were getting ever more creative as the pushrod cars often featured a long horizontal tub at the allowable height limit. Pat Fry in 2016 was now the Engineering Director at Manor, "To some degree I guess the chassis shape is coming down is a different set of constraints really. With the way the rules are written, I think a good example is if you look at Red Bull or the McLaren, it's done so you can get the pushrod actually out the top of the chassis. If you look at a picture of the McLaren, you've got the complete top of the chassis on the legal limit and with the rocker posts out the top and the pushrod connects above it so you can get your pushrod load ratio down that way."

Another commonly seen feature of the current generation Formula 1 cars, is the small vertical 'strakes' that one finds atop the sidepods, or in various locations around the bodywork. Pat Symonds was the Chief Engineering Officer at Williams and explains in 2016, "Yep, what are they all about? They are not really turning vanes at all, they are vortex generators. An awful lot of aerodynamics to the cars at the moment, goes on at the rear corner. So, if you look at the rear brake ducts, they are stuffed with winglets. They are remarkably effective… they represent 6 or 7% of the downforce on the car, so there is a really big focus in getting really high total pressure, down into that rear corner. If you let the flow go over the bodywork, as you know, you just pick up the boundary layer and you lose that high-energy air. So what you need to do is to trip it. Get the boundary layer down and get the energy in the air. The vortex generators increase the downforce on that rear corner of the car." As confirmed by Allison at Ferrari in 2015, the game is to simply get the air to reattach at the back of the car. Symonds confirms, "Absolutely. Even the front wing is all about the rear of the car, it's not about the front of the car."

The rear corner is the most critical aerodynamic area of a Grand Prix car, 2015 Mercedes shown (Norm DeWitt)

One often finds tunnels incorporated into recent front-wing designs. One might assume that to create an airstream that then attracts the surrounding airflow to it. Pat continues:

In a way, again, they don't produce a jet, they produce a vortex. We are really trying to get all the vortices to work together. We have three things if you like, that we'd like the front. One, to produce downforce at the front of the car. That's the trivial part, and that's relatively simple to do. Two, we are also trying to control the wheel wake off the front wheels. They have this horrible, really destructive wake. Of course it tends to leave the front wheels and dive in towards the rear of the car. Again, very destructive to those rear corners that we were just talking about. Even to the rear wing when it starts tumbling into the rear wing area, particularly when you have yaw on the car. So, we try to put vortices that push that wake out. We often do that by using counter-rotating vortices. So, if you get one vortex going like that [clockwise] and another going like that [counter-clockwise], it's rather like in the middle they tend to push the air out. A lot of the tunnels and the endplates are to do with trying to get rid of that wake. Three, then we use the Y250 vortex, which comes from the intersection of the non-lifted part in the middle of the wing to where it joins the other part. To some extent it does add to get rid of that wheel wake, but we are trying to use that to run that along the sidepod and seal that area and stop that turbulent wake from coming back under the car. It's trying to stop this bad air getting in more than anything else. The tunnel vortices, together with the cascade vortices and the endplate vortices, they are largely to do with just trying to get the wheel wake pushed out. You can see from the endplates that we use a lot of outwash. We are trying to get this outwash to drag the wake outboard."

An odd feature of recent design has been weird-looking finger-style terminations sticking forward on the rear-wing endplates. Symonds says, "We are all using these strakes and things. Again, more and more, because we are limited on the height of the diffuser quite severely, we are trying to get the expansion by expanding laterally, not just vertically." Opening up the endplates expands this laterally? "Exactly that. So we are using a lot of lateral expansion."

Perhaps even stranger than the slotted endplates is the concept of the Spoon wing. With most conventional sports car design, where the air comes over the greenhouse, the wing is twisted to a flat or lifting profile. That is because it reduces the drag, improves the airflow off the back of the greenhouse, and then the wing twists to a higher-downforce configuration near the sides of the car where it receives uninterrupted airflow. Formula 1 has often gone in the complete opposite direction, with spoon wings that dip down to create even higher-downforce

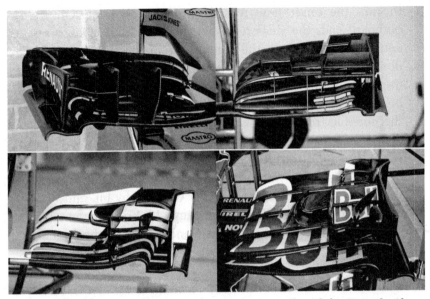

The critical front wing designs vary from race to race. Above left – Renault, Above right – Manor, Below left – Sauber, Below right – Toro Rosso (Norm DeWitt)

angles approaching the centerline of the airbox/wing. One would think the airflow cascading down above the airbox would be similar to the effect of the air coming down the greenhouse of the GT cars. It looks counter-intuitive, yet was run on both the Mercedes and Manor cars at Monza.

Pat Fry explains, "What is the design concept there? It's all about trying to even out the pressure distribution across the wing." And Pat Symonds adds:

> It's not as much of a polar opposite as you might imagine, actually. It is true to say that we get more downforce from the tips than we get from the center. The difference between Formula 1 cars and sportscars is that we run our wings very very very close to instability, because we are trying to get so much downforce from them. To give you an example, you may notice that at a pitstop we clean the rear wing because it is so critical. If we get insects on the leading edge, we'd stall the wing, that's how critical it is. Aircraft do actually suffer from the same thing. They don't get the stall that we would get, but they pick up drag and pick up fuel consumption. So we are trying to run the whole wing very close to the stability limit. Now of course

the air, the onset flow to that wing, is different. On the outboard ends the air is relatively clean air. The airbox is not entirely different from the greenhouse of the sports car. But don't think about cars going down the straight, that's not the important thing. Think about cars going around a corner. The big difference between a Formula 1 car and a Sportscar is that the horrible front-wheel wake is much more contained on a Sportscar. What we've got it that s**tty air coming onto our rear wing. We suffer from about that much more, which means the onset flow onto the wing is not everything we want it to be, so we try to manage the profile of the wing bit by bit, and what that has led to is things like spoon wings. It's a real fundamental difference between a closed-wheel and an open-wheel racing car.

Another feature of the modern Formula 1 car is the variety of chassis-rake solutions across the grid, with the Red Bull cars along with the Force India leading that trend with the most chassis rake. Force India's Robert Fernley explained in 2016, and again, it all comes back to the front of the car, "Whatever you do, it all starts at the front, the way you are channeling the air around the car. The rake is only about managing what is coming in from the front. Our aero philosophy is very similar to the Red Bull program, it's all about being able to pull the air out and funnel it through. It's always about how much flow you can get, all about increasing the flow and therefore the speed. You are allowing more air to get in."

Is the concept to accelerate the air under the nose through a tighter area? Fernley answers, "Within the design limits it is how you exploit those limits, it gives you a little bit of flexibility and the rake is a way to control it. When you look at the Red Bull as it's going down the straight, it starts to come down a little bit. I think it's only Red Bull and Force India running that much rake."

In motorcycle Grand Prix racing, winged aero developments of the 1970s would find a resurgence in 2015. Across the paddock, a variety of winglets began to appear on the fairings, and as it was explained to me by a Ducati engineer, they were there mostly to provide wheelie control. One assumes that such things were relatively unnecessary with the previously sophisticated software allowed. Now, with the basic Dorna software packages used across the field, the winglets help provide an aerodynamic aid to help maintain the proper attitude under maximum acceleration. In 2016 Jorge Lorenzo confirmed that Yamaha had not tried the unsprung downforce route, by putting winglets on the front fender, versus the 'sprung' downforce of putting them on the fairing. Lorenzo says, "I don't know what the rule is about that. We have not tried that." The story was the same at Repsol Honda, where Dani Pedrosa confirmed they hadn't tried that approach either. The fairing-mounted winglets weren't a huge advantage as seen from Dani's viewpoint, "If you have the correct one, yes. But we have had also crashes out of this because it wasn't in the correct point or giving the correct effect. Win win? Not really." As often happens with such things, the projecting winglets were banned for 2017 under safety concerns given the sharp edges of some designs.

Jorge Lorenzo says, "My riding style is different, I lean a bit more than the other riders and I am more on the throttle, so there is slightly more graining [of the tire]." Ducati had a more advanced aerodynamic development program with the winglets than the other manufacturers in 2016. Are you struggling with wheelie control? Lorenzo replies, "Not that much in wheelie, but more in front contact almost everywhere, so in some areas we have to close the throttle more, so we lose a little more than the other brands, but they decided to change the rules."

Has there been a significant difference in tire temps with the removal of the winglets for 2017? As Nicolas Goubert, Deputy Director, Technical Director, and Supervisor of the Michelin MotoGP Programme says, "To make a difference in tire temperature just from downforce is something we see in motorcar racing where they have quite a lot. Maybe... I didn't say they didn't have any effect." Obviously the winglets had a very minor, if even observable, effect upon the tire temperature. Andrea (Dovi) Dovizioso reports, "For sure the winglets make a factor, but in the two years I did the podium [here in 2014, 2016], the front tire condition was much better than the competitors." It is hard to draw any comparison here regarding tires, as the 2014 machine had no winglets, while the 2016 bike had large winglets with endplates.

Is the missing downforce from the removed winglets part of the problems Ducati is having with the 2017 front tire at Austin? Dovi says, "Nah, I don't think so. At the start... acceleration, for sure we lost a lot compared to last year." Have you been able to work within the software to get back some of wheelie control development. "Coming out of the turns we are better than our competitors, this is not a bad point for us."

By mid-season a series of winglets best described as a box kite, sprung out of the sides of Lorenzo's Ducati fairing, but did not find favor on his teammate's (Dovizioso) bike for a few races until the advantage was proven.

Aero plays an enormous role in Indianapolis car racing, given the complexity of recent aero packages in use through 2017. Chuck Sprague of Penske Racing explains:

Pack racing certainly teaches the guys respect, but when it goes wrong, it goes wrong big. I would defer to Rick Mears' idea, which is more power and less grip. If everybody is within a one-tenth of a percent of everybody, nobody ever has enough margin to pull off the pass. We tested at Mid-Ohio with Rick in '89 or '90, when the rear wings were basically unlimited as long as they fit in the box. We tried the low-drag setup, which was a short-oval setup, and then we tried the traditional road-course wing. Then we tried this huge monster they called a cascade wing, with two leading elements and two trailing elements like an airplane with all the flaps down. The lap times were within 0.15 seconds for all three setups, but the difference in gearing between the low-drag setup and the high-drag setup was three teeth, which was huge. If you looked at the profiles on the data, you could see huge differences in speeds, straightaways versus corner, yet the lap times were virtually identical. You would think that you'd want the low-downforce setup to pass, but here's where Rick Mears comes in. He says, 'Where do you pass people? End of the straight. Where do you need downforce? Under braking. But if you put the low-drag setup on it, you have to get on the brakes before the other guy. But if you put the high-downforce setup on it, you can't catch him to outbrake him. So I'll have the one that's not too hot and not too cold, but is just right.' So he took the middle setup.

So if the best setup is the Goldilocks theory of being just right, how much downforce should today's Indy spec car provide? Rick Mears describes it, "Downforce to a driver is like making money. I make $30,000 this year, and made $60,000 the next, then you tell me the third year that I have to go back to $30,000 I'm going to think, 'OMG, I'm dying. It's impossible, I'm going to die, it can't happen. I'm going to starve to death.' The danger card gets played about how you are going to hurt somebody." The reality is that there is nothing more dangerous than the open wheeled version of a NASCAR restrictor plate race, where the high-downforce cars are flat on the floor all the way around a superspeedway in a giant scrum. Mears concurs, "That's what I told them. This is where you are fighting different entities. We've already created the monstrosity of a pack race and the media has seen that. Now if you don't have a pack race, they say it's a ho-hum dull and boring race."

Indycar adopted a new concept for 2016, the domed skid to reduce downforce and keep the cars from getting airborne when turned around. It was a bit astounding that so many of the Chevy aero kits took off once going backwards at Indy. The ACO required the Delta Wing to go into the wind tunnel backwards to establish that it wouldn't take off at Le Mans. One can only hope they require something like that for Indianapolis. Scott Dixon explains:

> I don't think it was a one-manufacturer situation. I think the Honda was implemented to run a spine wicker as it was worse in yaw in some situations. There were only three spins that month with the Chevies and there were more significant things that were causing the accidents. The speedway sidepods moved the center of pressure significantly when you are turning, and with Helio's one they had missed on COP significantly and that caught them out. With one a tire went down, and with Ed's I think it was a similar version of the COP thing. I think if a Honda had been in a similar position... if you look at how open the rear pods were and how the back of it basically looks like a wing ready to take off, you'd think that if it went backwards it would go up in the air.

> The shape of the sidepod for the Chevy body kit changes significantly between qualifying and race trim. In race trim we have a wedge and a coke bottle. In qualifying it's just a big ramp. And when that is turning in yaw, it rotates the car a lot quicker. In the wind tunnel you don't see too much of that, which is how it caught everybody out. Say if your race COP was a 36%, to qualify you would need to lose maybe 3% or 4% to get the same amount of rotation. The Chevy version between qualifying and race trim was a lot of pieces, and it was right on the edge of

efficiency. With the domed skid it put us into an area where when the track heated up, it was really hard to get it into that sweet spot. From team to team you could see that we thought it was more efficient running the winglets like that, and then Penske thought about running it on the right side. That's just how it is.

It gives us all something to talk about. Scott laughs, "Exactly! It will be interesting if they keep the body kits around to see what they are going to do. It's hard as they are super expensive and there are a lot of pieces that you have to buy. When Texas moved to become a day race, you had to put on all the downforce you could find. Track temperature is the biggest culprit. Some teams were running road-course winglets opposed to superspeedway winglets to get more downforce. Others weren't running them as they were about ten grand to buy and they'd assumed, 'Who would need road-course winglets at a superspeedway'? It's expensive, even for the manufacturers."

Sebastian Bourdais on the domed skids, "It is no equalizer, it is a safety feature that raises the car and takes the downforce away. It's not designed to make the car any nicer, but it worked. We've had a lot of cars that obviously took off going sideways in the last couple of editions of the 500 with the DW12. Now, all of a sudden we've had a lot of cars spinning and hitting the wall, but not taking off, so… mission accomplished. Is the car nicer to drive? Absolutely not! We lost a ton of downforce, and we've got enough wings that we can find the downforce back, but then the CG is higher so the car is not as nice. In the 2016 race, everybody was running more downforce than we used to, to try and make up for the loss of mechanical grip from when the CG was raised." The qualifying setup is another thing entirely. Sebastian continues, "In qualifying, with these aero kits the rear wing is tiny and is basically just a balancer. I still say more power would be great, but you need the downforce to keep the car under control. It needs the downforce to stay drivable."

Scott Dixon shares his opinion:

The domed skid was just for safety. From what we've seen at Indianapolis and in the crashes we've seen there, that and the area where the flip-up things helped significantly. I don't think anybody got enough praise for Indycar for doing that. Driving-

-wise it made it more difficult for sure across the board. I think it took away maybe 150 pounds of downforce by raising the car by 9mm, but then you had to put on almost double or three times that to make the car work the way you did in the past. Putting on more wing to get it to work. You'd try to improve the mechanical balance, there is still a very big emphasis on mechanical grip because the tires degrade so quickly if you can get a few extra laps versus your competitor it helps so much. It did feel like the car was quite high, the front feels like it's floating quite a bit and it's more affected by being in traffic with wash-out.

Downforce reduction might be the goal, but getting to that result is not as simple as unbolting wings. Scott continues:

It's an easy thing to say, applying it is quite different. Taking off a bunch of downforce is one thing, but you've got a tire that has half the load, it's got no grip. As we've seen, there have been more pack races, but Texas has become much more of a driving track than a typical mile and a half where you are flat out. In the early days you had the pack races where you just followed the white line and kind of got crazy. 2015 at Fontana was the fine line where it probably wasn't going to be a pack race, but I guess the simple answer is that it's more complicated than just taking off downforce. We went to Iowa because we've had significant loading at Iowa in the corners the past few years for an extended period of time, pulling six G's and we've seen suspension failures too. At that test we took thousands of pounds of downforce off the car and it really didn't change the loading, but in that scenario I think it would have made the racing a lot worse. You took off downforce, but kept going flat because there is so much banking and that becomes the most significant part. You'd take off 10,000 pounds of downforce and then it would change the actual loading by 200 pounds.

For 2018 Indycar is coming with a new single aero package less dependent upon wings and more dependent upon the underbody. Rick Mears explains:

Part of the new package is that it's more dependent upon utilizing the bottom and not so much upon the top, so there is less disturbed air and it is better in traffic. Downforce – I think it's less, and for me I don't think you can go too much [that way] but I think it's going the right way. Until now, it's been a steady drip, drip, drip in the other direction, and at least now we are seeing what we need to do and we are working hard at it. You've got to keep trying stuff, and if you don't change anything, nothing ever changes. In today's world everybody is afraid to make a change because if it's wrong, they get beat up. It takes the enthusiasm out of making changes, but until you try, you never know. Now we spend more money on trying to find things that are such a small difference. If you had more power and less downforce you wouldn't even notice that. The whole point is competition.

Yes, the whole point IS about competition, tempered with safety concerns for competitors and fans alike. Hopefully the sanctioning bodies will be able to balance the endless advances in aerodynamic wizardry with that fact.

Rear view of a Mercedes Grand Prix car. Note the emphasis upon the winglets in the brake duct area (Norm DeWitt)

CHAPTER 20

THE TECHNOLOGY OF SPEED

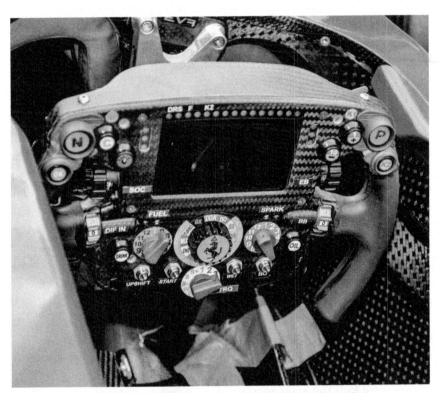

The steering wheel of Sebastian Vettel's Ferrari shows the complexity of operating a Grand Prix car (Norm DeWitt)

In every era there are inventive solutions to resolve the issues of the day. This chapter examines a number of technical developments that resolved these challenges. An example of one would be the age-old compromise of going through a series of corners where one needs to upshift somewhere while under lateral load. The downside of the upshift is that it destabilizes the car, and the ensuing wiggle (or worse) costs speed. For generations of racers the solution was to go into the corner one gear higher and if you had a powerband of reasonable width, let the torque pull you through the first part of the corners so that the corners could be completed without shifting. This technique works particularly well in a downhill series of corners such as Laguna Seca's corkscrew, and probably not at all in an uphill series, unless you have a huge powerband akin to a Can-Am car, where it probably doesn't much matter which higher gear you select.

Imagine, then, the complication of having a seamless-shift gearbox where the next gear is triggered instantly, with the attendant 'kick in the seat' that comes from the shift. One marvels watching Formula 1 cars accelerate through a series of bends, snicking through the gears without a hint of a chassis twitch. The answer to this riddle lies in the technology of the modern Grand Prix car, where advances have been made to soften the seamless shift when it is advantageous to do so.

Paddy Lowe explains, "There are two aspects of that. There is the seamless shifting that we now do, there is no interruption of torque. A full-on seamless shift without any torque reduction will of course give you positive results, because you engage the next gear before you disengage." Is there a minute time where there is no drive? "No it's quite the reverse. The drivers say its full throttle throughout and if you do nothing except the straight forward seamless shift, you will get an increase in torque transiently because of the energy that comes out of the crank as you decelerate it when it drops 1,500 revs or whatever. That energy goes into the transmission as an increase in torque, there is no reduction that you have with a conventional shift. That's the full-on, full-energy power shift, which is what you would do down the straight, where that energy is useful. You can soften that a bit by inducing a little bit of clutch slip during the shift to just momentarily dampen out that deceleration energy from the crank. We now have the techniques to create a very smooth shift without interruption."

Hence, with the Mercedes Grand Prix car, there is a G-load accelerometer to determine if the car is in a high-G corner where the chassis might be more sensitive to that crank energy upon shifting, and thereby trigger that bit of clutch ship. Paddy adds, "Yeah, that's one aspect. The second aspect is the power unit. First of all the torque is anyway much stronger at all revs than the old V-8s and it has a much smoother torque curve by the nature of the engine being a low-revving turbo, the turbo engine filling in all those holes. The torque of the engine is much more consistent at all revs, and where it isn't we can smooth it out using the MGU-K electric motor, so putting that all together we have engines that have far more consistent and smooth torque. So the drivability of these engines is far better than the old engines. When you put all that together, it can give you a car that is much more able to run with stable torque through a gearshift."

Paddy Lowe with the most dominant team of his long career,
Nico Rosberg, Lewis Hamilton, and Mercedes (Norm DeWitt)

This could be highly useful technology for hybrid vehicles on the street, a reason for the manufacturers to be involved. Current hybrid vehicles are a stone axe in comparison to the sophistication of the system used in Formula 1. One wonders in this world of kinetic energy recovery, does the programming have to deal with the driver using brakes as well as throttle at the same time, an age-old rally driver technique. Paddy answers, "Those environments do overlap from time to time. Lewis tends to do it more than Nico, but I'm not completely sure as to what causes it."

Maintaining traction through a seamless-shift gearbox can be far more complicated given the small contact patch of the typical Moto GP machine. Honda typically had adopted a 'point and squirt' technique, where the bike squares off the corner and then explodes out in a straight line. The Yamaha has traditionally worked best using a smooth, arcing radius style of cornering, which is why Jorge Lorenzo was able to start his first Moto GP on pole position with the Yamaha, riding it as if it were a really fast 250 Grand Prix bike. Where this tradition went upside down was upon the introduction of the Honda seamless shift box in 2011. One wonders how this could be possibly be beneficial when the shock load from the crankshaft torque can make the bike squirrely on each shift.

Dani Pedrosa was riding the Repsol Honda in 2011, and every year from 2006 to 2017. He describes the effects of the technology:

> At the beginning, every time when we were flat out, not when we were at 50% of the throttle, the shift chop was very strong. You felt a super strong chop with some kind of wheelie. That upset a bit, so we had to turn down the effect to have more control and have constant acceleration. Actually the seamless was created not to have so much benefit on the full straight-line power, but to have stability in the corners when we shift because our main problem at the time was the rear grip. When we shift and have the cut from the ignition and then get back the power on, we even lose more grip and we had a lot of shock movement, so what we wanted to achieve was a more constant automatic feeling on the gear when we shift up just so that we don't lose some of the grip that we have in the corner. Then we also see benefits on full power in a straight line.

On the Respol Honda, the riders didn't need to go in a gear higher to reduce shifting shock under lateral load in the corner. Dani continues:

> I was doing the same gears. Before the shift shock was cutting off a lot and the timing was too big, the feeling was too mechanical, like "Whack!" The target was to improve the stability on the shift-up. It's more about how the gear engages. Of course we have electronic control, but I am not an engineer, I cannot go that deep into the technical, because I just give my feedback. I do not know what actually is happening and how they do it.

> One of the biggest improvements was the shift-down. Before, we used the clutch to shift down gears, and that clutch affects a lot the engine brake. So if you'd go from sixth to third, you'd have the engagement and it would go 'wuh', release and go 'wuh, wuh, wuh' and what we wanted to have was a more constant grip going 'wuuuuuuuuuh' and that was by creating a different clutch so we don't need to use the clutch into the turns, and it is full time, 100% of the time. We developed this two or three years after the seamless shift was first used. One of my strongest requests was how on a normal bike you have first gear, then neutral, then second, third, like that. I always had problems between second and first, many times I'd touch the lever and get neutral and maybe I'd miss the corner or give me like a highside almost. I said, 'I don't understand why it is like this, I would like to have all the gears – second, first, and then neutral.' We developed that and then a few years later all the other manufacturers have done it too. It's quite positive because you can go from second to first and you do not get the neutral in-between. One of the issue of this for me was going on the upshift from first to second because I have really short legs and when you are leaning on the right side, my leg would not touch and maybe sometimes I would not be so precise and just touch a bit and go into neutral instead of second and make a big mess. This was quite an interesting upgrade in 2013.

At Yamaha, their gearbox technology was far behind the factory Honda. The seamless shift Yamaha originally would upshift seamless, but wouldn't downshift seamless. Jorge Lorenzo rode the Yamaha Moto

The seamless shift Honda Moto GP machine of 2012 was an early successful integration of the concept (Norm DeWitt)

GP bike from 2008 to 2016. Does the rider go into a corner a gear higher to avoid the shock load of the gear change with the Yamaha? Lorenzo says, "Yes, sometimes. The shock is quite big. About three years ago it was [turns his body sideways and makes an alarmed face]. We have reduced it some." Closing out the discussion on the seamless shift transmissions of Moto GP, the different experiences between the Honda and Yamaha riders speak to the importance of technological advances in power-delivery management. While seamless-shift technology tamed the explosive Repsol Honda, at the same time it turned the flowing smooth Yamaha into something that could bite.

With the hybrid powerplants, or 'power units' of Formula 1, one wonders if there are issues with regenerative braking not fully charging up the power unit battery when it's a wet track and the brakes aren't used as hard as they would normally be. Paddy Lowe assures, "There are no situations where wet weather has affected apart from playing here and there being a little bit down." With improved development of the power units, coasting techniques were also typically no longer in use for 2015 versus their first year of use in 2014. Paddy adds, "The biggest factor is the circuit layout. Sochi is one, due to its nature, where it is very difficult to make it on fuel unless there is a safety car. So, we are still in a situation where at part of the circuit there has to be a reasonable amount of conservation. Sochi is the worst. It's not the length of the straights."

This seems counter-intuitive as one would think that fuel strategy would primarily come into play where you've got your foot to the floor at Spa for twenty-two seconds at a time. Yet, the highest demand for fuel economy is at tracks like Sochi. This is something you find throughout the paddock consistently, and Sochi always heads the list. The consensus is that it is because drivers are flat to the floor for a higher lap percentage than at any other circuit, a start/stop/start/stop circuit with medium-length straights.

Pat Fry says, "Tracks like Monza, I can't say exactly why they aren't and I'd have to go have a think about it." Maximizing the available technical combinations is very complicated. Pat continues:

In reality if you've got fuel when you want to charge the battery back-up, keep the throttle down and drive the electrical motor the other way and charge it up and convert fuel into electrical energy going down the straights. If you look at our speed profile, we'll come out of a corner once you've gotten out of the grip zone, you've then got a choice of 'do you open the wastegate' to get another 15-20hp, driving the turbo with the electric motor that is on the turbo. So then you are using electrical energy to open the wastegate. Then you close the wastegate and carry on pushing with the engine and with the K [Kinetic], with the electric motor that connects to the drivetrain. Then at some point, depending upon the energy you've got and the fuel you've got, you'll stop the K running, and carry on just running on the fuel. Then if you've got to save fuel, you lift and coast at the end. All the techniques are similar to the endurance cars these days. I was chatting with Allan McNish about how you aren't driving hard much of the time and you are lifting and coasting, saving fuel. I think if you've got a certain amount of energy, the best place to deploy it is at the start of the straight. It's a lot more efficient that having a few kilojoules here or there. I think they [in the WEC] are in more severe fuel savings than us, anyway.

The 2016 championship dominating Mercedes Grand Prix car
(Norm DeWitt)

Another item of note in the current Formula 1 car is how supple the dominant Mercedes appears to be. Johnny Herbert lends his opinion:

Look at the Mercedes now, and the amount of roll it has and the amount of grip it produces. It's soft, and it's very clever how it rolls and gets an aerodynamic seal on the endplate. It is so controlled, like where we were with the active cars. The engineers in 2017 are going to be loving it, but I'm worried that it's not going to be as much of a spectacle or an aero game. Are you going to see three seconds a lap? Are you going to see that? I understand that Ferrari and Mercedes are a brand trying to promote their technology through Formula 1… okay no problem. But to almost allow them to dictate where the rules are going and how the cars are going to be because they think that 'if we do this' it will give them an advantage, it's not about the engineers getting a perfect car. You want to see the drivers wrestling the cars around. It's quite fun with Max [Verstappen], as he's on the edge and that's fine, it's what racing is all about. He's very good at the moves he's doing and everyone wants to stop him. In Japan 2016, Lewis couldn't get past him because he was, I think, defending perfectly fine. We don't want to stop that, to say that you can't defend, we should want that to happen much more. Stop having rules for what you can and can't do. Are the people who are trying to implement rules, do they actually know what it's like in the cockpit? Do they know what factors Max and Lewis or Fernando are using? Look at Valentino Rossi with Marquez. There's a bit of a push and a shove. Yes!

As it turned out, Herbert was right to be concerned for 2017 as when the following car got within a second of the car it was trying to pass, the speed differential of the DRS often wasn't enough to compensate for the loss in aerodynamic efficiency. The aero became so disrupted that unless the straight was very long, it was nearly impossible to pass, and you end up with a parade such as what happened in Hungary 2017. In that example, the faster car on rare occasion got to within a few tenths, only to immediately fall back again due to the aero turbulence.

Fuel and lubricants are free (no spec manufacturer), and friction management is a huge part of contemporary race-engine design. One wonders how that works for Mercedes engines allowing different lubricant companies' products in their engines, such as with the Mobil 1 McLaren-Mercedes versus the lubricants used in the factory Mercedes. Paddy Lowe says, "Well you shouldn't and you try not to. It puts a far bigger load on the development program if everything you do has to be type approved on more than one type of oil or fuel. All the Mercedes engines are now running Petronas products. The exception was with McLaren having Mobil products, one of their major sponsors. In that case McLaren had to pay extra fees for extra development to use their lubricants in the engine."

When asked for specifics in 2017 as to how such comparative testing is achieved, the head of the Mobil 1 worldwide program essentially said, 'No comment.' Although he did mention there were over fifty different parameters. One might assume there are a few more parameters in play now, given the oil-burn horsepower increases that are now legislated as to the oil burn amount allowable.

Other revisions across the past few seasons were the changes in the clutch launch sequence, being changed to a single-paddle system that doesn't allow for engineering-staff-guided bite point manipulation after leaving the garage. Much of this started with the hand clutch pioneered at Renault, and by 2004 the combination of Trulli and Alonso ruled the start lines of the world. Pat Symonds recounts:

It all started with Launch Control with the 1992-93 active cars, and they started studying exactly what it was about launch that made for a good launch versus not a good launch. So what they came up with was some parameters, banned for 1994. They tried to figure out how to make this work mechanically, so they came up with a dual-paddle hand-actuated clutch system. They had really developed the good hand-clutch start system and that mechanical launch advantage at Benetton/Renault was to carry over for at least a decade. One of the things we had was that we used to run the car with much more of a rearward weight distribution than anyone else, and so as soon as the car launched, it meant we had so much more grip. It meant that we had to develop our entire setup around the weight distribution. Because we were much more involved with the tires then, we could tune the tires to what we required and we were able to tune them to our weight distribution. It also had the twin-paddle system similar to what we had in 2014.

In 2015 quite a few things changed. As the years have gone on, everyone was doing manual starts, and as always when you have stable regulations, people tend to gravitate to a good solution. What we are having to deal with at the moment is some significant changes to that. No longer are the engineers allowed to examine data from a practice start and advise the driver exactly how to adjust his clutch to make a better race start. The driver has to live with what he's got and he has to modulate his start. This year we are still able to use the two paddles to maintain control. What you do essentially is that both pedals have the same effect on the clutch, but the control system will take into account the one that is most disengaged. So you fully disengage the clutch with one paddle and hold the other one to where you think the bite point is. What we used to do was to adjust that bite point with a switch, which we are no longer able to do. The driver now is trying to hold the bite point on one paddle and disengage the clutch with the other. When the lights go out, he lets go of the paddle that kept the clutch completely disengaged, and the control system then starts reading the other one, and in an ideal start the driver will hold that paddle completely still, precisely on the bite point. As the car starts to accelerate he will feed more clutch in and eventually let go of it.

In reality, you are accelerating at over 1G, so it's quite difficult to hold the position, so there is a little bit of modulation going on with it. The big advantage of having the two paddles is that you can concentrate on holding one of them on the bite point and not have to worry about the car moving. For 2016 we had to go to the single paddle and that really makes it much more like a road car. The driver has to find the bite point at the start and when the lights go out, he has to feed in just like you would do with a manual transmission profile. You can't just hold it at 10,000rpm and just drop the clutch. You really can't do that these days. If you've got a nice big V-8 with a big flywheel on it, you can do that sort of thing because there is a lot of inertia in the engine, it won't bog down. The inertia of a current Formula 1 engine is really small. If you over-engage the clutch, it will tend to pull the engine down if the grip is high. If the grip isn't enough, it will spin the wheels. You are effectively connecting a low-inertia engine to a high-inertia drivetrain and that is not an easy thing to control. As far as the driver is concerned, the change has not had a real effect. Before the driver could look at the data from the practice start from our grid position. The engineer could not transmit data to the car, but the engineer was able to look at the data and advise the driver to select 'clutch position 5', which might change the clutch by one-tenth of a millimeter, or in your language, "four thou." You have to get the clutch to the ideal temperature for the ideal launch, and we are not entirely sure exactly where that temperature is. It is such a multi-variable problem. The ideal launch depends upon the temperature of the tires, the grip of the circuit, and the slope of the circuit. It is difficult to optimize and it's not an easy problem. It is not a controlled experiment, and it is not easy to do testing on it.

How does the current testing protocol work during the typical three-day Formula 1 weekend? Pat continues:

On Friday [or Thursday] the circuit is changing a lot, especially on a temporary circuit like Monaco. The evolution in grip between the start of P1 and the end of P2 is massive, so you can't test anything. On top of that, you've only got two sets of tires for the first session. Friday is all you've got, but very often on Friday morning there are things you can do. People are looking at flow visualization on new aerodynamic parts that have come from the wind tunnel. The wind tunnel is a simulator, so you are checking reality. Tire models are pretty good, but they aren't perfect. So, we might think something is a better way of doing something, but Friday morning we can see whether it is. More of a concern to me is not the performance of the car, it is about the reliability. Now that we don't have testing, if we bring new parts for the car we want to run them on Fridays for two or three weekends on the car, before we take them for the races. You can only do limited running with parts on Saturday, as it might be difficult to get them changed in time for qualifying on Saturday afternoon. The minute you start qualifying, then the car goes into parc ferme.

Gian Paolo Dallara once mentioned that maybe 80% of aerodynamic modifications that show promise in computational fluid dynamics (CFD) actually show improvement on the car. Nick Wirth says it is closer to 99%. Pat Symonds wryly explains, "Yeah, Nick Wirth has told me lots of things, so I had to sack him twice. There are two distinct categories, and those are mechanical parts and aerodynamic parts. For aerodynamics it's not high, but the truth is I don't know the answer to that. Let's try to figure it out. There is a race every two weeks. We do eighty runs a week in the wind tunnel, so let's say 160 runs in the wind tunnel, and we if we are lucky we might bring three or four parts to the next race, and those are all ones that have gone through CFD, so it's pretty low."

Along with the Hybrid era of Formula 1 arrived the electrical actuation of the rear brakes, as it is virtually impossible to make a seamless braking experience with hydraulic brake actuation and the increased amounts of kinetic energy recovery on the current cars. This has similarly been a major issue in the WEC hybrid cars. Previously things were a lot more complicated when it was being done mechanically, and the end of hydraulic rear braking in Formula 1 was during the KERS era with V-8 powerplants (there is still a hydraulic back-up system in case of electronic failure). With the increase in kinetic recovery in 2014, these older mechanical/hydraulic systems were no longer able to offer seamless braking. Pat Symonds says, "They also developed a mechanical preload linkage control system that preloaded the balance bar with springs of varying capacity to get proper control over the right progression of braking from front to rear. It was a very complex series of pre-load springs at the balance bar to get a hard pedal, but now with brake-by-wire they have eliminated a lot of that nonsense and there is a hugely firmer brake pedal that is noticeable to the driver in the cockpit."

The mechanical pre-loaded balance bar concept is further complicated by the desire to avoid adding drag to the pad/disc interface. Brembo are the suppliers of braking systems to a number of the Formula 1 teams and all the Indianapolis 500 competitors. Mauro Piccoli is the Racing Director at Brembo. He discusses braking technology:

> These were the first attempts at controlling the balance of the brakes. The balance bar can be adjusted by the driver, a fixed value between front and year. This was a bit different. They tried to understand how to control the brake characteristics through the stop. It requires to use maximum brake at the rear and move from rear to front during the stop, in order to use the grip in the best way and avoid wheel locking. This concept has been translated with the help of some springs that act upon the balance bar, and from a certain point on, are basically adding force to the balance bar in order to change the balance distribution during the stop. You have the two master cylinders and the balance bar. The force distribution between the two is higher for the one that is closer to the pivot point. If I add another force, you now have three forces that are acting upon the balance bar. Changing the pre-load or where you start to use that spring [the third force], you can control the balance distribution through the stop.
>
> The concept behind that was trying to control the brake application through the stop, not only to fix the value and also to leave the driver with using modulation, the possibility to control the front and rear. This obviously translates into a performance advantage, especially with the Pirelli era when the control of the wear of the tire was resulting in winning the races or not, from the effect of being able to control the brake bias during the stop. It would put the driver in a position to better control the brakes and avoid wheel locking resulted in a

pure advantage of performance. This was when the effect of the KERS was minimal compared to the full brake power of the car.

In 2014, Formula 1 went to a system that doubled the regenerative braking of the cars and reached the point where it needed electric actuation of the rear braking to permit it to result in a seamless braking as it is felt from the cockpit. The previous hydraulic systems were borderline acceptable with the 2013 spec KERS systems, and doubling the regenerative braking no longer could even approach a seamless result when combined with hydraulic braking. Were there pitot tubes that analyzed the speed of the car and that adjusted the balance at the brakes? Mauro explains:

The old system was not an active system. You cannot make something that is analyzing the behavior of the car and reacting this way. The challenge is to predict the behavior of the car, previously by brake balance, and from brake-by-wire now.

Now the pedal is acting upon the master cylinder, but the rear cylinder is not connecting to the line, but with a simulator that is the ECU of the car that is receiving the temperature of the front brakes and pressure of the front brake, is understanding how much the KERS is recharging, and commanding the rear brakes. In reality the driver is not anymore connected to the brake system, but he acts upon the simulator and that is actuating the rear brakes. You have a hydraulic power of the car that is the same as what is controlling the gearbox and parts of the engine. With a valve you can modulate the high-pressure signal down to control the brake pressure. If you have a failure, you might be giving too much pressure or not enough pressure and the system recognizes if there is a failure and it opens the valve that re-opens the circuit between the rear master cylinder and the rear brakes, and puts the driver straight back in connection with the rear brakes [hydraulically]. Obviously in this case, you cannot get the best use of the rear brakes, because how the KERS is charging or not, the feeling of the pedal will be completely different, it will change the feeling of the rear brakes. This is all for the safety of the driver.

In the pre-2014 KERS era, a KERS failure played absolute havoc with the braking system. Mauro continues:

> Mercedes had a failure of the KERS in Canada, but were both so much in an advantage over the others they were able to continue. But when you have a failure of the KERS, most of the time the car gets so slow it is not worth continuing. The rear brakes now are dimensioned upon the torque that is required with the KERS that is regenerating. The problem is that if you don't have the KERS that is acting on the rear axle, the torque you have to develop with the rear brakes is much higher. Obviously the way to design the rear brakes is to optimize the mass. Part of the torque goes to recharging the batteries. If this component is missing and just the rear brakes are stopping the car at the rear and the full torque goes onto the rear brakes, which are now smaller brakes with smaller calipers, you can have trouble. Because it is not an active system, it is not measuring the rear torque requirement on the axle and commanding the pressure on the rear. The goal for us was to have material that was very flexible in pressure and temperature for working range and to provide the teams with a map of our carbon in order to predict a certain point with pressure and temperature, the Mu that is generated by the brakes. This is something our customers appreciate a lot.

In the WEC, the Toyota was unique in having kinetic energy recovery front and rear, adding yet another level of required sophistication to braking technology. Alex Wurz describes their development:

> Under braking the rear brake balance should never change from what the driver uses on a conventional system. The first year when we went to these systems, you saw a lot of drivers suddenly spinning off, and that was when the calibration was off. The carbon fibre discs do not heat up linearly, so you have to be very prepared for when it blends over into the brakes. It has to be seamless, as if it isn't I either spin or run off. I can't have any margin for a braking move of 5 meters in case the handover is not done correctly. I think we [Toyota] are ahead of the game on this subject, but it is one of the key areas of race setup and

tuning now. We still have things to learn, especially when we are out of our normal operating zone with the temperature of the brakes. In our case we always want to recover more energy than we could. It makes no sense to brake more than your motor can recover for it is wasted in heat by going to the disc brakes.

The revolutionary 6 wheeled Project 34 Tyrrell of 1976 won the Swedish Grand Prix (Norm DeWitt)

Another of the many clever innovations that are no longer allowed, is changing the number of wheels as desired for traction purposes. This was pioneered by the six-wheel Tyrrell P34 of 1976, which did win one Grand Prix. Six-wheel variants were also tested in the 1970s by March and Williams, and they both had four drive wheels in back, all of similar size to the front wheel/tire package. Williams first tried this concept on the FW-07, but then shelved it for a few years. It emerged again when the Cosworth-powered Williams of 1982 was clearly being outclassed, and that called for a drastic solution if the team was going to be able to defend Keke Rosberg's driver's championship for the following year.

Dickie Stanford remembers, "It got banned before we could race it in 1983. It actually started life on a '07' previously, and the idea of it was to increase traction. They transferred it from one car to the other, to the FW-08. We were one of the last teams that was left with the Cosworth DFV when everybody else had gone turbos. It actually worked, and we broke the track record at Paul Ricard in testing, but then literally a week later it got banned. It is called the FW-08B, and was never raced. Keke Rosberg and Derek Daly drove it." There was only one six-wheeler rear assembly ever manufactured and it has twin mechanical differentials with the gearbox between. The six-wheeler ran in the FW-08B configuration at Silverstone for the fortieth-anniversary celebration of Williams Grand Prix Engineering, although Sir Frank's presence as an entrant in Formula 1 predates that by eight years as Frank Williams Ltd (see Chapter 12).

The only 6 wheeled Williams gearbox and dual differential assembly that was used in the FW-07 and -08 (Norm DeWitt)

With rare exception, if you had a normally aspirated car in 1982, you'd better hope you'd found some sort of advantage to have a chance at a win. Eddie Cheever had only raced Cosworth-powered Grand Prix cars up through 1981, when he signed to drive the Ligier-Matra V-12 for 1982. This was the car in which he was to get his best Grand Prix finish. Eddie recalls:

> The Ligier had no torque and used a lot of fuel. It had good horsepower, probably had more than the Cosworth did, but it was really good around Detroit because you could just put your foot in it, as you didn't have to worry about wheelspin. Detroit was tough to put the power on the ground. Normally they would give you a set of tires where the fronts were the same as the backs. For the first time ever, one of the drivers used a certain back tire, and then changed to a much softer front tire, but then on the restart f**king John Watson did exactly the same thing. All of a sudden the rumor was going through the pits that this was the way to go and then John's got it. He won and I finished second. He passed four of us on one lap, he was like a hot knife through butter and Watson was as good as anybody I'd ever seen.

In this case the cars that finished 1-2 had a couple of advantages, those being either Watson's soft front tire selection versus being on the one track that rewarded the Matra V-12's lack of torque. As it always has been and always will be the case, having the perfect tire choice for the conditions is a tough combination to beat.

Danny Sullivan knows all about the disadvantages of having a normally aspirated engine in a turbocharged world, as he drove alongside Alboreto in the Benetton-sponsored Tyrrell-Ford during that 1993 season. Sullivan remembers, "We were still somewhat competitive at certain tracks, but ultimately it was a DNF. That's really what happened to Tyrrell, he lost the sponsorship because he had no turbocharged engine and Benetton wanted a turbocharged engine and was going to go. I tried to broker and get them both to put money into Brian Hart so that at least we would keep the sponsorship and I would keep the drive. Brian had a good little engine and with about 100 grand in development money it would have been really good. But Ken always believed that Ford was going to bail him out, and so he lost the sponsorship. I said it's got to take some money to bring money and it was a pity. As you know Benetton went on to buy Toleman and went on to win World Championships. That could have been Tyrrell."

The Tyrrell was at its best as a street fighter, and with the Linden dip tearing the Formula 1 cars apart, Ken Tyrrell responded that his cars were built to race under these conditions and that they shouldn't change the track. Sullivan says, "The Linden dip in 83… I never had a problem, I always thought that Detroit was ten times rougher than that. With the manhole covers you couldn't run it too low. The Tyrrell was a strong car, it was really well built. Ronnie Peterson said back in the day, 'They have the wrong thing about making the tracks smooth, they need to make the cars stronger.' I was pulling away from [eventual winner] John Watson until something failed on the car, and at that stage I was in front of them. We lost the potential of a really good result." Eddie Cheever points out, "Every generation of monocoques in that era were getting better and better, lighter and stronger with more honeycomb."

Cheever had his first taste of what he had been up against at Tyrrell and Ligier-Matra when he was signed by Renault for 1983. "When I was at Tyrrell it was my first encounter with Renault. I went to Renault one winter and they had V-6s lined up next to about fifty gearboxes and I don't know how many mechanics. I thought, 'This is unbelievable.' For 1983 I was the teammate to Prost. I was number two, and that was the winter when they took away the ground effects. Renault had gone to Le Castellet with the car [RE-30] the way it finished the season with the ground effects [1982] and it was unbelievable. It was like 1,000 horsepower and so much downforce, it was physically difficult to drive. It was the first turbo I drove, but then they took the ground effects off and that was that. I drove the RE-30 in Brazil and Long Beach."

At Renault, Cheever quickly realized it was not going to be a level playing field. Eddie continues, "My engines were from the same group as Lotus' engines." Welcome to driving for a French manufacturer with a French superstar teammate. Alain Prost nearly won the '83 World Championship with the RE-40 Renault, and much of that success was due to his talent in developing a car. Cheever recalls the focus that Prost (who was nicknamed 'The Professor') had in debrief sessions. "A vein would pop out of his head like a finger when he would get focused, and he would go on for three or four hours. After about an hour my brain would turn to mud. He won every race he ever won the hard way. I'm not ready to call [Simon] Pagenaud 'The Professor' just yet."

Drivers always could play a major role in hiding their advantages, such as when drivers don't put in a whole lap and reveal the potential of the car or particular part they were trying out. Pat Symonds explains, "Alonso was very good at it. He wasn't just hiding it from teammates, he was also hiding it from his competitors. He wasn't hiding it from the team [engineering staff], as we could see the data and could see what he could do." Telemetry has exposed the driver's secrets, which now are most often shared with everyone within the team.

In the world of Indycar, Rick Mears was initially at the cutting edge of this new age of telemetry. He recalls, "When we first started using the telemetry, I was one of the first drivers that started using a computer at the hotel. You couldn't even find a laptop back then. I drove all over Indianapolis trying to find one, there were so few and they cost an arm and a leg back then. There were three or four channels... steering, throttle, speed was all we were reflecting at that point. They'd give me a floppy at the end of the day and I'd study it at night. When we'd start collecting more data, then it was, 'Now we've got to figure out the best way of doing this to make it work with the team and on the track.' How to make use of it better than somebody else does."

As the technology advanced, the ability for an inspired designer with a few good fabricators to show up and change the world has ended. Rick Mears concurs:

I agree with spec cars in today's world. Once we got away from aluminum and steel, you can't go home and think outside the box, you've got to have an autoclave. That's when the car numbers went away. Penske, Lola, Reynard, March... why did they all go away? Here's an example, you've got March and Lola. Say we find out in testing that the March is hot this year and has the advantage, so Lola doesn't sell any cars. So they spend the whole year in design and development so they get the advantage next year and everybody jumps to their cars. But if you don't have the advantage that second year, can you afford to go do that for another year? How long can you afford to do that? There are reasons why things cannot go back to where they were. I think we are all headed in the right direction. I've been spotting for Helio at Indy every year and I can see it. In the last six to eight years it's been steady growth, the grandstands [at Indianapolis] are always full, but you can visually see it by the number people in the infield. Last year [the 100th running] it was absolutely packed.

Thankfully in Formula 1, by decree there are a large number of chassis manufacturers, versus relying upon a handful of vendors, or in the case of Indycar... one. This fact is why the large number of technology junkies prefer Formula 1 as they devour the latest aero tweaks of the week. The mechanical tweaks aren't always as noticeable. Scott Dixon explains:

Typically on a street course the rear roll center is quite high, and depending upon the driver in the front it can be at zero [ground plane] or above, or at zero or below. It's a big tuning thing for us, not so much for the other guys, but we use it. Dario and I were quite similar, especially on developing in one direction. He typically didn't like the car as neutral as I did, but did with a lot of things such as roll centers and weight distribution. TK and I are on the two different ends of the spectrum. He's very aggressive, very hard on brakes, he doesn't like a loose car and always runs a front bar. I don't run a front bar, I run a rear and he won't run a rear. We are kind of the maximum split you can have. Early on we tried to go in one direction, but it just didn't work. He exerts a lot of energy in the car, but it's just his style.

I kind of feel the brakes, using finesse feeling what the car is doing and being accurate when trail braking. When Dario and I worked together we were very lucky that we were quite similar.

In Formula 1, it appears the rear roll center is similar to the current Indy car. Studying the rear roll center from photographs and connecting the dots, it appears that the rear roll center is a couple of inches above the ground plane. In the front the anhedral layout appears to drive the roll center really high in the front. Pat Symonds says, "In the rear it is up a little bit. In front, just because the suspension is pointed up like that doesn't necessarily mean that the roll center is high if they are parallel."

True enough, as if the wishbones remain perfectly parallel, then the roll center is at the ground plane. Of course, if they don't remain perfectly parallel, then the roll center is then bouncing all over the place. Pat continues, "The roll center at the front of the car is not where you'd like to put it for kinematic reasons, the design at the front is from aerodynamics. Then try to make the best kinematics out of it that you can. The same is true at the rear, as we would not run the lower wishbone as high as we do were it not for aerodynamics. It's not ideal from a kinematic point of view and it certainly isn't ideal from a stiffness point of view, and hence it's an incredibly heavy solution. Mounting the lower wishbone up at driveshaft height compared to mounting it at the bottom of the gearbox, is probably making you carry 7 kg extra weight perhaps. Aerodynamics just drives so much of it."

On a superspeedway, it can be a challenge to get one's head around spring/damper packages that are different at every corner of the car. There is no set ratio, although they are closer from right to left than one might expect. Is there a given percentage higher on the right side spring rate? Michael Cannon, an engineer at Dale Coyne Racing, explains:

We look at diagonal loads in the car, as it gets compressed into the banking and the car starts to build load. We have a certain range we like to stay in as far as the left rear and right front contribution of stiffness. We spring and bar to try and hit those targets, and then we adjust the damping. We run several different permutations on the seven-post rig so we have a solution for every spring. If you are squatting on the RR, you are lifting off on the LF. So as a result you start statically with right front weight, which has a lot of left rear weight as well. The car is sort of like a three-legged bar stool, if you will, and then once you are in the corner then the load sort of settles down. Then the majority of the load is carried by the right rear, and then you are trying to plant the left rear as well by having a lot of right front cross-weight.

When the team arrives at the track they know which spring they want, they know which damper package to throw at it. Cannon says, "This is the third year that we have run this basic car and so we know the neighborhood we are living in. We know we are going to stay in this damper setting range for this particular spring."

Perhaps the most surprising of all was when I was sitting next to AJ Foyt's paddock rig, waiting to chat with AJ in the Fontana paddock in 2014. A team mechanic had partially disassembled the team's golf cart and obtained a large part from it. Foyt emerged from the transporter, picked a particular hammer, and studiously whacked the offending part repeatedly. Handing it back to the mechanic, it was quickly reassembled and upon completion, the golf cart started and ran flawlessly. Foyt's reaction to my stunned expression was a hint of a smile… and a wink. For all of you who have seen Foyt pounding away with a hammer on Indycars across the decades, be reassured that the studiously applied 'whack' can work.

At the opposite end of the spectrum from AJ's hammer collection is the story of Harry Robertson's contributions to racing safety at Indianapolis. As with so many other racing technologies, from carbon brakes to the monocoque chassis, fuel safety technology came directly from aviation applications. Robertson spent much of his life developing this fuel system technology and proper installation protocol to the extent that the US Army estimates his work saved more than 8,000 lives in aircraft applications. His first installations were in Bell 'Huey' helicopters used in Vietnam, with an immediate beneficial effect. His tank and fuel assembly integration systems became known as 'Robbie tanks' and he is in the National Aviation Hall of Fame and Army Hall of Fame. Harry recalls:

I went to a lot of Midget races after the war, as my dad would take me to them. I was at a Midget race and I saw a guy have a wreck and burn to death, he got cremated in the car right there. When I built my own drag car, I had an early crash-worthy fuel system in the car with a container that could change shape and not leak. It contained a bladder that was highly flexible and somewhat tear-resistant. The plumbing line had a link to extend itself rather than break off and leak. I grew up not very far from Clint Brawner's shop and would hang out there when I was a young man. I worked over there pushing a broom, cleaning the cars and parts, and being the kid that was getting in their hair. At one time Brawner was known as the winningest mechanic at Indy. I knew Roger McCluskey like a brother almost.

When I got out of the Air Force in 1961, I approached the Flight Safety Foundation and requested funding for fuel system crash-worthiness research. They applied to the Army for funding and I ended working for their Phoenix division, the Aviation Crash Injury Research group. I went there with the intention of seeing if there was some way that I could resolve the system for aircraft, working mostly with helicopters, but not exclusively. Anything where there were rapid acceleration changes, and how you could safely control the fuel in a way that could control the spillage, and keep it from being easily ignited. The Army agreed to fund that if I would go to work for the Flight Safety Foundation, and we started with my ten-year treatise to determine how it could be done. In nine years, we crashed forty-three aircraft with instruments and a lot of high-speed cameras to understand how when these accidents occurred, what forces did what to injure or kill the people, and also how it released fuel that could catch on fire.

We started the program with the concept that you can't fix stuff if you don't know what broke and why. We had to understand the problem, so we spent the first year primarily doing that, and a pattern emerged in what was happening to the fuel systems that would cause fuel to leave and catch on fire. Obviously if the bladder ruptured, it would create a mist cloud, and that will catch fire easily whether it is alcohol or gasoline. The pattern was that if the bladders are punctured, you do not want the rip to propagate.

There were similar issues with racing car applications. Robertson continues:

Most bladders being used in racecars of that era in the early 60s, were somewhat like an inner tube, but when it was punctured and the container changed shape rapidly, you had a flame thrower as the fuel would eject out rapidly. You needed to have the bladders that could change shape in a way that would allow the container to deform, and you'd want the fittings not to pull out of the bladder. You'd look at test crashes and you'd find that the filler cap and filler neck was still attached to the vehicle and the bladder was some distance away with a hole where the filler neck used to be.

Materials were developed that were extremely cut-, penetration-, and tear-resistant. Once the materials were perfected, it was ten to fifteen times more resistant than anything that had ever been used before. Then you needed to stabilize where the metal fittings were for the fuel cap, and where lines came out of the bladder for the fuel control systems at the engine. We had to get involved in the fuel system plumbing, using frangible breakaway valves and fittings that could be pulled apart and not leak.

Interest in their work from USAC came in the wake of the horror that was the Dave MacDonald/Eddie Sachs crash during the 1964 Indianapolis 500. "They knew of our research work and were hot to go on it, asking if our crash-safe fuel systems could be installed, or to come up with kits for Indycars. We told them, 'We are research guys, we are not a factory that can make kits that we can sell. We are just not there today.' Our capability was building prototype stuff, the certification process for military use was coming, but it was not there yet, so the program did not go through. They went ahead with some tougher bladders for about nine years. They called them crash-resistant bladders, but they were a long, long ways from what it took to do the job."

In 1970, Harry Robertson left the research group and started Robertson Research Engineers:

I saw on TV [in 1973] when they had the fire with Salt Walther spilling the fuel that got thrown up into the stands, and Savage was killed. I wrote them a letter that night, how the parts could now be bought and tweaked accordingly. Ironically the next day after I mailed the letter, Frankie DelRoy, USAC's safety guy, came to Phoenix and contacted me on the spot. He said, 'We've had to solve this problem because the insurance company can't let the fuel be scalding people in the stands,' and how that wasn't going to fly. The insurance companies had given them an ultimatum to get it solved or there wouldn't be insurance for the track.

I went back to Indy and talked with them. They wanted me to sign a document that as I tell them what to do, that I couldn't be feathering my own nest by selling the products myself. I would write the specs for the parts, and inspect the cars and tell them what is right and what is wrong. The first thing of the many components we did was to create an integration policy where we mandated a self-sealing breakaway valve at the firewall so if the engine became displaced, the fuel would be trapped on both sides and not leak. We got that mandated for later that year. Goodyear was the only manufacturer who could pass the new bladder-strength specification. I wasn't a Goodyear man, I wasn't on anybody's side. I wrote all the USAC specs for the bladders, fillers, caps, hoses, and fittings. Then came the handbook as to how you install it all, for the mechanics, published in 1975. McCluskey was a real big help with the parts integration, as he was one of the few guys that understood how it was going to work and why, as he had been at some of our crash testing.

Harry Robertson's active involvement in tech approval for the fuel system integration went from 1975 until 1998. Thanks to his efforts, fire is seldom a concern, other than from turbocharger, engine, or gearbox oil. Harry adds, "When Gordon Smiley got killed it was a horribly violent crash, and the fuel tank was completely out of the car, sitting there on the track and not one drop of fuel leaked out. A crash-worthy fuel system has to be hooked up right or it won't do what it is supposed to do. The cars were all different, but certain things have to be incorporated or it won't work. It's a case of how you apply the load and all of the pieces are designed to perform at various load applications."

There are a virtually endless number of technical details to ferret out about past or current Grand Prix and Indianapolis cars. These can range from head-scratchers to sheer appreciative awe. May it always remain so.

The 1.5 litre supercharged 8-cylinder Grand Prix Alfa-Romeo of 65 years ago, the state of the art (Norm DeWitt)

The 1.6 litre turbo-supercharged 6-cylinder Red Bull of 2016. Imagine what the next 65 years might bring (Norm DeWitt)

CHAPTER 21

THE FINAL FRONTIER

Nelson Piquet Jr. won the Long Beach Formula E race of 2015,
35 years after his father won there (Norm DeWitt)

At the 2011 Pau Grand Prix, the support race for the Formula 3 main event was a race featuring electric cars, using sports coupes of reasonable pace in a race remembered mostly for the giant pileup at the exit of Foch corner. By this point, hybrid cars had reached the mainstream and by 2012 had begun to dominate the 24 Hours of Le Mans, first with using carbon flywheels spun up by the electrical energy derived from regenerative braking, or with Supercapacitors (see *Making it FASTER* for details on these 2012-13 cars). The Formula E car was introduced at the 2014 Consumer Electronics Show (CES), with Lucas di Grassi (2016-17 Formula E champion) giving demonstration runs that left a mediocre impression at best. Lucas told me the following day that they had been required to turn down the motor power for the exhibition, due to the proximity of the crowds.

Lucas di Grassi explains the control and operation of the new
Formula E car to me at the Consumer Electronics Show

The system used in Formula E's energy regeneration is a manual system, quite unlike the computer-guided systems of the World Endurance Championship (WEC) Hybrid P1 class. However, the cars were unable to maintain reasonable racing speeds over enough laps to complete the envisioned race distances, so the solution was to have two cars, and all drivers must make a pit stop to switch cars sometime near mid-race. Of course, this did nothing but institutionalize the impression that electric car racing was not yet ready for prime time, given that the cars were incapable of impressive speeds for long enough to complete a reasonable race distance. Even with this two cars per race solution, with approximately 220hp, the result is something akin to a garden-variety Formula 3 car (240hp).

Another issue is the lack of sonic stimulation, as Formula E cars sound like louder radio controlled toy cars, except for the occasional scrabbling of tires that sometimes are heard over the motor. To compensate for this, during the race, the organizers have installed large speaker boxes all around, blasting high-energy music to try to somehow make up for this sensory deprivation with sonic annoyance. The effect is not unlike that of the Burning Man festival, with 12,000 different mixes being 'shared' simultaneously through boom boxes to create a pulsing din.

The one thing that the current edition of Formula E does offer is a technical exercise by drivers having to manage their energy in the most efficient manner. This explains why the series is dominated by the top drivers from the WEC Hybrid P1 class, as that is a way of life for those seeking success in the big prototypes. There is the obvious strategy of using energy coming onto the longest straights, as you get more of a lasting benefit from the energy usage. Most drivers run the race dividing the number of laps by two between the two cars, to spread out the available power evenly between segment one and two.

Sadly, there is the social media-driven aspect of the competition as well. "Fan Boost" is an invention by the FIA and the organizers of Formula E that allows fans to vote for their favorite drivers on the FIA Formula E website, three of whom then receive a temporary boost in power for the race. This has created the absolute antithesis of a fair and equitable competition, and Lucas di Grassi was the only driver to finish on the podium at Long Beach who didn't have the benefit of it. After seeing that first Formula E race at Long Beach, I approached di Grassi at Le Mans. When he asked me what I thought of Formula E, I replied, "I think Fan Boost is the biggest bunch of bulls**t I've ever seen in my life." Lucas replied, "It's a good way of seeing it."

For the race at Monaco, they eliminated the long uphill climb from St. Devote to the Casino Square for the Formula E race. Lucas notes, "It could be uphill or downhill, you have to understand that the power output is the same, so the consumption of energy is the power output times the amount of time you are flat out. They took that section out because they don't want Formula E running the same course as

Formula 1." That was probably wise as the lap times of the Formula E cars would be glacial in comparison to the modern Grand Prix car.

Tesla uses lipstick case-sized batteries by the hundreds for their vehicles. Lucas di Grassi explains, "It is the same, similar, it is a lithium cell. With this giga-factory they are doing, the costs of these batteries are going to go much lower. I fly model planes also, and the fastest electric planes are as fast as the combustion planes. For the first time in history, four electric planes are in the top ten in the World Championships, the races are ten minutes."

However, Formula E races are not ten minutes, so pure performance has been sacrificed due to the current limitations of the technology. Lucas adds, "The problem is the battery, as the power to weight ratio of an electric motor is much higher than a combustion engine. You could put one motor on each wheel, the motor is 16 kilos, it is nothing. So you would have 1,200hp with one motor at each wheel, total weight of 64 kilos." As a result the electric racing car could be the ultimate setup for a short race where anything goes. He continues, "Eventually yeah, the Pikes Peak Hill climb is that. As soon as the batteries develop, we will see the cars going much faster and much more reliable. When the manufacturers are joining and building their own cars with their own systems, we will see the cars making a big difference. The series is young. In five years' time it will be a massive change and it will start to get at a really high level."

There is little doubt that Lucas is correct, and when the manufacturers can bring their own chassis, motors, energy recovery, and energy storage devices, this could be a technological circus of great relevance and interest, much like the Zero TT at the Isle of Man where the only rule was that it be 'carbon free' (see *Making it FASTER* for the story of MotoCzysz). But for now, Formula E remains a momentum-conservation and energy-management game between spec cars, where although the motors used are now open spec, they are regulated to provide equivalent power. Lucas concurs, "Like Formula 3, it is like that. The more you slow down the more energy you consume. Maybe a lap in the wet uses more energy than a lap in the dry, because if you carry the momentum around, you usually are quicker because your apex speed is quicker, plus you use less energy." In late 2017, Formula E decided to prolong the spec battery until 2025, taking what could have been the single biggest sphere of technological improvement away from the series.

Formula E Champion Lucas di Grassi in the hairpin at
Long Beach, 2015 (Norm DeWitt)

A big part of the game is the regenerative braking, with drivers sometimes 'backing it in' coming into the corners. Scott Speed says:

If somebody is backing it in, it had to do with the brakes being far over-braked for the car. Especially in the first year, all the teams were fighting through a lot of problems with the braking. Because they were carbon brakes, you'd get one that would accidentally start working and it would generate some heat and then it would really start working and locking up... you would lock up the tire for the rest of that session and there was nothing you could do about it. Front to back, side to side, everything, and there were all sorts of problems. I know it's better now, I just tested the new car, but in the first year there were lots of issues in how to get the brakes to operate consistently. They didn't have proper cooling, it was just a big hassle. If one of the corners happened to gain a little bit of temperature then it would exponentially skyrocket, and then you were done, which was a big thing in 2015.

It's a little bit of a chess match doing it because you have to think during the race, but the strategy is not rocket science. The drivers are the cream of the crop and there is a lot of really good talent there. Regardless of where your track position is, if you are faster than some guy, you can use a little bit of extra power and get around him. There is a brake and a gas, there are levers on the wheel that allow you to regen or not. We are always regenerating 100% of the time and so we didn't even use the levers, it was always 100%. When I touch the brake it was always as much as we could possibly get.

Strategy plays in for perhaps the first two laps as you don't get the regeneration that you would at full, so because of that lack of regeneration you are going to miss a certain amount of rear-brake bias. So to start the race you would start with two turns of brake balance to the rear and each one-half a lap or so you would go one-quarter turn in to match the feed-in of the regen. At the start of the race when the tank [battery] is full, you didn't have that extra regenerative braking at the rear so you need to compensate for it. After two laps the energy would come down to a point where you would get full regen. Normally you have a tiny bit of power still in the battery at the end of the run, as when it goes, you stop. It was found to be better to run with as much power as it would let you, and coast more. There is a small amount of regeneration when you lift, but we tried to make it nice and neutral so you wouldn't be slowing down too much while you were coasting. For example, if I needed to lift 200 meters before the corner, I just slowly started braking into the corner and then tried lifting 200 meters before, coasting all the way in and then at the last second braking as hard as I can. That is way faster, so much faster than slowly decelerating. I was shocked at how much it didn't cost you by coasting into the corner. It didn't really cost you too much lap time, lifting and coasting.

The 2016-17 season three saw the predictable battle between di Grassi and Buemi, two of the superstars of the WEC Hybrid P1 era. The title went to di Grassi, in no small part helped by the absence of Buemi when his Toyota WEC commitment took priority, whereas di Grassi's Audi Sport R18 team had withdrawn from WEC racing at the end of 2016, eliminating his scheduling conflict. Regardless, both have been dominant drivers in the series and are worthy champions.

Formula E has been pushing the connections to gamers, with an overwhelming presence at the Consumer Electronics Show while having less emphasis upon traditional motorsport media. Is the future path to Formula E going to come from the gamers in simulators? Scott Speed calls it, "I think that is total horses**t because the simulator is so different. Jean-Eric Vergne was the Ferrari test driver, and he sat in a simulator all the time. And every time before a Formula E race we would go sit in the simulator and I'd be at least a half second faster than him on it because I'm a video gamer. I wouldn't say that I'm a better race driver than Jean Eric Vergne, but I'm a better sim driver by a lot. I attack a simulator like a hacker, trying to figure out what it wants me to do because it's not real."

It appears that the one thing that Formula E has provided is the option of choice for German car companies wishing to pay penance for their seemingly endless revelations of vehicle emissions gaming. It's understandable, coming from that background, that those companies would make any attempt to somehow repaint their tarnished brands in green.

Scott Speed was also the last American driver to have a regular drive in Formula 1, in the early years of post-Minardi Toro Rosso. The technical highlights can be summed up in a single word. "Electronics. That was 100% of it, and our cars were really fast in the rain. In 2007 it rained one time and that was for practice at Monaco and I was P1 for almost all of that session. My most memorable moment in Formula 1 actually was my last race in the Nurburgring. I drove from eighteenth to ninth on the first lap in the pouring rain and then we came in for pit stops. They told me that they got the tires mixed up and we spent a minute in the pits kind of sorting that out. Once we got going again in a monsoon, the first turn was a big river down there and we hydroplaned off the track. So it was only a two-lap race for us, but it was definitely the most memorable as it was a lot of fun to pass so many cars in a Formula 1 car."

Speed continues, "I think open-wheel racing is going to die. For one, it's still much easier to get into a stock car racing program. For relatively cheap I can put you in my Rallycross car and guarantee you way more views, especially with the right demographic that matters, which is the 18-24." Speed is one of the few with Formula 1, Formula E, and Global Rallycross experience. He gets recognized away from the track a great deal more since moving to Rallycross racing with Andretti than he ever did from his Formula 1 days, which probably explains much of his concern. Scott reminisces:

I remember my very first F1 test in '05 vividly, because I tested with three other GP2 [now Formula 2] level drivers. I kind of considered us all to be at the same level of ability, everybody was good at that level. But when we all went to test in the Formula 1 car at Barcelona, there was a huge gap between us in lap times. Myself and Neel Jani were close, I was a few tenths faster than Neel, but after him there was a HUGE gap, we were both almost three-quarters' of a second faster than the other two. It had to do with that extra amount of grip and feeling the edge. Feeling the edge at four G's lateral in a corner is a lot different than three G's. An Indy car or a GP2 car is about three G's but a Formula 1 car is over four G's, and that makes a big difference. To feel the edge of grip at that level is much more of a knife edge and it was a big separator. You could put someone in a Formula 1 car and you would know within a day if they were going to be able to do it or not.

I have a lot of friends in the WEC from that Red Bull Junior program. Brendon Hartley is doing great now, Neel, and Buemi, and we'd all play ping pong in between trainings in Austria. There was a big camaraderie as we were given opportunities because we were good. There's a big difference these days as those programs don't exist anymore. We had such mutual respect for each other as we were all selected based upon ability, not because we brought 100 million dollars or whatever it is to race. There's a big distinction between those two types of drivers. I think that disparity is like a grand canyon between them now. The chance for a kid to go from the go-karts up through some kind of ladder series without money is almost non-existent now. Before, there were junior teams. Renault had one, Mercedes had one, that's how all of us got our chances. They don't exist anymore, and if they do exist they are very small. The landscape has changed a lot.

Does the transition also apply between a P1 Hybrid and a P2 prototype? Neel Jani explains, "Look at [Nicolas] Lapierre, he's super quick in LMP2 and they kicked him out at Toyota. They took him back now, but somehow he's not able to transit. At Spa, Sarrazin was on the same lap time as me, they had a chance to go for pole. He completely f**ked up and they were lost as it was his chance on the start and he went straight with locked up wheels. But in P2 he won everything ever since he got kicked out from Toyota. So the answer is, yes, I think it can happen. No one doubts that he's quick, it seems he's having a problem transitioning. The speed he has, he's quick wherever he goes, but at Toyota it never worked out for him. But, at the end of the week we may be saying, 'Oh s**t, now he's won Le Mans.'"

1996 World Champion Damon Hill OBE says, "I myself decided that I wanted to race, and that I didn't want to go to University. I bought a bike, I bought a van, and I went and raced. It was possible even on the meager amounts of money that I was earning. Those were the building blocks that helped me get all the way into Formula 1. Now it's almost prohibitively expensive. Karting in the UK, it's easy to spend 100 thousand US dollars to be competitive in just the junior categories of racing. And then when you get into GP2 we are talking millions of dollars in budget. Where is the return, what is the benefit? Costs in sports are complicated because people want to win, and there will always be somebody with more budget that wants to win."

Scott Speed adds, "Formula 1, how do you get there? Look at the GP2 field, how do you pick a driver, it's just rich kids, that whole system is broken and I don't see any sign of it getting fixed. How can you justify spending that amount of money?"

Isn't it upon the sanctioning body to try and do something? Damon answers, "Well, you would think so, and that somebody would say, 'What are we trying to create here?' Maybe natural selection is some people's philosophy, and there will always be a bigger dinosaur out there to scare everyone off and dominate. Maybe that has become the pure essence of the sport. It's getting to the point where the gambit is so huge that the people look at it and it can be seen as obscene expenditure in a world where not everybody has so much money to burn. That could go against the sport, it could be seen as profligate and irresponsible. After all, what are we trying to do here? We are trying to demonstrate our skills, and it could be sustainable in a cyclical way. The people who love it, pay for it. The money is recycled and we live off the insane asylum."

It isn't just the aspiring drivers that have to deal with financial inequities and soaring budgets. The Haves and the Have-nots on the Formula 1 team ownership side were never more obvious than in the Friday team-management press conference at COTA's USGP in 2014. Two of the teams, Caterham and Marussia, had recently gone into admi-

*The contentious and forever memorable press conference for
team management at the 2014 USGP (Norm DeWitt)*

nistration (the British term for bankruptcy), so neither had showed up for the US round. In the most riveting press conference I've ever attended, the Haves in the front row were furiously trying to change the subject from the financial inequities of Formula 1, to what was happening that day on the track. The Have-nots, Sauber, Force India, and Lotus, were literally seated in the back row of the bus (stage). These three teams were having no part of attempts to change the subject, and it was quite possibly only the memory of the carnage wrought on the American Formula 1 scene after the 2005 Indianapolis USGP fiasco, that kept all three of the teams from packing up and leaving the event. Hence, the race had eighteen starters, when it was looking dangerously close to only having twelve for a while that Friday.

Damon Hill reflects, "There is a Bob Dylan line about how, 'Those who are first shall soon be last.' It's all very well when you are in the strong position to want to stay there, but for anything to have a sporting element there has to be acceptance that you can take on the competition. The whole idea of sports is that you think you can beat other people given a fair chance. It's not very fair when you know under the rules, that there is no way that you can beat them." That describes the current Formula 1 in a nutshell, as there are automotive factories (Mercedes, Ferrari, Renault, McLaren) competing against smaller independent teams such as Force India, Williams, Sauber, Toro Rosso, and Haas, with Red Bull–TAG Heuer Racing being the one massively funded team that defies categorization. As of 2017, every one of those non-factory teams are leasing engines from the currently competing factories.

As Jenson Button told me in 2014, "Obviously the great pity with Marussia and Caterham not being here is for the people who are working for those teams. It is a very tough situation for them as we found at the end of 2008 [with the sale of the Honda Formula 1 team to Ross Brawn for one English Pound]. It is sad to see that teams cannot stay in this sport and obviously you would say that things need to change to make for a full grid and it makes the sport. On the other side of it, 135 teams have been and gone, so it's happened before and it's going to happen again."

As if the struggles for an equitable sharing of the spoils between competitors wasn't enough of an issue, as Scott Speed pointed out, the Formula 1 ladder system is well and truly broken for the drivers as well. With the WEC shedding Hybrid P1 teams left and right, and those orphaned drivers finding Formula E as the best port in a storm, let's wrap this up on an optimistic note.

There was an electric vehicle in development fifty years ago. It provided stellar performance and incredibly compact packaging, incorporating all-wheel drive, four-wheel steering, and its own solar panel array, all with a total operational weight of eighty pounds. It is the one electric vehicle that everyone can agree was both a technological achievement as well as something we'd all love to experience.

The Lunar Rover had only been used on three Apollo missions: 15, 16, and 17, which means that there are only six humans who have ever driven on the moon. It was a packaging marvel, given the space constraints in the Apollo missions. Al Worden was the Command Module pilot for the Apollo 15 flight. He shares these details:

> We had a lot of difficulties getting the thing qualified for the flight. As a matter of fact, the lunar rover might have been scheduled for an earlier flight, but it couldn't because it wasn't ready. We actually had to send it to a proving ground up in Santa Barbara to get it certified for flight, because Boeing was having a little bit of trouble with the electronics on the rover. We had to take it to General Motors to sort of straighten it out, and that's how we got the rover finally cleared for flight.

> It was about a 480-pound package, and the wheels folded up in the middle. There was a chain hoist that they would take out of a compartment and drop it down to the surface and unfold the wheels and put pins in to hold them. You know, the thing worked pretty well. The problem with walking around on the

surface of the moon is that if you don't have a rover, you've got a certain amount of oxygen that you carry with you, so you can't go beyond the point where you can't walk back without running out of oxygen. With the rover, which was a little Jeep, you could carry all kinds of stuff. They actually got out about 1.7 kilometers from the lunar module, and that's a pretty long walk if you had to walk it. The lunar rover made it possible for them to look at both Hadley Rille and over to the base of Mons Hadley. So the rover was extremely valuable and it was a great thing for the last three flights.

I think other flights might have brought back more rocks than we did, which was about 170 pounds. But I think Dave [Scott] and Jim [Irwin] got the limit of the number of rocks we could carry back on our flight. I was by myself in lunar orbit while they were on the surface. I probably had more to do in lunar orbit than they did on the surface. They would work eight-hour shifts and then they took the rest of the day off. I would work twenty hours a day, photographing the moon, running remote sensors on the lunar surface that could measure X-rays, microparticles, microwaves, and those kinds of things.

While Scott and Irwin were on the lunar surface exploring in the Rover, they weren't the only Apollo 15 astronauts that had an amazing new tool. Al adds, "I could take some very low-light pictures for the first time, and carried a camera that had perfect focus with a special Nikon lens that was f1.01 and had film that was ASA5000. I could take very low-light level pictures of the universe and certain phenomenon that we'd always wanted to take pictures of. It was a special Nikon camera, I carried the first Nikon. All the others were Hasselblads, as they had the deal with NASA. I asked for the Nikon as I wanted to take pictures of the Gegenschein, which is the ring of unconsolidated rock out past the planetary system that never consolidated into a planet. You can see it at certain time on a perfect night in a place like Chile, you can see the sunlight reflected off the rocks. I was able to photograph that."

So much for being 'the loneliest human in the universe.' Al laughs, "That was the best time of the flight, as I got rid of those guys for three days! Not a bad deal at all. Also, I was in the vehicle that could get home, so if they weren't good to me…"

Milt Windler was Flight Director for the Apollo Missions. "The Lunar Rover, it was Marshall [NASA Marshall Space Flight Center] that did that, and I thought it was incredible. The thing was all folded up on the lunar module. You pulled the lanyard and the whole thing fell out and expanded and then they drove it off. It was incredible, it worked so well, and it gave us so much. It really was amazing how they were able to package that into a relatively small space. The design was pretty much the same, I don't remember too much being different about it other than maybe the tools and the cameras on it. Once you've got something that's working, if it ain't broke, don't fix it. It gave us a whole lot more capability, even the little rickshaw thing [cargo trailer] made a difference."

John Young salutes the flag. The Lunar Roving Vehicle is parked in front of Lunar Module Orion (Charlie Duke – NASA)

Charlie Duke was lunar module "Orion" pilot for Apollo 16, later riding shotgun while bouncing along on the lunar surface, and recounts the adventure:

John Young would not let me drive, he said, 'I'm driving the Rover, you navigate us.' It was a rough ride, it was really cratered everywhere, up and down rocks, and the suspension system on the Rover makes it really bouncy. It ended up fishtailing back and forth, I guess because the steering was real sensitive as it had four-wheel steering. You were really glad you had your seatbelt on. The Rover weighed about eighty pounds up on the moon and so it was springy. Each wheel had its own independent suspension.

I had a movie camera on the Rover and when I turned it on you could see this thing bouncing across the moon was really a rough time. It's called the Cayley Plains but it wasn't plain. It was a valley about ten kilometers across and the last day we drove north about four kilometers and the second day we went south and actually climbed a place called Stone Mountain and we went up maybe 300 meters or so. It was really steep, and the Rover would climb a twenty-five-degree slope. The objective was to go to the furthest part on the traverse, so when you got there you had maximum oxygen, cooling water, and electrical power to get back if it broke down out there and you had to walk. We felt that about four kilometers was the most we could walk.

It was ten feet long by five feet wide, but the guys that designed the Rover only had a five by five spot to fit it on the Lunar Module. The chassis folded up, behind the seats was a hinge point, and in front of the instrument panels was a hinge point. The wheels came in and everything rolled over on top of the seats, and the back wheels came up and turned over, so you had the wheels butting up against each other and that was just bolted up to the side of the spacecraft. I don't think we took a picture of it while it was on the side of the vehicle [Lunar Module]. We had pulled the pins on the top hinges, and then on the side of it there were some cables you pulled hand over hand, and that turned a jack screw that pushed it out. When it got to about forty-five degrees, it just kind of unfolded.

The only time that the whole Rover was photographed was by me on [Apollo] 16. We did what we called the Grand Prix, and it was to show the engineers what the whole vehicle actually looked like, everything else was mounted on the vehicle. So I took about ninety seconds worth of film of it going out and coming back, rooster tails everywhere with the tires bouncing around. One time it was off the lunar surface with the right front tire. It was remarkable.

Charlie Duke, Lunar Module Orion pilot and
Lunar Rover passenger – Apollo 16 (Norm DeWitt)

Apollo 17 was the final moon landing, and it carried the last of the Lunar Rovers. Gene Cernan drove the Rover on the moon, as well as being the last man on the moon (the title of his memoir). He reports, "It didn't have wings but at one-sixth gravity, I tell ya, you were on three wheels a good part of the time. It didn't go that fast, but I broke the speed record going downhill (his Lunar LSR is credited at 18km/hr). It had four-wheel drive and Ackermann steering fore and aft, so you could do anything with it. It allowed us to go places to do things that we could never have done without it."

As with Charlie Duke on Apollo 16, Jack Schmitt didn't get to drive the Rover either. Gene continues:

Well, you know that's the Commander's choice. I didn't let Jack Schmitt drive it either, and well… he didn't ask because he knew what the answer was going to be. I would have been very uncomfortable in the right seat. We covered some thirty-six kilometers, and we had landed in a valley that had mountains higher than the Grand Canyon is deep. Believe it or not, in one-sixth gravity it's very difficult to walk up those mountains, and going down you've got to be in control of your own center of gravity. With the Rover, we'd go on the side of hills, up hills, and down hills. We covered a big portion of that valley that we wouldn't have had time to on foot, but we probably couldn't have hiked up some of those hills. It had four wheel-independent electric-drive motors, so if you lose three of them, you've still got one. The plan was that we would go out to the furthest point we were planning to go, so just in case it broke down we were always on the way home. Time [away from the Lunar Module] was the real issue.

Cernan had also been awarded the Distinguished Flying Cross. "I guess I've earned it. I say 'guess' because those of us who went to the moon were getting our pictures in the paper while our buddies who were Naval or Air Force aviators were getting their butts shot off in Vietnam, so those are the guys who earned them. I'm proud to have it as that's a pretty significant group of people. Flying is a part of my life, and it's been my life since that dream as a kid. I was a World War 2 kid who was dreaming about flying airplanes off aircraft carriers and that dream eventually led me to the moon. I tell the kids, 'Never count yourself out, because you just never know.'"

Gene Cernan's final words as he stepped from the moon still echo across the years, "As I take man's last step from the surface, back home for some time to come – but we believe not too long into the future – I'd like to just [say] what I believe history will record: that America's challenge of today has forged man's destiny of tomorrow. And, as we leave the Moon at Taurus–Littrow, we leave as we came and, God

Gene Cernan, the last man on the moon, driving the
Lunar Rover – Apollo 17 (Harrison Schmitt – NASA)

willing, as we shall return, with peace and hope for all mankind. Godspeed the crew of Apollo 17."

In a world that often celebrates mediocrity, remember that we stand on the shoulders of giants. The technological gains that can be had through competition, be it with the Soviet Union or the pit next door, are real, relevant, and potentially revolutionary. The racing series of the future should encourage advances in technology with real-world potential uses. This is what makes racing relevant, versus the endless parade of spec car series, and something that the rule-makers should never forget.

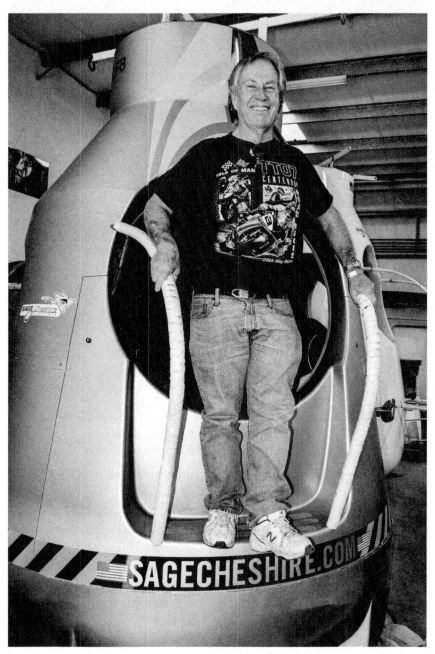

*That's one small step for Norm... exiting the Red Bull
Stratos capsule at less than record altitude (Art Thompson)*

CHAPTER 22

PENSKE VERSUS GANASSI

*Mike Hull, Managing Director of Chip Ganassi Racing,
chats with Roger Penske (Norm DeWitt)*

In 2008, while delivering the race cars to Sonoma Raceway (Sears Point), the Penske transporter carrying the primary cars of Helio Castroneves and Ryan Briscoe caught fire and burned, extensively damaging the two Penske Dallaras inside. It was the Wednesday before the race and Team Penske had to scramble to fully prepare two backup cars in time for Friday practice.

Rising to the occasion, Castroneves qualified in pole position, with teammate Briscoe alongside. Championship rival Scott Dixon in the Ganassi car qualified fifth. It was a stunning recovery for Penske.

Standing on the grid next to Chip Ganassi's pit box, Roger Penske wandered over. Ganassi, never one to miss an opportunity, turned to Penske and said, "You know, I tried everything. I burned your transporter down, it didn't make any difference..."

Five years later at the same track, the battle for the lead was between Will Power (Penske) and Scott Dixon (Ganassi), who were pitted next to each other on pit road. At the final pit stop, the right-rear tire changer for Power was clipped by Dixon as he left from the pit box behind. A penalty was issued to Dixon, which essentially ruined his race. Yes, the right-rear tire changer had used a wider than normal path around the back of the car, but that's his call to make. The incident stoked the Internet-conspiracy-theory crowd like none in recent memory.

As has often been the case in recent years, that 2013 championship came down to Penske versus Ganassi in the series final race, at Fontana's AAA Auto Club Speedway. Once again, a Penske car (Castroneves) was pitted directly in front of Scott Dixon. Before qualifying, Roger Penske came strolling over to the Ganassi pit. I assumed he was going to offer something along the lines of "Best of luck." As Roger reached out to shake Chip's hand, he said, "Well, I've got the guy with the wheel ready to go." Everyone within earshot was immediately laughing out loud.

Despite being the fiercest of rivals, there is camaraderie and a lighter side to these proceedings that doesn't get enough recognition or appreciation. When it is all said and done, racers are all part of a gypsy caravan that camp together in the paddocks of the world. Perhaps at times a dysfunctional family, but a family nevertheless.

*Roger Penske drops by the Chip Ganassi pit at Fontana 2013.
The camaraderie is obvious to see (Norm DeWitt)*

CHAPTER 23

DEATH BY SHOE

*Mark Webber in the Red Bull, winning the 2010 Monaco Grand Prix,
in a world before the "shoey" (Norm DeWitt)*

In 2016, Australian Moto GP racer Jack Miller won his first premier class Grand Prix at Assen, Holland, and celebrated by drinking champagne from his shoe. It wasn't long before Daniel Ricciardo got into the act in Formula 1, first drinking from his shoe at the German Grand Prix in 2016. Soon after, Ricciardo moved on to bigger challenges, getting the podium guest of honor or interviewer to drink from his shoe. His first victim, at the Belgian Grand Prix, was fellow Aussie, Mark Webber.

Me: Why do Australians drink champagne out of shoes?

Mark Webber: No idea. It started, I think, in Australia with touring cars, where one guy did this. It must be a barbeque thing. Then I had nowhere to go, obviously when Daniel asked me on the [Belgian GP] podium. But he has fresh boots every race so I was okay.

Me: It was at the end of the race.

Mark Webber: Spa is always cold so there is no sweat.

Soon after, at the United States Grand Prix in October, Daniel's latest victim was Gerard Butler, the actor. True to his craft, after drinking a hearty swig from the shoe, he collapsed and acted out his death scene. More recent victims of "the Shoey" have included Sir Patrick Stewart, who survived the experience intact. As I'd always suspected from watching endless episodes of Star Trek, The Captain of the Enterprise is harder to kill than King Leonidas of Sparta.

Gerard Butler the actor (King Leonidas of Sparta in 300) does the "shoey" as Daniel Ricciardo celebrates (Norm DeWitt)

Death by shoe. Showing his acting skills, Butler acts out his death scene (Norm DeWitt)

CHAPTER 24

SECOND IS FIRST LOSER

The agony of defeat, by 6/100ths of a second. Helio Castroneves collapses in disbelief (Norm DeWitt)

It was about an hour after the end of the 2012 Indianapolis 500 where the wild battle between Dario Franchitti and Takuma Sato ended with Dario in Victory Lane, and Takuma in the wall. On the last lap, Takuma had gotten up the inside of Dario going into turn 1. Pinched down to just above the white line and slightly over the limit, his car began to rotate while maybe a foot ahead mid-corner. It was a miracle that he didn't take Dario with him, as the cars were in constant contact as Takuma's car rotated 90 degrees. There were many venomous questions being asked about Takuma Sato's move in post-race interviews, which seemed grossly unfair – journalistic grandstanding at its worst.

Dario and I were in the elevator going up from the winner's interview on the ground floor, with me sharing the sequence of photos that I had taken of the last lap in turn 1. He pointed out Takuma's hand position early in the sequence, saying, "Look, his hand is up on the wheel, it's already gone."

Getting out of the elevator at the cafeteria level I grabbed a table and began to shed camera gear. Robin Miller wandered by on his way to the press room.

Robin Miller: What did you think?

Me: I think that if you've got a chance to win the Indianapolis 500 on the last lap and don't take it, you don't belong here.

Robin (smiling): Exactly.

Congratulations to Dario Franchitti, winner of the 2012 Indianapolis 500, and to Takuma Sato, winner of the 2017 Indianapolis 500.

*Dario Franchitti's final pit stop as Takuma Sato drives by during the
2012 Indianapolis 500 (Norm DeWitt)*

*Sato with his hand up on the wheel as he takes the lead into the first
turn on the final lap (Norm DeWitt)*

Contact! As Takuma Sato begins to rotate, he makes contact with Dario Franchitti (Norm DeWitt)

Somehow, Franchitti did not get collected as Sato did a rotation alongside in Turn One (Norm DeWitt)

Dario Franchitti, 2012 Indianapolis 500 winner
(Norm DeWitt)

Takuma Sato, 2017 Indianapolis 500 winner
(Norm DeWitt)

INDEX

O

Offenhauser (Offy) xx, 20, 25, 32, 74, 88, 117, 118, 125, 141, 142, 145, 182
Oliver, Jackie 48, 225, 228, 230, 234
Olson, Warren 4, 5
Olvey, Gary 327
Owen, Alfred 58, 72

P

Pace, Carlos 236, 243-245
Parnell, Reg 3, 124, 227
Parrott, Pete 196
Patrese, Riccardo 257, 258, 305, 322
Patrick, Pat vi, 144, 185, 205
Pedrosa, Dani 363, 372
Pellatt, Ron 149, 150, 243, 256
Penske PC1 183, 184
Penske PC3 184
Penske PC4 183-185
Penske PC5 151, 184
Penske PC6 184, 186, 190, 280
Penske PC7 186
Penske PC9 PC9B 190-193, 198, 309
Penske PC10 PC10B 194, 195, 197-201, 309, 333
Penske PC11 PC11B 199-201, 288
Penske PC12 201
Penske PC15 201, 202
Penske PC16 202-204
Penske PC17 198, 204, 205, 212
Penske PC18 205-207
Penske PC19 208
Penske PC20 208, 209, 214
Penske PC21 209
Penske PC22 209-211, 213
Penske PC23 PC23B 212, 214
Penske PC24 213, 214, 215
Penske PC25 215
Penske PC27 216, 217
Penske, Roger 107, 143, 179, 180, 193, 199, 204, 220, 269, 278, 297, 347, 413-415
Penske, Roger Jr. 199
Pescarolo, Henri 228, 236, 237
Peterson, Ronnie 55, 67, 145, 273, 386
Peugeot xix, , 128, 156
Philippe, Maurice 51, 102, 135, 168, 171, 172, 173
Phillips, Jud 96, 97, 105, 124
Piccoli, Mauro 381
Pilbeam, Mike 71
Piquet, Nelson 151, 152, 253, 256, 257, 258-260, 272, 277, 302, 395
Politoys FX3 236, 237
Pryce, Tom 245

CPSIA information can be obtained
at www.ICGtesting.com
Printed in the USA
LVOW13*2125140318
569872LV00004B/7/P